WITHDRAWN FROM LIBRARY

D1592764

Integrating Emergency Management and Disaster Behavioral Health

Integrating Emergency Management and Disaster Behavioral Health

One Picture through Two Lenses

Edited by

Brian W. Flynn

Ronald Sherman

Butterworth-Heinemann
An imprint of Elsevier
elsevier.com

Butterworth-Heinemann is an imprint of Elsevier
The Boulevard, Langford Lane, Kidlington, Oxford OX5 1GB, United Kingdom
50 Hampshire Street, 5th Floor, Cambridge, MA 02139, United States

Copyright © 2017 Elsevier Inc. All rights reserved.

All opinions, interpretations, conclusions, and recommendations contained herein are those of the authors and should not be construed as representing the positions or policies of an author's institution including, but not limited to, the Uniformed Services University of the Health Sciences, the Veterans Affairs Administration, the Federal Emergency Management Agency, and the United States Department of Defense.

No part of this publication may be reproduced or transmitted in any form or by any means, electronic or mechanical, including photocopying, recording, or any information storage and retrieval system, without permission in writing from the publisher. Details on how to seek permission, further information about the Publisher's permissions policies and our arrangements with organizations such as the Copyright Clearance Center and the Copyright Licensing Agency, can be found at our website: www.elsevier.com/permissions.

This book and the individual contributions contained in it are protected under copyright by the Publisher (other than as may be noted herein).

Notices

Knowledge and best practice in this field are constantly changing. As new research and experience broaden our understanding, changes in research methods, professional practices, or medical treatment may become necessary.

Practitioners and researchers must always rely on their own experience and knowledge in evaluating and using any information, methods, compounds, or experiments described herein. In using such information or methods they should be mindful of their own safety and the safety of others, including parties for whom they have a professional responsibility.

To the fullest extent of the law, neither the Publisher nor the authors, contributors, or editors, assume any liability for any injury and/or damage to persons or property as a matter of products liability, negligence or otherwise, or from any use or operation of any methods, products, instructions, or ideas contained in the material herein.

British Library Cataloguing-in-Publication Data
A catalogue record for this book is available from the British Library

Library of Congress Cataloging-in-Publication Data
A catalog record for this book is available from the Library of Congress

ISBN: 978-0-12-803638-9

For Information on all Butterworth-Heinemann publications visit our website at https://www.elsevier.com

Working together to grow libraries in developing countries

www.elsevier.com • www.bookaid.org

Publisher: Candice Janco
Acquisition Editor: Sara Scott
Editorial Project Manager: Hilary Carr
Production Project Manager: Punithavathy Govindaradjane
Cover Designer: Matthew Limbert
Cover Image Credits: Michael Rieger/FEMA; Steve Zumwalt/FEMA; Andrea Booher/ FEMA News Photo

Typeset by MPS Limited, Chennai, India

This book is dedicated to those who have
experienced unthinkable loss in disasters.
And to those who help them recover
from the unthinkable.

Contents

List of Contributors xvii
Acknowledgments xix
Introduction xxi

Section I
Context

1. **Where Emergency Management and Disaster Behavioral Health Meet: Through an Emergency Management Lens**   3
   Nancy Dragani and Valerie L. Cole

   An Emergency Management Perspective   3
   Nancy Dragani

   National Incident Management System and the Incident Command System   4
   Threat, Hazard, and Risk Assessments   7
   All Hazards Planning   8
   Whole Community   8
   Resiliency   9
   References   12

   A Disaster Behavioral Health Perspective   13
   Valerie L. Cole

   Foundation: Incident Command Structure   13
   Understanding the Difference Between Typical Community Mental Health Care and DBH Services   15
   Chasing Damage   17
   Mass Casualty Events   18
   Preparing for the DBH Effects of Disaster   19
   DBH in the Recovery Phase   20
   Workforce Behavioral Health Protection   21
   Recommendations for the Future   21
   References   22

   Making Integration Work   24
   Nancy Dragani

2. **Where Emergency Management and Disaster Behavioral Health Meet: Through a Disaster Behavioral Health Lens** 25
   Rachel E. Kaul and Ronald Sherman

   A Disaster Behavioral Health Perspective 25
   Rachel E. Kaul

   Introduction 25
   The Formation of the Field 28
   The Ideal of Integration 28
   Emerging Forces Driving Policy 30
   The Focus on Resilience 32
   Central Challenges to Address 33
   The State of State Behavioral Health Systems 34
   Standards and Plans 34
   Culture 34
   Future Directions 35
   References 37

   An Emergency Management Perspective 40
   Ronald Sherman

   Meeting the Needs of Disaster Survivors 41
   Meeting the Needs of Disaster Workers 42
   Opportunities for Promoting Integration 43
   References 45

   Making Integration Work 46
   Rachel E. Kaul

   References 48

3. **Why Is Integrating Disaster Behavior Health Essential to Emergency Management? Challenges and Opportunities** 49
   Albert Ashwood, Steven Moskowitz, Brian W. Flynn, and Ronald Sherman

   Through an Emergency Management Lens 49
   Albert Ashwood

   An Emerging Profession 49
   Who are Emergency Managers and What Do They Do? 51
   Looking at Clouds from Both Sides: A First Person Account
      From an Emergency Manager 53
   Sunday, April 16, 1995 53
   Wednesday, April 19, 1995 54
   Saturday, April 22, 1995 56
   May Through June 1995 58
   Lessons Learned 59

Through a Disaster Behavioral Health Lens   61
Steven Moskowitz

Wellness in Chaos   64
References   66

Making Integration Work   67
Brian W. Flynn and Ronald Sherman

Identify Behavioral Health Resources   67
Explore Scope and Limits of Practice and Availability   68
Orientation to the World of Emergency Management   68
Assure BH Focus on Both Survivors and Workers   68
Establish and Support a Positive EM Culture   69
Include DBH Professionals in Exercises and Drills   69
Focus on Shared Needs and Challenges   70
Monitor, Evaluate, and Revise the Nature and Success of Integration   70

4. Why Is Integrating Emergency Management Essential to Disaster Behavioral Health? Challenges and Opportunities   73
Anthony H. Speier and Ronald Sherman

Through a Disaster Behavioral Health Lens   73
Anthony H. Speier

The Culture of Emergency Management   74
The Culture of Disaster Behavioral Health   76
A Brief Review of Disaster Behavioral Health Literature (1977–2016)   78
Integration Through the Assimilation of Two Cultures   80
Systemic Considerations   81
References   81
Web Links   84

Through an Emergency Management Lens   85
Ronald Sherman

Why Does Disaster Behavioral Health Need Emergency Management?   85
How Can Integration Be Facilitated?   87
Factors Hindering Integration   88
Strategies to Help DBH Gain and Sustain Integration   89
Culture and History   89
Mission   91
How Can DBH Become Indispensable to EM?   91
How Does All This Look at the Various Parts of the Disaster Cycle?   92

Making Integration Work 94
Anthony H. Speier

Empowerment at Work: An example of State System Integration 95
The Promise of an Integrated DBH and EM 95
References 96

Section II
Key Areas of Integration

5. Integration in Disasters of Different Types, Severity, and Location 101
   James M. Shultz, Marianne C. Jackson, Brian W. Flynn, and Ronald Sherman

   Through a Disaster Behavioral Health Lens 101
   James M. Shultz

   Complex Systems Thinking for EM and DBH Integration in Disasters 101
   Different Levels of Disaster Response 102
   The Event Dictates the Level of EM Response 102
   The Event Defines the Level of DBH Response 104
   Concluding Comments and Take-Home Lessons for EM and DBH Professionals 112
   References 117

   Through an Emergency Management Lens 119
   Marianne C. Jackson

   An Emergency Management Perspective 119
   Integration in an Urban Evacuation: Yonkers Mudslide 119
   Integration in a Multicultural Mass Violence Event: Binghamton New York Shootings 120
   Integration in a Human Exploitation Case: Forced Labor of Deaf Mexicans 122
   Integration in a Terrorist Event: The Boston Marathon 123
   Understanding Variations in Local, State, and Federal Disaster Authorities 124
   References 125

   Making Integration Work 126
   Brian W. Flynn and Ronald Sherman

6. Not All Disasters Are the Same: Understanding Similarities and Differences 129
Daniel W. McGowan and James Siemianowski

Through an Emergency Management Lens 129
Daniel W. McGowan

Nature of Preparedness 131
Nature of Foundation Elements 134
Nature of Current Services 138
References 140

Through a Disaster Behavioral Health Lens 141
James Siemianowski

Predisaster Strategies 147
Develop Local and Statewide Disaster Behavioral Health Plans 147
Develop DBH Annexes in State All Hazard Plans 147
Develop Local and Statewide Disaster Behavioral Health Assets 147
Provide Basic Psychological Training to Emergency Personnel 148
Include Behavioral Health Representatives in Emergency Planning Efforts 148
Incorporate Disaster Behavioral Health into Exercises and Drills 148
Formalize the DBH Role Through Legislation 148
Postdisaster Strategies 149
Integrate Senior Mental Health Officials into the Command Structure 149
Use Mental Health Experts in a Consultative Role 149
Link Disaster Behavioral Health Leadership and Experts to Other Decision-Makers 149
Require the Lead Mental Health Agency to Develop a DBH Incident Action Plan 150
Develop a Disaster Behavioral Health Coordinating Committee 150
The Future of Integration 150
References 151

Making Integration Work 152
Daniel W. McGowan

7. What Can DBH Actually *Do* To Make Emergency Managers Jobs Easier? 155
April J. Naturale and Lesli A. Rucker

Through a Disaster Behavioral Health Lens 155
April J. Naturale

Consultation to Leadership 155
Survivor Stress 157
Responder Stress 158

Understanding the Body and Brain Responses to Extreme Situations: What It Means for Integration	159
Risk Communication/Media Information	165
Training and Education	171
Program Evaluation, Measurement, and Monitoring	173
Tracking, Documenting, and Projecting Behavioral Health Consequences	173
References	174

Through an Emergency Management Lens — 177

Lesli A. Rucker

Introduction	177
Topic Areas	178
Awareness and Understanding	178
Messaging	181
Operations	182
Sheltering	182
Strategic Locations	184
Program Implementation	185
Benefits to Emergency Managers	186

Making Integration Work — 188

April J. Naturale

Reference — 189

8. **Expanding the Tent: How Training and Education Partnerships with Other Professions Can Enhance Both EM and BH** — 191

    Laurence W. Zensinger, Gerard A. Jacobs, Brian W. Flynn, and Ronald Sherman

    Through and Emergency Management Lens — 191

    Laurence W. Zensinger

The Course Syllabus: Disaster Behavioral Health for Emergency Managers	192
Who Is in Charge?	194
Do Expectations Reflect Reality?	194
What Is Your Approach to the Media?	196
Bringing Allied Professions into the Tent	197
Emergency Management Training and Education and Disaster Behavioral Health	199
Summary and Further Thoughts	201
References	202

Through a Disaster Behavioral Health Lens		203
Gerard A. Jacobs		
National Biodefense Science Board (NBSB)		206
Education/Training Strategies in Disaster Behavioral Health		207
Educating the Education Community		208
Educating Criminal Justice and Law Enforcement		209
Psychological First Aid Training		209
Learning about Each Other		210
Mental Health Professionals		211
Emergency Managers		212
Disaster Drills		212
Obstacles in Cross Training		214
Stigma About Receiving Psychological Support		214
Costs		215
Time		215
References		216
Making Integration Work		218
Brian W. Flynn and Ronald Sherman		
9.	**Linking with Private Sector Business and Industry**	
	Diana Nordboe and Susan Flanigan	221
	History and Overview	221
	Role of Private Sector and Nongovernmental Organizations (NGOs) in Emergency Response and Recovery	226
	Evolution of Disaster Complexity and Private Sector Collaboration	228
	The Power of Water	229
	Case Example: Sandy Hook School Shootings	230
	Case Example: Omaha Metropolitan Medical Response System	232
	Missouri's Evolving Practice: From Floods to Ferguson	234
	Case Example: Ferguson, Missouri	234
	Case Example: Virginia Responds to the 9/11 Attack on the Pentagon	237
	Community Support	239
	Integrating Volunteers into Disaster Response	240
	Strategic Steps Toward Accomplishing Emergency Management, Disaster Behavioral Health, and Private Sector Integration	241
	Step One: Educating Individuals and Organization	241
	Step Two: Preparing Individuals and Organizations	242
	Step Three: Reach Out to Key Partners	242
	Step Four: Live Emergency Management for Life	243
	Conclusion	243
	A Nation Responds: Diana Nordboe's Reflection on 9/11	243
	Susan Flanigan's Reflections: Ferguson	245
	Making Integration Work	248
	References	249

Section III
Special Opportunities to Enhance Integration

10. **Integration in the Emergency Operations Center (EOC)/Emergency Communications Center (ECC)**   255
    John J. Brown, Jr., Chance A. Freeman, Brian W. Flynn, and Ronald Sherman

    Through an Emergency Management Lens   255
    John J. Brown, Jr.

    An Emergency Manager's First Hand Account of Encountering Behavioral Health Effects   256
    Emergency Operations Centers   257
    Emergency Communications Centers   257
    Linking the EOC, ECC, and Field Operations   258
    Integrating Behavioral Health in the EOC and ECC   258
    Understanding the Stresses   258
    Addressing the Stress: What to Do?   260
    Conclusion   261
    References   261

    Through a Disaster Behavioral Health Lens   263
    Chance A. Freeman

    Why it is Important for DBH Personnel to be Located in an EOC and What Roles Can They Play?   263
    Situational Awareness   264
    Briefing   264
    Assure Provision of Behavioral Health Services   271
    Challenges in Integrating Within the EOC   272
    Defining Terms and Mutual Education   274
    Establishing the Partnership   274
    References   275

    Making Integration Work   276
    Brian W. Flynn and Ronald Sherman

11. **Risk and Crisis Communications**   277
    Brian W. Flynn and John P. Philbin

    Through a Disaster Behavioral Health Lens   277
    Brian W. Flynn

    Mutual Understanding of Roles and Skills   278
    Mutual Respect and Trust   278
    Working with Emergency Managers to Increase Understanding of Victim Priorities   278

Working with Emergency Managers to Develop Anticipatory
 Guidance 279
Fostering Emergency Managers' Communication
 with Victims 281
Assisting in Crafting Messages 282
Assisting in Monitoring and Managing Stress of Emergency
 Management Personnel 283
Conclusion 284
References 284

Through an Emergency Management Lens 286

John P. Philbin

Today's Communication Environment 286
A Systems Approach to Effective Communication During Disasters 288
Summary 292
References 293

Making Integration Work 294

Brian W. Flynn

12. **How to Navigate External Factors: Legal, Ethical,
 and Political Issues** 297

Berl D. Jones, Jr. and Daniel Dodgen

Through an Emergency Management Lens 297

Berl D. Jones, Jr.

Some Important Context: Speed and Privacy 297
Evolution of the Disaster Assistance Application Process 298
The Evolution of Disaster Behavioral Health Services 301
The Intent and Complications of Providing Assistance 301
Fraud and PII 303
In-Person Contact 305
The Current Process: What DBH Needs to Know About
 Information Sharing 305
References 306

Through a Disaster Behavioral Health Lens 307

Daniel Dodgen

What are the Relevant Current Laws, Policies, and Ethical
 Guidances for Disaster Behavioral Health Services? 308
The Law 308
Policy 310
Ethics 312
How Does the Disaster Context Impact Ethical, Legal,
 and Policy Questions? 314

What are the Long-Term Ethical and Policy Concerns
for Impacted Individuals and Communities? 315
References 317

Making Integration Work 319
Berl D. Jones, Jr.

13. Sustaining Integration: A Way Forward 321
Brian W. Flynn and Ronald Sherman

Why Integration Breaks Down 321
Change of Leadership 321
Change of Authorities 322
Political Landscapes 322
Changing Evidence and Practice 323
Visibility of Impact 323
Competing Demands and Priorities 324
Four Pillars of Sustained Integration 325
Mutual Trust and Respect 326
Demonstrated Benefit 326
Resources 327
Adaptability 329

14. Conclusion/Summary 331
Brian W. Flynn and Ronald Sherman

Key Findings 332
Foundational Agreements 332
Exceptional Opportunities 333
Significant Challenges 334
The Path Ahead 336
Conclusion 337

Index 339

List of Contributors

Albert Ashwood, B.A.
Oklahoma Department of Emergency Management, Oklahoma City, OK, United States

John J. Brown, Jr., M.S., B.S.
Office of Emergency Management, Virginia, VA, United States

Valerie L. Cole, Ph.D.
American Red Cross, Washington, DC, United States

Daniel Dodgen, Ph.D.
US Dept of Health & Human Services, Washington, DC, United States

Nancy Dragani, B.A., M.A.
Federal Emergency Management Agency, Denver, CO, United States

Susan Flanigan, CBCP
Missouri Department of Mental Health, Jefferson City, MO, United States

Brian W. Flynn, Ed.D.
Uniformed Services University of the Health Sciences, Bethesda, MD, United States

Chance A. Freeman, B.S.
Texas Department of State Health Services, Austin, TX, United States

Marianne C. Jackson, B.A., M.A.
Federal Emergency Management Agency (Retired), New York, NY, United States; NYC Emergency Management, New York City, NY, United States

Gerard A. Jacobs, Ph.D.
University of South Dakota, Vermillion, SD, United States

Berl D. Jones, Jr., M.P.H., CEM
Federal Emergency Management Agency, Washington, DC, United States

Rachel E. Kaul, LCSW, CTS
U.S. Department of Health and Human Services, Washington, DC, United States

Daniel W. McGowan, B.A., M.A.
McGowan Enterprises, Inc., Helena, MT, United States

Steven Moskowitz, M.S.W.
New York State Office of Mental Health, Albany, NY, United States

April J. Naturale, Ph.D., M.S.W
ICF International, Fairfax, VA, United States

Diana Nordboe, M.Ed.
Emergency Manager, Independent Consultant, Carter Lake, IA, United States

John P. Philbin, Ph.D.
Crisis1, LLC, Reston, VA, United States

Lesli A. Rucker
Cenibark International, Inc., Richland, WA, United States

Ronald Sherman, B.A., M.A.
Independent Consultant, FEMA Federal Coordinating Officer (Retired), United States

James M. Shultz, M.S., Ph.D.
University of Miami Miller School of Medicine, Miami, FL, United States

James Siemianowski, M.S.W.
CT Dept Mental Health and Addiction Svcs., Hartford, CT, United States

Anthony H. Speier, Ph.D.
Louisiana State University Health Sciences Center, New Orleans, LA, United States

Laurence W. Zensinger, B.A., M.S.
Director of the Recovery Division for the Federal Emergency Management Agency (Retired), United States

Acknowledgments

We are both humbled and gratified by the many people who have given so generously or their time, effort, and expertise to make this book possible. It is only through your good work that these important topics can be shared so widely and comprehensively.

We would like to first thank the many contributors to the book. You were carefully selected for this unique project by virtue of your real-life experiences and your demonstrated ability to conceptualize what you do and to effectively communicate your knowledge and wisdom. Each of you have walked the walk as well as talked the talk. This is a powerful combination that is seldom found.

We would like to thank our guardian angels at Elsevier who have supported and guided us through this project from start to finish. Sara Scott was the first to approach us with the idea for such a book. Somehow she knew that neither of us could resist such an opportunity. She cast the line and set the hook. Hilary Carr worked with us throughout the project and was ready at a moment's notice to guide us, support us, and help us stay organized. She was both our coach and cheerleader.

Michelle Herman served as research and editorial assistant on the book. Her contributions contributed immeasurably to making the content clear and accurate.

We both would like to thank our wives, Donna Flynn and Carol Sherman, for their support, encouragement, and tolerance throughout more than the past year as we poured ourselves into this effort. We realize it distracted both of us from our time with, and attention to, you. We also realize that both of you had every reason to expect that once we gave up direct disaster response leadership roles, we would be more physically and psychologically available. Neither of us can say no to an opportunity to support the work of those who help survivors who have suffered so much. We cannot tell you enough how much your love and support continues to mean to us.

Introduction

Preparing for and responding to disasters and large-scale emergencies make for strange bedfellows. These tragic events require that individuals, groups, and governments work in harmony if victims and survivors are to be optimally served. As disaster response has become more standardized and formalized over the past several decades, the complexities, challenges, and promises of integrating critical partners has become clearer. One of the most challenging and promising integrative opportunities is between emergency management (EM) professionals and behavioral health (DBH) professions involved in disaster preparation and response. As the chapters of this book will reveal, there are immeasurable advantages to be gained from the integration of these two professional domains. At the same time, this is not an easy match. These are two rapidly evolving areas of theoretical, legislative, and practical formalization grounded in conceptual structure, research, and real-life experience.

Both the structure and content of this book are designed to guide the reader through key areas of important integration, articulate the challenges and opportunities involved, and provide practical guidance for implementation and application. The structure is designed to model an integrated approach to the topics presented. Chapters will provide perspectives on the topic from both professions as well as case examples or suggestions for accomplishing integration.

The editing of this volume is also intended to model integration. Each of us comes from a different profession and has designed the book based on not only our own experiences. We have fully incorporated the suggestions from both EM and disaster behavioral health communities to identify the most relevant topical areas and contributing authors.

Fundamental to the foundation in designing and editing this book, both us have walked the walk. Combined, the two of us have more decades of hands-on involvement in disaster preparedness and response that we like to contemplate. Our experience spans significant governmental, legislative and policy development and implementation, consultation to national and international leaders, knowledge development and dissemination. Perhaps most important, is our real-time experience in disaster preparation and response in some of the most complex and difficult situations the United States has faced, such as in the aftermath of both natural and human-caused disasters.

I, Brian Flynn, have lived primarily in the DBH and science world. During my 31 years in the U.S. Public Health Service (USPHS), in addition to other responsibilities, I worked in, managed, and supervised the federal government's domestic disaster mental health program. In that role, I served on-site with EM professionals at many, if not most, of the nation's largest disasters. Since I retired from the USPHS in 2002 at the rank of Rear Admiral/Assistant Surgeon General, I have directed nearly all of my professional efforts toward advancing the field of preparing for and responding to large-scale trauma. I have provided training and consultation to both public and private entities throughout the United States and internationally. I currently serve as Adjunct Professor and Associate Director of the Center for the Studies of Traumatic Stress, in the Department of Psychiatry at the Uniformed Service University of the Health Sciences in Bethesda, Maryland.

I, Ron Sherman, spent almost 29 years as an EM specialist with the Federal Emergency Management Agency (FEMA). I worked on over 200 federal disaster operations, including some of this nation's most devastating events. Many times I was in the role of Federal Coordinating Officer (FCO), the on-site official in charge of all federal response and recovery efforts. After Hurricane Katrina, I served as the FCO in Alabama before becoming the Senior Housing Official responsible for disaster housing operations for the entire Gulf Coast. After retiring in 2007, I continued EM involvement by starting a Citizen Corps Council in my hometown and am now the leader of a Community Emergency Response Team. I provide EM consulting services to communities and emergency response training for volunteer groups. I successfully integrated the operations of a Community Emergency Response Team with a new Medical Reserve Corps team and made Psychiatric First Aid a requisite part of the training curriculum.

We have both "been there and done that." We have lived first-hand the enormous opportunities that emerge where EM and DBH professions understand each other and integrate our experience and expertise. We have also witnessed situations where this was not the case. Failure to understand, respect, and value the perspectives and responsibilities of the other field have compromised preparedness and response. The stakes are high. In the end, our ability to work together and integrate our professions makes a difference in the lives of countless disaster victims and survivors. We owe them no less than our best.

Contrary to what the prior few paragraphs may have implied, this book is not all about us. We and our supporters at Elsevier are providing the gateway for the almost incalculable knowledge and experience represented by the contributors to this volume. An introduction to this book would be incomplete without a discussion of who they are and what they have so generously brought to this book.

First, all contributors were selected because of not only their status and credibility in their professional domains. A defining criterion for selection was not only a conceptual grounding on specific topics, but their real-life experiences in operationalizing their expertise. Every contributor has been in a position to prepare for and/or respond to real disasters. This is not just theory for them. They too have walked the walk. Think about the magnitude of what this means. It means that you, the reader, hold in your hands almost 400 person years of collective experience, wisdom, and advice. We are proud to have had their willing and eager participation in crafting this unique book.

Yet, for us, that is only part of the picture. Both of us know these contributors well. Certainly as important as the knowledge they bring are the values they represent. The writers of this book have spent much, if not most, of their careers in service to protecting their national and global neighbors, giving their all in helping others rebuild their lives and communities in the darkest hours, and have brought comfort to the vulnerable, frightened, and displaced. They have healed the broken. For their service, many, if not most, along with their families and colleagues, have paid a price. One does not do this type of work, no matter how noble, without testing the limits of health and relationships. Yet, we know they would tell you, the reader, as they tell us—they would not want to do anything else. In these pages, they serve once again by sharing with you the lessons they have learned trusting that you will continue to build on both their work and their values. We all hope that you will use what you take from these pages to "pay it forward."

TOPIC SELECTION

Together, pooling our collective decades of experience, we identified topics that we felt were the most central to facilitating meaningful and practical integration of EM and DBH. To assure that the content reflects the needs and priorities of both fields, we distributed a draft of the book's content and structure to individuals and groups with credibility in both professional domains, asking for their input both on topic and potential contributors. The topics and contributors identified through that process are contained on these pages.

STRUCTURE

The format of this book is a bit different than one might be used to. From the start, we were determined to avoid a structure that would keep the two professions talking only to the reader and not to each other. We have attempted to model integration as not only a goal but as the foundation for this book. We also were driven by a commitment to assure that content was not only theoretical, conceptual, or practical—but rather, all of these, combined.

As a result, in each chapter, readers will find a primary contributor representing either EM or DBH. This is followed by a commentary on the chapter topic by a contributor from the other profession. Finally, each chapter will contain a case example or practical advice to implement or make operational the topical content in terms of integration.

TERMINOLOGY

Words matter. One of the challenges of a volume such as this is the inclusion of different professions that have different terms, frames of reference, and common acronyms. Assuming that most readers will read specific chapters as their interest and responsibilities dictate rather than read from start to finish, we have done our best to include and often repeat terminology and key references throughout the book.

We should mention from the start that we have chosen to use the terms *behavioral health* (BH) and *disaster behavioral health* (DBH) in this book. As Dr. James Shultz describes in more detail in Chapter 5, Integration in Disasters of Different Types, Severity and Location, DBH is not a familiar expression throughout the rest of the world. Instead, "mental health and psychosocial support" (MHPSS) is the phrasing that is recognized and used worldwide by the World Health Organization, United Nations agencies, and numerous organizations involved in disaster and humanitarian response.

We have used the term behavioral health instead of mental health because it is more inclusive, places a high value on behavior, and is rapidly become the preferred term at least in the United States.

It is our hope that readers will understand that language is evolving, and evolving differently in various parts of the world. We hope that these differences will not be distracting. Optimally, readers will see this as an ongoing dynamic in the development of shared understanding.

Section I

Context

If integration of emergency management (EM) and behavioral health in disasters is to occur, it must rise on a foundation of mutual understanding and respect. In practice, many in each profession often have little understanding or awareness of the other. In preparation for each profession, there is little exposure to the other field. When behavioral health experts find themselves participating in disaster preparedness and response, they seldom, at least initially, know much about the field of EM. Likewise, when emergency managers first encounter behavioral health experts while preparing for and responding to disasters, they seldom have a comprehensive understanding of roles behavioral health professionals can play.

In this section, readers will gain an in-depth understanding of what each profession does. The authors provide examples where attempts at integration have succeeded or fallen short. These examples show how an enhanced understanding of each other's roles can help each profession complement the other's efforts.

Chapter 1

Where Emergency Management and Disaster Behavioral Health Meet: Through an Emergency Management Lens

Nancy Dragani[1] and Valerie L. Cole[2]
[1]*Federal Emergency Management Agency, Denver, CO, United States,*
[2]*American Red Cross, Washington, DC, United States*

An Emergency Management Perspective

Nancy Dragani

Every day, somewhere in the United States, someone is recovering from a disaster. The disaster may be small in scope, such as a house fire or localized tornado or a major event that affects thousands like Hurricane Katrina or the attacks on September 11, 2001. Regardless of size, the lives of those who are impacted have changed. Each of these events leaves families in turmoil—homes ripped apart and people forced to piece their lives back together. For most survivors, recovery will take place—maybe not as soon as they would like, but eventually. Time will dim the terror, dull the pain, and ease the memories of their struggles to regain the life they had before the disaster. On the other hand, some impacted by disaster will experience psychological or emotional trauma that will change their behavior and lead to potentially damaging outcomes.

In his work, *A Treatise Concerning the Principles of Human Knowledge*, 18th century philosopher George Berkeley posed a question, which is commonly paraphrased, "If a tree falls in the forest and no one hears it, does it make a sound?" (Berkeley, n.d.). Berkeley's theory posited that perception creates reality—a concept that has application within an emergency management (EM) and disaster response framework. One could argue that, if a physical disaster occurs and there is no impact on individuals, it is not really a disaster. Of course, there are other serious situations that impact individuals and governments, such as ecological or financial emergencies. But, in the world of EM, a disaster has an inherent and inextricable link to its effect on people. The impact may be direct such as loss of housing, personal property, or a job. It can also be indirect when essential government services like transportation systems, utilities, or public buildings are damaged or destroyed. Regardless, when people are impacted, effective preparedness, response, and recovery must take into account the whole human—physical, psychological, and emotional.

A sequence of events, including wildfires in the 1970s; Hurricane Hugo; September 11, 2001; and Hurricane Katrina, led the EM profession beyond a military-based, civil defense approach to disaster response and recovery. Terms and processes accepted as the standard today, such as the National Incident Management System (NIMS) or Incident Command System (ICS), all hazards planning, resiliency, and whole community are relatively recent evolutions in the field of EM (Federal Emergency Management Agency (FEMA), 2004).

NATIONAL INCIDENT MANAGEMENT SYSTEM AND THE INCIDENT COMMAND SYSTEM

Following widespread and deadly wildfires in Southern California in 1970, an interagency fire group in southern California determined a better system was needed to coordinate operations, particularly when multiple agencies and jurisdictions were engaged in the response. This group, led by the U.S. Forest Service, identified two key areas for improvement: the first was the need for a standard terminology, operating procedures, and command structure; and the second was a way to prioritize and coordinate resources during a multiagency, multijurisdictional response (FEMA, 2004). In 1972, Congress allocated $900,000 to the U.S. Forest Service to develop a system that addressed these deficiencies. The system evolved into the ICS and the Multiagency Coordination System (MACS) (Neamy & Nevill, 2011). However, for nearly 30 years, its use was limited primarily to the fire community, even though recognition was growing that ICS could be effective for any response, regardless of cause, setting the stage for its use in an all hazard environment. In response to the attacks on the United States in September 2001, President George W. Bush issued Homeland Security Presidential Directive-5, commonly referred to at HSPD-5. This directive was released on February 28, 2003, with a subject line that read, "Management of domestic incidents," and had a single, clearly stated purpose: "To enhance the ability of the United States to manage domestic incidents by establishing a single, comprehensive national incident management system" (Department of Homeland Security, DHS, 2003).

The NIMS, based largely on the fire service ICS, differs only in the addition of an Information and Intelligence Management function, which can provide analysis and sharing of intelligence during an event.

DHS launched NIMS in March 2004. When the federal preparedness grants in 2006 rolled out, DHS made NIMS compliance a grant requirement (FEMA, 2004). Ten years later, NIMS and ICS have been largely institutionalized in EM agencies across the nation and are beginning to be used in planning efforts outside traditional EM areas, such as school safety plans and major retail center emergency response planning.

So what are the core principals of NIMS and ICS? NIMS has five components:

- Preparedness
- Communications and Information Management
- Resource Management
- Command and Management
- Ongoing Management and Maintenance

What NIMS and ICS Meant to Disaster Behavioral Health (DBH)

NIMS/ICS Element	Relevance to DBH/EM Integration
Preparedness	Integration begins in the preparedness phase. DBH practitioners should take basic ICS and NIMS training, look for additional opportunities to learn about the preparedness process, work with the EM community to establish a seat at the table, and become culturally competent in understanding EM.
Communications and Information Management	Communications is a behavioral health intervention. Chapter 11 "Risk and Crisis Communications" describes in detail integration factors and strategies regarding communications and public information activities.
Resource Management	DBH resources are among the many resources of interest to EM. By integrating EM and DBH efforts, DBH type, timing, nature, and duration of DBH resources can be more accurately and efficaciously determined. Chapter 7 "What Can DBH Actually *Do* To Make Emergency Managers Jobs Easier?" describes how behavioral health practitioners can make emergency managers' jobs easier.
Command and Management	In disaster situations, DBH, like all other elements of response operations, will generally function under EM command and control functions. However, there may be situations when an Incident Commander wishes to incorporate DBH into an active operation, either to support responders or to identify critical concerns with survivors. DBH resources must understand these structures and be prepared to function within either of these frameworks. Few DBH professionals are familiar with these systems until they enter the world of disaster preparation and response. Chapter 10 "Integration in the Emergency Operations Center (EOC)" describes in detail how DBH can integrate in various EM facilities and operations.

(Continued)

(Continued)

NIMS/ICS Element	Relevance to DBH/EM Integration
Ongoing Management and Maintenance	There are DBH needs and considerations in preparedness, response, and recovery. The nature of the DBH needs change over time and the types of DBH expertise needed will also vary. Ongoing DBH needs and requirement should be an integral part of EM ongoing efforts through the event cycle.

ICS is one of three functions under Command and Management; the other two are MACS and Public Information. According to the FEMA ICS Resource Center, ICS is a scalable, standardized management tool that can be used for emergency and nonemergency events. To illustrate its adaptability, ICS trainers often use, only half-jokingly, the example of ICS as a management tool for planning a wedding or graduation party.

ICS has 14 core principals organized into six areas. The first focus area, Standardization, only contains one principal. However, it is arguably one of the key elements of ICS—common terminology. Using plain language, rather than agency-specific codes or acronyms, is critical to a successful multi-agency response. Imagine the challenges that would occur if two agencies came together in an active response and used different codes. One agency may use a 10–99 for "officer down" and another may use the same code for "temporarily out of service."

The second focus area, Command, has two essential functions. Establishment and Transfer of Command address the question of "who is in charge" and how command is transferred in such a way that all the essential information is provided to the incoming commander. Chain of Command and Unity of Command identifies how the lines of authority flow within the incident management organization and stipulates that each person has a designated supervisor.

The third area is Planning and Organization and includes four features. The first feature, Management by Objectives, establishes specific, measureable objectives for a defined incident period and then focuses efforts to achieve the objectives. Modular Organization, the second feature, simply refers to the scalability and flexibility of the ICS, or the ability to scale up or down depending on the size and scope of the incident, as well as any specific hazards. Incident Action Plans communicate the overall incident objectives, addressing both operational as well as support activities. Manageable Span of Control recommends the span of control or line of authority of any single individual should be from three to seven directly reporting individuals.

The fourth focal point is Facilities and Resources and includes two features: Incident locations and facilities encompass the various operational and support facilities such as Command Posts, Bases, Camps, Staging Areas, and Mass Casualty Triage Areas; and Comprehensive Resource Management, the processes for categorizing, ordering, dispatching, and tracking resources.

Communications and Information Management, the fifth area, has two features. The first, Integrated Communications, addresses the need for a common communications plan and interoperable communications processes. Information and Intelligence Management, added to ICS as part of HSPD-5, allows for the gathering, sharing, and managing incident information and intelligence.

The final focus area, Professionalism, is the largest with seven features but largely reinforces several of the preceding features, including Incident Action Planning, Unity of Command, Span of Control, and Resource Tracking. Three other areas of attention include ensuring Personnel Accountability, by requiring Check-In before receiving an assignment and reinforcing that personnel only respond when requested or Dispatched/Deployed by an appropriate authority.

THREAT, HAZARD, AND RISK ASSESSMENTS

The first step in emergency planning is identifying threats, vulnerabilities, and risk. A threat is anything that can cause harm to people, property, or the environment. Vulnerability is a weakness that is exposed when faced with a threat. Risk, then, is the combination of threat and vulnerability. An urban environment faced with a threat that may require evacuation may be at greater risk due to reliance on public transportation than a suburban environment where most families own personal vehicles. In this case, reliance on public transit systems creates an increased vulnerability in an evacuation scenario. In another example, an area with an active wildfire threat but little population may be at a lower risk based on the limited vulnerability of a population.

In 2013, FEMA released a new tool to measure risk—the Threat, Hazard Identification, and Risk Assessment or THIRA. The THIRA is a four-step process that helps the community and planners understand risk and identify capability requirements. Communities can then begin to map their risks to the capabilities needed to achieve their desired outcomes and the resources required to achieve their targeted capabilities. FEMA has identified 32 capabilities that are common and grouped them into five mission areas: prevention, preparedness, response, recovery, and mitigation of disasters. Some core capabilities are only in one mission while others can be found in multiple mission areas. The capabilities run the gamut—from planning to mass care services, search and rescue to supply chain integrity, cybersecurity to interdiction and disruption. The result of the THIRA process is that communities know what they need to prepare for, what resources (either owned by the community or available through mutual aid) are needed to meet the required capabilities, and what actions can the community take to avoid, limit, or eliminate a threat or hazard. The risk assessment feeds into the planning process.

ALL HAZARDS PLANNING

Until President Jimmy Carter created FEMA in 1979, there was no single agency in charge of coordinating the federal response to disasters. In 1988, when the Robert T. Stafford Act became law, FEMA was required to develop a federal response plan and each state was directed to develop a state emergency operations plan. The Federal Response Plan (FRP) was released in 1992 and heralded the advent of a new type of emergency plan. Up to that point, emergency plans focused on specific events, such as earthquakes or nuclear attack, or were written by individual agencies focused on their activities. The FRP was the precursor to the National Response Plan, and the current federal plan is called the National Response Framework (NRF).

Each of these plans or frameworks is based on the assumption that there are core activities that do not fundamentally change, regardless of the cause of an event. For instance, if a building collapses, search and rescue must occur, debris must be cleared, public information must be disseminated. While the way a search proceeds, debris is cleared, or the content of a message may change, the act of search and rescue, debris clearance, or public information does not fundamentally change, regardless of whether the building collapsed because of a tornado or a bomb. This is the core precept of all hazards planning.

In FEMA's planning guidance, *Developing and Maintaining Emergency Operations Plans*, released in November 2010 (Federal Emergency Management Agency (FEMA), 2010), several key planning principles are outlined. Planners must use a logical, analytical approach to work through complex problems. Plans must consider all threats and be able to address traditional as well as catastrophic events. An effective all hazard plan should identify actions and the resources required to implement a response. It will clearly define roles and responsibilities as well as how other levels of government can support the primary jurisdiction.

Planning must include participation from all the stakeholders in the community and be inclusive in its approach. In other words, the whole community must be at the planning table.

WHOLE COMMUNITY

So what does the "Whole Community" mean? In 2011, FEMA Administrator Craig Fugate launched a new concept that advocated ensuring the whole community is involved in planning and response. FEMA (2011) released a document describing the concept, "A Whole Community Approach to Emergency Management: Principles, Themes, and Pathways for Action." This document outlines the intent of the program and offers suggestions for implementation. Aimed primarily at the state and local EM community, the program recognizes "the importance of bringing together all members of the community to collectively understand and assess the needs of their respective communities and determine the best ways to organize and strengthen their assets, capacities and interests" (FEMA, 2011). Through this approach, FEMA hopes to build a more effective path to a secure and resilient society. It also provides an impetus for EM to seek out and engage with DBH practitioners to really fill out the whole community approach.

The whole community concept suggests a philosophical shift away from a reliance on a standard set of actions. Instead, there is in an increasing recognition that in order to build a prepared society, we have to engage all members of the community, understand the needs and motivations of our citizens, build on what currently works in community engagement, and "move beyond the easy to looking at the real needs and issues a community faces" (FEMA, 2011, p. 7).

RESILIENCY

Much has been said about the need for personal preparedness and a more resilient population. In fact, one could argue that the more prepared individuals are, the more psychologically resilient they may become. They will have positioned themselves to be a survivor; as such, they will be more resilient.

A resilient population is one that has the ability to bounce back, to adapt to adversity, and return to a state of normalcy. However, resiliency is more than being prepared for a disaster; it includes understanding the risk, accepting some measure of personal responsibility, and then proactively engaging in solutions.

Likewise, one can understand a risk and yet choose not to accept personal responsibility for that risk. Most drivers understand that having a vehicular accident is a realistic risk of traveling by car. Yet, according to a press release issued by the Insurance Information Institute, 12.6% of drivers are uninsured (Insurance Research Council, New Study Reveals a Declining Trend in Uninsured Motorists, 2014). Drivers may understand the risk, but choose not to, or are not able to, accept personal responsibility and engage in a solution by purchasing insurance. The challenge for emergency managers is to develop messages that not only clearly convey risk but motivate personal responsibility as well as advocate multiple options to engage in solutions.

Positioning theory, which looks at the roles and rights, duties, and responsibilities that individuals assume based on their perceived or actual position in the world, is one way individuals can choose, or be placed in, the role of victim or survivor. A 2009 article published in *Theory and Psychology* called "Recent Advances in Positioning Theory," by Harre and colleagues, explored new applications of the use of positioning theory to explain interpersonal encounters (Harre & Moghaddam, 2009).

According to the website Changing Minds (http://changingminds.org/), people define roles based on what they have learned from social interaction and education. Once the roles have been defined, people develop expectation about those roles and then encourage others to behave accordingly. According to Davies and Harre (2007), in their article "Positioning: The Discursive Production of Selves," the challenge with role theory is that evaluating actions identified by roles is static and based on formal, ritualistic aspects. They contend that how people position themselves may be a more reliable precursor to their actions than their static role. A key element in positioning theory is belief that language not only communicates but shapes the way individuals act based on how they position themselves relative to the language used (Davies & Harre, 2007, p. 2). Craig Fugate understood this when he reframed those impacted by disasters from "victims" to "survivors," first, as director of the Florida Division of Emergency Management and then as the administrator of FEMA.

Discourse, or dialogue, creates a framework in which one positions one's self or others. The assumption of position will determine where an individual is placed in the narrative. Davies and Harre (2007, p. 4) assert there are two ways individuals get positioned into the narrative: interactive positioning and reflective positioning. In reflective positioning, the individual defines his or her own position. In interactive positioning, the individual's position is defined by someone else. In both cases, the positioning may or may not be intentional. However, just as conversations shift and change based on input, an individual's position can flex based on how the dialogue changes the narrative, and, therefore where each player fits in into the storyline (Davies & Harre, 2007, pp. 2−3).

FEMA Administrator Craig Fugate's attempt to reframe victims of disasters to survivors is consistent with the concept of positioning theory. By positioning individuals impacted by disaster as survivors, FEMA is helping them assume the responsibility to be accountable for their own survival, as well as an implied duty to help others. By assuming the role of survivors, they become not only enablers in their own response and recovery, but also allow government resources to focus on those who are positioned, regardless of cause, as victims. In the role of survivors, people may even feel a responsibility to provide for others more impacted than themselves or their families.

> Language matters. It matters from both psychological and operational perspectives. DBH professionals should be aware of the history, importance, and evolution of the use of terms like victim and survivor and other EM terminology.

Individuals placed in the role of "victim" may feel justified in abdicating any responsibility for preparedness, survival, or recovery, which requires the government to fill that role. In this case, they will also likely feel they have a right to certain provisions and benefits, such as ice, food, water, and shelter, in the immediate aftermath of an event as well as government grants or loans during recovery. As victims, they may feel these rights are inviolable and can feel betrayed if their needs are not met to the standard they believe is warranted. The position an individual is placed in or places himself in is not static. The environment, other individuals involved in the encounter, and outside stimulus all change the framework of the interpersonal relationship and the position of the individual relative to role, rights, responsibilities, and duties. Consequently, an individual may assume the role of a survivor for an event that is either familiar or planned. For example, many individuals who live along the Ohio River in southern and eastern Ohio are relatively sanguine about the flood risk from the river. To them, the risk is part of river life, and they have plans in place to move their possessions and their families to higher elevation when river flooding is predicted. When the water recedes, they clean up their homes and property and resume their daily activities. The most responsible residents maintain flood insurance and understand that in the event of flooding, flood insurance will provide the greatest assurance of recovery. Within the framework of positioning theory, they have demonstrated a closer nexus to survivor than victim. They accept the risk of flooding as well as the responsibility to prepare and recover from that risk. However, since positions are not static, rights and responsibilities shift as new patterns emerge. In the event of a less common disaster, such as a tornado or chemical event, they may be less prepared, and therefore more likely to assume the role of victim. In this case, they may be more inclined to wait for outside assistance and may feel betrayed if that assistance is not readily offered.

Individuals can choose, or be placed in, the role of victim or survivor. They can choose or be placed in role of responder or casualty, which will then influence their perception of their rights as well as responsibilities. If messaging is reflective of personal perception of self and personal perception is defined by language and dialogue, then it is critical that those who prepare the messages use appropriate language to foster positive positioning. Craig Fugate was spot-on when he reframed victims as survivors. He used words to inspire a different context for those who have lived through disaster, and by doing so, encouraged a stronger, more resilient self-image of disaster survivors.

> DBH professionals are acutely aware that individual perception often trumps facts and heavily influences how individuals behave. Positioning theory is a good example of a shared conceptualizing between DBH and EM, even if terminology may vary. Finding common ground is an important first step in integrating DBH and EM.

REFERENCES

Berkeley, G. (n.d.). Immaterialism. Philosophy Pages. Retrieved from <http://www.philosophy-pages.com/hy/4r.htm>.

Davies, B. &. Harre, R. (n.d.). Positioning: The discursive production of selves. Retrieved from <http://www.massey.ac.nz/~alock/position/position/htm>.

Department of Homeland Security (2003). Homeland security presidential directive-5. Retrieved from <https://www.dhs.gov/sites/default/files/publications/Homeland%20Security%20Presidential%20Directive%205.pdf>.

Federal Emergency Management Agency. (2004). NIMS and the incident command system. Retrieved from <http://www.fema.gov/txt/nims/nims_ics_position_paper.txt>.

Federal Emergency Management Agency. (2010). Developing and maintaining emergency operations plans: Comprehensive planning guide 101, version 2. Retrieved from <http://www.fema.gov/cpg_101_comprehensive_preparedness_guide_developing_and_maintaining_emergency_operations_plans_2010-2.pdf>.

Federal Emergency Management Agency. (2011). A Whole Community Approach to Emergency Management: Principles, Themes, and Pathways for Action. Retrieved from <https://www.fema.gov/media-library-data/.../whole_community_dec2011_2_.pdf>.

Harre, R., & Moghaddam, F. (2009). Recent advances in positioning theory. *Theory and Psychology, 19*(1), 5–31.

Insurance Information Institute. (2014). New study reveals a declining trend. Retrieved from <http://www.iii.org/fact-statistic/uninsured-motorists>.

Neamy, R. & Nevill, W. (2011). From FIRESCOPE to NIMS: How NIMS developed out of the earlier FIRESCOPE program. Retrieved from <http://www.firefighternation.com/article/command-and-leadership/firescope-nims>.

A Disaster Behavioral Health Perspective

Valerie L. Cole

The premise of this book is that there is much that disaster behavioral health (DBH) can bring to the EM table. This section will present a brief overview of how DBH fits into the world of emergency management (EM) from the perspective of the behavioral health lens.

There are many misconceptions about DBH that are held by emergency managers, first responders, and others involved with disaster response. Some of the misconceptions are:

1. The Office of Mental Health for the jurisdiction can and will handle any behavioral health issues that may arise due to the disaster. Any mental health practitioner is qualified to practice DBH.
2. The magnitude of a disaster is measured solely by number of homes damaged or destroyed. If there has been minimal property damage, there is very little need for DBH.
3. DBH is only needed when there have been fatalities or acts of terror.
4. DBH is only needed during the *response* phase, or it is only needed during the *recovery* phase.
5. The only function of DBH practitioners is to recommend self-care or suggest that the responder is not fit to handle the situation and should leave. For responders who want to continue working on the disaster, it is dangerous to talk to a DBH specialist.

These misconceptions will be addressed one at a time. Before we delve into these misconceptions, we will look at how DBH fits in into the structure of EM.

FOUNDATION: INCIDENT COMMAND STRUCTURE

In a typical Incident Command System (ICS) model, the focus is on a unified command that will allow for rapid response, interoperability between agencies, and an emphasis on saving lives as well as protecting the community. A typical command table of organization has five main sections: Command, Operations, Planning, Logistics, and Finance/Administration (FEMA, 2008). Within this command structure, there is not necessarily a specific place for behavioral health. Following are examples of the multiple positions that behavioral health specialists can assume under this command structure.

> The FEMA model taught to emergency managers has no specific place for behavioral health. There is a Medical Unit in the Logistics section, but this placement reflects the need to provide services to responders, not the impacted community. This Medical Unit is responsible for staff force health protection. Behavioral health is certainly part of the overall effort to ensure a fit workforce but is meant to protect the emotional well-being of the community rather than focusing on the wellness of the unit.
>
> In California's Hospital Incident Command System guidebook (California Emergency Medical Services Authority, 2014), the Behavioral Health Unit Leader works in the Operations section under the Medical Care Branch Director, while Employee Health and Well-Being is situated in Logistics. This placement allows for a logical connection to other medical units whose mission is to provide care to patients.

During a mass casualty when a large-scale DBH response is indicated, the Hospital Incident Command System (HICS) Guidebook calls for a modification of the typical Hospital Incident Management Team structure (California Emergency Medical Services Authority, 2014). In that scenario, the Guidebook recommends that DBH be a Branch with a Branch Director reporting to the Operations Section Chief.

In some cases, a DBH specialist could be located in the Planning section as a technical specialist. For terror incidents or incidents involving community panic, fear, hysteria or anger, it may be appropriate to have a DBH responder at the Emergency Operations Center (EOC) to advise EM leadership on typical responses, risk communication, or management of community support. As mentioned in the HICS Guidebook, behavioral health specialists should be called upon to assist in planning for support to patients and their families, as well as staff and their families.

> *DBH Myth 1*: The Office of Mental Health for the jurisdiction can and will handle any behavioral health issues that may arise due to the disaster. Any mental health practitioner is qualified to practice DBH.

It is important here to provide some context for how disaster response is structured nationally. There is an overarching structure, The NRF is established by the Department of Homeland Security DHS (http://www.fema.gov/national-response-framework). Within that document, there are specialized functions identified as Emergency Support Functions (ESFs) in which the scope of responsibility as well as governmental entities responsible for leadership are identified (DHS, 2008, https://www.fema.gov/pdf/emergency/nrf/nrf-overview.pdf). It should be noted that disaster preparedness in the United

States is based on what is known as an *All Hazards* model meaning that preparedness and response occurs based on identifying core principles and procedures common to all types of disaster events and then response becomes customized through selective applications of relevant ESFs. Typically, states use this framework as the architecture upon which to structure their own response planning and strategies.

However, a challenge arises if the planners are focusing on the areas of responsibility known as ESFs as the organizing principles when incorporating DBH into emergency planning. ESF 6 describes the mass care response during a disaster whereas ESF 8 is concerned with the medical response. Hospitals generally request resources under ESF 8 in the Operations section, including behavioral health resources. However, for some response agencies that often provide disaster mental health resources, such as the American Red Cross, the agency liaison will be focused on ESF 6 services and DBH may be overlooked as an asset by the agencies responsible for the medical response.

UNDERSTANDING THE DIFFERENCE BETWEEN TYPICAL COMMUNITY MENTAL HEALTH CARE AND DBH SERVICES

When planning for EM, often state or local mental health agencies are included in planning and response. Emergency managers often believe that this has covered the issue of DBH. However, the differences in mission and focus between the behavioral health as it is practiced day-to-day and how it is addressed in disasters have implications for EM's ability to protect and serve the community from the psychological effects of disaster.

DBH responders need to have specific knowledge and skill sets that are not generally found in mental health practitioners in state or community mental health systems (King, Burkle, Walsh, & North, 2015). In the All-Hazards Planning Guidance distributed by the Substance Abuse and Mental Health Services Administration (SAMHSA) in 2003, the Agency recommended that all personnel in a disaster response be trained in appropriate DBH interventions, typical disaster reactions, and effective interventions (U.S. Department of Health and Human Services, 2003).

One difference between DBH and community mental health is the population being served. The community mental health system is designed to diagnose, treat, and monitor community members with mental illness. This most frequently includes those with serious and persistent mental illness or substance abuse disorders. While those populations are at increased risk for impairment due to a disaster, DBH providers serve the whole community. Disasters affect everyone in a community: from children to adults, from professionals to workers in the service industry, and includes all ethnic and socioeconomic groups. People who are not in "the system" need care, and some even need high levels of care. Individuals and families who were not receiving services previous to a disaster often resist accessing behavioral health services that are offered by the health and social services systems after the event. Lebowitz (2015) stresses the need for planning and collaboration among community mental health agencies but does not mention collaboration or communication with EM. The two systems of community service are not well-coordinated.

In some states and municipalities, the community mental health system is composed of contracted agencies instead of the traditional public or nonprofit entities. In these cases, representatives of the contract agencies are less likely to be at the planning tables and generally will not have providers who have been specially trained in DBH. Because of contract content and obligations that do not address services in disaster situations, these agencies may be even less likely to consider making practitioners available in a disaster than agencies that are government entities and directly control their personnel. There is an increasing need for contracts to clearly address expectations and understandings in times of emergency and disaster such as training, participation in preparedness activities, and deploying staff. If these issues are not addressed, especially the fiscal implications, there is increased likelihood of delayed, misdirected, and compromised response.

In addition, community mental health systems are frequently functioning at capacity or beyond before the disaster and have little, if any, room to expand or redirect their level of service after a disaster. EM needs to plan for and mitigate the effects of disaster on the community mental health system. Some examples include: natural disasters may impede routine service delivery, the workers at community mental health centers may be affected by the disaster themselves, or mass casualties may overwhelm current systems of care. In a system that is typically overworked and overloaded on a good day, a disaster can act as a breaking point that will disrupt the normal provision of care and impact the whole community (Lebowitz, 2015). The inclusion of community mental health agency representatives at the EOC allows for the thoughtful and effective planning needed to minimize disruption to vital services (see chapter: Integration in the Emergency Operations Center (EOC) for more discussion of EOCs). This inclusion may help fill a necessary community need, but does not adequately fulfill the need for DBH services for the entire community.

On the other hand, DBH is the provision of support and care to the whole community to mitigate the psychological effects of a disaster in both the short term and the long term. When a disaster hits, members of the community that have been functioning adequately will find their coping mechanisms challenged. Even such simple stressors, such as school closings due to weather, can place emotional burdens on a family that a DBH intervention may alleviate. Planning for the types of behavioral and psychological responses that can be expected in a community after a disaster is a very different task than mitigating the impact of disaster on the community mental health system. When only one type of system is considered during planning and response, the community suffers as a result. Consideration of the collaboration among agencies, such as fire departments, is especially important for disaster response with vulnerable populations, such as school-aged children (Bergstrand, 2008; California Emergency Medical Services Authority, 2014).

> DBH Myth 2: The magnitude of a disaster is measured solely by number of homes damaged or destroyed. If there has been minimal property damage, there is very little need for DBH.

CHASING DAMAGE

Disaster response agencies, such as FEMA and the American Red Cross, estimate the amount of assistance that a community will need based on the amount of damage that has been done to a community. Usually that estimate is determined by assessing the number of homes and other structures that are damaged or destroyed. EM also takes into account damage to infrastructure, such as utility service, roads, hospitals, and schools. The difficulty for DBH is that the level of structural damage does not correlate directly with the psychological impact of an event.

For example, an event that causes widespread property damage and destruction, such as a slow-rising and long-lived flood, may have a low to moderate psychological impact. If the residents know in advance of the threat, have time to prepare to collect their most precious belongings, and evacuate well before the flood reaches their homes, there will still be a sense of loss for the houses and the community but the event will not be perceived by many as a trauma. Response agencies will classify the disaster as a Major Disaster (a higher level, or Level 4–7 in American Red Cross terms) which then leads to resource allocation based on that assessment. However, from a behavioral health standpoint, the likelihood of long-term psychological effects is low, especially if the response is handled competently, and the need for DBH is minimal.

In other types of events where property damage may not be great, the traumatic effect of the event may be pronounced. For example, in the spring of 2011, several southern and Midwestern states experienced a very active and deadly tornado season. Tornadoes are typically fickle, shifting from one area to another during their path, with very little warning. Because of the recurring incidence of tornadoes, residents experienced the fear of being caught in the tornado's path repeatedly. Even when the tornado missed a block of homes, the residents who had been hiding in their bathrooms or basements hearing the winds and the debris being blown about only a block or two away were traumatized. From an EM resource allocation perspective, the residents in those homes that the tornado missed had very little need. However, from a behavioral health perspective, the risk of psychological consequences of the disaster is high, based on type of exposure and previous experiences. In order to minimize that risk, DBH assets should be allocated to that area to provide immediate support and assess needs for long-term services.

Emergency managers are unlikely to be aware of, or to appreciate the behavioral health implications of, experiencing "near misses" and of multiple disasters in a short period of time. If the goal is to prevent or mitigate the impact of property destruction and damage, an emergency manager may choose to ignore those areas where damage has not occurred and to allocate resources, energy, and attention to only areas with significant damage. In order to assess and respond to both types of needs, it is crucial for a DBH specialist to be in the EOC from the beginning of the disaster threat.

Public health emergencies, such as the Flint water crisis, may have long-term psychological impacts that may not be obvious to emergency managers. Without having a behavioral health expert at the planning table, the community may not have the opportunity to mitigate and address the emotional impact of those events. As another example, a small-scale active shooter event or terrorist event may not activate a "typical" EM response, but could have long-term psychological implications.

> *DBH Myth 3*: DBH is only needed when there have been fatalities or acts of terror. In those cases, EM should call on the "trauma experts" for help.

MASS CASUALTY EVENTS

The one type of event in which most emergency managers will readily recognize the need for DBH is a mass casualty event. Especially if the casualties are due to intentional acts by humans, the level of trauma that is likely to be experienced by the families of the casualties, the survivors, the bystanders, and the responders is generally acknowledged as being potentially quite severe.

Many behavioral health practitioners consider themselves "trauma experts" and may come to the forefront in such a situation. While their intentions are generally honorable and emergency managers have the community's best interests in mind when consulting with them, often the "experts" do not understand the nature and structure of an active response situation. Psychologists, social workers, and others who have been trained to treat post-traumatic stress disorder (PTSD) have generally been taught to administer a program consisting of hour-long sessions in a controlled environment, like at a therapist's office, which can span a period of weeks to months.

Working with trauma victims at the time of the disaster is quite different than a trauma intervention in the months after an event. The appropriate response at the time relies on a strengths-based model which encourages the survivor or victim to access their support system and their preexisting coping skills and strategies. Reliving or retelling the details of the event and their reactions to it is not helpful for prevention of long-term psychological consequences and can actually be harmful to the survivor. This approach is in stark contrast to generally accepted best practices for PTSD treatment—many of which include repeated exposure to the trauma.

Emergency managers may not understand or appreciate the need for specially trained DBH practitioners to provide services during a response to a mass casualty event (King et al. 2015). Without this awareness, they may choose to take advantages of services that are offered by well-meaning behavioral health providers from the community. Not only might those services be ineffective, they may also be harmful by re-traumatizing individuals who need comfort, support, and safety, rather than consolidation of memories of the traumatic or life-threatening event.

The multiagency nature of the response to a mass casualty may create unforeseen difficulties for emergency managers in the coordination of the DBH response. Behavioral health practitioners from the community may not be accustomed to working with law enforcement agencies or in a highly structured incident command scene. Emergency managers need to clearly communicate to the DBH responders the hierarchy of reporting and command, limits to access to survivors, and confidentiality challenges present in a mass casualty response. Behavioral health practitioners need to respect and acknowledge these conditions. Without mutual understanding, behavioral health practitioners may overstep their authority or create confusion and conflict within a response effort. All parties are likely to become frustrated to the detriment of service provision to survivors and their families.

> *DBH Myth 4*: DBH is only needed during the response phase, or it is only needed during the recovery phase.

PREPARING FOR THE DBH EFFECTS OF DISASTER

There are many opportunities to strengthen a community's resilience and prepare for the DBH effects of a disaster—just as emergency managers convene, plan, exercise, and drill with first responders to prepare for rapid, effective response. Emergency managers should similarly work with DBH experts to build community capacity to anticipate and respond to likely community reactions that often occur in the aftermath of disasters.

A survey was conducted with Kansas mental health agencies to assess the degree of disaster preparedness found that respondents felt that the state was not prepared for the mental health consequences of disasters (Hawley et al., 2007). However, more surprisingly, the study authors did not question emergency managers about their knowledge of mental health resources or their plans for including mental health in recovery efforts.

There is a current emphasis on identifying and addressing the disaster-related needs of various populations, such as those with disabilities, functional and access needs, mental illness, or substance abuse (Bergstrand, 2008; Institute of Medicine, 2015). These initiatives need to include vulnerable populations such as the elderly and children as well. Specific recommendations regarding the need to prepare for meeting the needs of children have been developed by the National Advisory Committee on Children in Disaster—http://www.phe.gov/Preparedness/legal/boards/naccd/Documents/healthcare-prep-wg-20151311.pdf.

DBH IN THE RECOVERY PHASE

The line between response and recovery is often quite fine and crossed very early in the disaster response. For DBH, an effective response leads immediately into recovery from the traumatic effects of the disaster and optimally reduces the need for long-term psychological care. When attention is paid to the likelihood of long-term consequences of disaster, individuals who have been most severely exposed should receive priority of care.

DBH responders should remain on the scene of a disaster even after the immediate response has been concluded. For example, in October 2015, there was a building in Brooklyn that exploded, killing two and injuring more than a dozen residents (Crook, 2015). The New York Fire Department responded quickly, and the Greater New York chapter of the American Red Cross arrived on the scene to provide mobile canteen support (hot and cold beverages and snacks), disaster mental health, and disaster spiritual care DSC. The DMH and DSC responders remained at the scene for 50 h while the firefighters continued to look through the rubble for another victim (Ryan, 2015). All other emergency responders had left, but because of the nature of the incident, both spiritual and emotional support was still needed in order to assist the survivors in their efforts to begin their recovery.

> *DBH Myth 5:* The only function of DBH practitioners is to recommend self-care or suggest that the responder is not fit to handle the situation and should leave. For responders who want to continue working on the disaster, it is dangerous to talk to a DBH specialist.

In addition, DBH experts in the EOC can provide support to the emergency manager and the other leadership in the EOC. During the long days (and often nights as well), a colleague who is charged with monitoring stress levels while being active in the response can intervene to prevent and mediate conflict, offer another perspective, provide distraction from the high-stress environment, or simply provide comfort. If the DBH expert is in the EOC from the beginning of the response, trust is developed that allows supportive interactions to occur in a natural, spontaneous manner without threat to the emergency manager's sense of integrity and competence.

The DBH expert in the EOC can also prevent a slowdown or poor decision-making due to an emotional crisis by offering guidance on managing the environment and the response. When needed, a DBH staff member will provide much-needed support to other EM personnel in order to allow the continuation of vital work. The intention is to minimize a need to release anyone from duty throughout the response.

WORKFORCE BEHAVIORAL HEALTH PROTECTION

A DBH expert in the EOC can also provide consultation to emergency managers on stress management and strategies to avoid conditions that are likely to lead to burnout, compassion fatigue, or vicarious traumatization. Attention needs to be paid to the effects of structural, organizational, and event-related stressors on all responders. A DBH expert can advise emergency managers on the conditions that could be improved to reduce stress on the responder while still providing needed services to the community.

RECOMMENDATIONS FOR THE FUTURE

There is an urgent need for understanding and shared knowledge as well as shared experiences between EM and DBH specialists. The community will benefit from dispelling the myths described above and facilitating a dialogue between the two professions. Rather than being seen as a hindrance and drag on response efforts, DBH efforts should be recognized as facilitating and improving the emergency response for the whole community (Amaratunga, 2007).

In May 2015, New York State convened emergency managers and disaster mental health agencies and subject matter experts from around the state to develop recommendations for moving forward to prepare the state for any type of disaster that may arise. Among their recommendations were clear mandates to both emergency managers and DBH representatives to communicate more effectively and comprehensively with each other, train and drill together, and include DBH at EOCs when disaster strikes (Hawley et al., 2007). The value of these recommendations is not limited to the state of New York. There is broad applicability in all cases where the goal of effectively integrating DBH and EM is being pursued (Fig. 1.1).

Love this. After reading, I was visualizing a nice txt box here labels something like "Guide to Using DBH Experts". Might include very practical advice like identifying them beforehand, having them do exercises with you to evaluate their utility, have range of specialist experts identified (e.g., child, elderly, medical related, etc.), nature/duration of availability, etc You could come up with such a list I know.

Again, you call to add or not but, to me, this just cried out for a nice table/graphic or something to really capture the reader.

GUIDE TO WORKING WITH DISASTER BEHAVIORAL HEALTH EXPERTS

Pre-planning

- Create a list of *disaster behavioral health* experts in the area
  - American Red Cross
  - Medical Reserve Corps (may have a behavioral health expert on team)
  - Corporate support companies
    - KonTerra
    - Crisis Care Network
    - Others
  - State Disaster Behavioral Health team
- Identify *disaster behavioral health* experts in special topic areas
  - Children
  - Elderly
  - People with functional and access needs, including disabilities
  - Ethnic/cultural concerns
- Invite them to drills and exercises
  - Include them in planning the drills and exercises
  - Add injects related to behavioral health
  - Identify behavioral health impacts to unique populations
- Consultation related to special events, e.g., anniversaries of tragedies or community commemorations

During an event

- Include DBH expert at the EOC and on leadership teams
- Bring in specialists for under-represented community groups
- Utilize DBH to plan for staff care
- Consult with DBH on mitigation of long-term psychological impacts

After an event

- Work with DBH to plan for anniversaries and memorials
- Use DBH to develop community resilience-building strategies
- Include DBH in after-action reviews and evaluations

FIGURE 1.1 Guide to working with DBH experts.

REFERENCES

Amaratunga, C. (2007). Mental health emergency preparedness. *Prehospital and Disaster Medicine*, 22(3), 205–206. Available from http://dx.doi.org/10.1017/S1049023X00004660.

Bergstrand, G. (2008). Working together on emergency preparedness. *Minnesota Fire Chief*, 64–65.

California Emergency Medical Services Authority. (2014). *Hospital incident command system guidebook* (5th ed.). Rancho Cordova, CA: EMSA.

Crook, L. (December 5, 2015). Fatal blast thought to be gas was arson, FDNY says. Retrieved from <http://www.cnn.com/2015/12/05/us/new-york-brooklyn-explosion/>. Accessed 08.06.16.

FEMA. (2008). *ICS review material*. <https://training.fema.gov/emiweb/is/icsresource/assets/reviewmaterials.pdf>.

Hawley, S. R., Hawley, G. C., Ablah, E., Romain, T. S., Molgaard, C. A., & Orr, S. A. (2007). Mental health emergency preparedness: The need for training and coordination at the state level. *Prehospital and Disaster Medicine, 22*(3), 199–204.

Institute of Medicine (2015). *Healthy, resilient, and sustainable communities after disasters: Strategies, opportunities, and planning for recovery*. Washington, DC: National Academies Press.

King, R. V., Burkle, F. M., Jr, Walsh, L. E., & North, C. S. (2015). Competencies for disaster mental health. *Current Psychiatry Reports, 17*(3), 1–9.

Lebowitz, A. J. (2015). Community collaboration as a disaster mental health competency: A systematic literature review. *Community Mental Health Journal, 51*(2), 125–131.

New York State. (2015). New York State disaster mental health summit: Review and results from meeting held on May 29, 2015.

Ryan, D. (September 19, 2015). Government liaison, community partnerships and disaster mental health, Greater New York Chapter, American Red Cross (V. Cole, Interviewer).

U.S. Department of Health and Human Services. (2003). *Mental health all-hazards disaster planning guidance*. Washington, DC: HHS.

Making Integration Work

Nancy Dragani

Although some progress has been made in terms of integrating DBH into the EM system, much work remains to be done. The first hurdle to overcome is differences in culture between the two. EM is a child of civil defense—a military-based civilian operation. Many of the senior leaders in EM continue to be drawn from the military or the first responder community. Within those groups, there is often an unspoken bias against seeking out DBH or mental health practitioners. The attitude too often continues to be, "if you can't stand the heat...get out of the kitchen." This personally held belief can lead some emergency managers to turn a blind "policy" eye towards involving any type of mental health support, especially DBH, in their planning, response or recovery efforts. Recognizing that DBH is a component of community healing is the first step in making integration work.

> Behavioral health practitioners should automatically be inserted into the Public Health and Medical core capability subset, or even better, the core capabilities should be expanded from 32 to 33 with a specific capability focused on DBH. This action would formalize the inclusion of DBH in the EM structure.

DBH must be at the table for all phases of EM—from prevention to mitigation, preparedness to response. All hazard planning already incorporates the whole community into the planning process, but a specific call for behavioral health professionals to participate in every phase will ensure they are part of the dialogue. Any discussion of a resilient population cannot occur without recognizing the part individuals play in determining their role—victim or survivor.

Chapter 2

Where Emergency Management and Disaster Behavioral Health Meet: Through a Disaster Behavioral Health Lens

Rachel E. Kaul[1,*] and Ronald Sherman[2]
[1]U.S. Department of Health and Human Services, Washington, DC, United States,
[2]Independent Consultant, FEMA Federal Coordinating Officer (Retired), United States

A Disaster Behavioral Health Perspective

Rachel E. Kaul

This comprehensive review of the intersection between emergency management (EM) and disaster behavioral health (DBH) uses a historical perspective, as well as first hand field experiences, to identify challenges, successes, and opportunities for future growth within disaster response coordination.

INTRODUCTION

Disaster experts and practitioners have emphasized the need for comprehensive integration of behavioral health into emergency preparedness, response, and recovery in articles, books, and policy documents (Pfefferbaum, Schonfeld, et al., 2012; Reissman, Reissman, & Flynn, 2007; U.S. Department of Health and Human Services, 2003). In the aftermath of broad scale or particularly traumatic events such as 9/11, Hurricane Katrina, and the Sandy Hook Elementary School Shooting, planners, responders, and health professionals began to

---

*Ms. Kaul is a Senior Policy Analyst with the US Department of Health and Human Services' Office of Policy and Planning, but this chapter reflects her personal opinions and do not necessarily represent the views of the Department of Health and Human Services or the United States.

recognize the important role behavioral health plays in overall health and in disaster response and recovery. However, consistent behavioral health inclusion into broader EM operations and effective collaboration between the two fields has remained challenging and inconsistently achieved at state and federal levels (National Biodefense Science Board, 2010). There are deeply ingrained differences in professional culture between the two. A lack of a systemized approach to DBH makes psychological support strategies a difficult fit into an EM paradigm. The limited attention emergency planners are able to pay to the inclusion of behavioral health into preparedness activities limits coordinated integration of behavioral health during response and recovery phases.

Personal Experiences From the Field:

A seasoned emergency management professional recently commented to me that, although we had worked for the same agency for a number of years and had a positive professional relationship, he did not really understand what I do for a living. I am currently a disaster behavioral health subject matter expert and policy analyst for a federal response agency. Before this, I functioned as one of the original state disaster mental health coordinators hired through a Substance Abuse and Mental Health Services Administration (SAMHSA) grant established to encourage behavioral health all hazards planning (an approach that incorporates planning for many types of events in a single plan with annexes) across the country. Even after years of effort to establish effective collaboration between those in the field of emergency management and those in the disaster behavioral health community, I realized my colleague's question points to an ongoing disconnect and lack of understanding of roles and function between the two arenas.

Rachel Kaul, Disaster Behavioral Health Professional.

Many emergency managers report they rarely interact with mental health and substance abuse practitioners, and that they feel uninformed about behavioral health practice. In addition, most policy guidance and operating procedures make little mention of behavioral health and, even if they include it, offer no concrete direction on how integrations should occur. As a result, emergency managers may be unsure about how to identify the need for behavioral health personnel, uncertain about the appropriate time to engage them, and in possession of little knowledge about what exactly they do once they are involved in response or recovery activities. The DBH community continues to struggle to find effective methods of encouraging participation in and consistent inclusion into preparedness activities on the part of funders and from EM. This is further complicated by a lack of standards in terms of practice in the DBH field and a limited evidence base to support assessment and intervention approaches (Andrew & Kendra, 2012; North & Pfefferbaum, 2013).

Planners at the state and federal levels are not in the habit of routinely incorporating behavioral health professionals into medical response teams, often considering behavioral health concerns an aspect of response to be addressed after the initial life-saving and medical response activities have begun to wind down. Opportunities to connect people with much needed information, emotional support and assessment, or resources early—which literature suggests is valuable (Ruzek, Young, Cordova, & Flynn, 2004) — are missed in this approach. One contrast to this occurred during the Super Storm Sandy response, in which federal health and medical response teams deployed with an embedded mental health capability along with medical assets. The response was further augmented by mental health teams deployed through the US Public Health Service. Anywhere medical services were available; there was also a behavioral health capacity. Both responders and leadership highlighted this aspect of the operation as very successful during unpublished after action reviews. The presence of professionals to assist with psychological triage, referral, and resource provision enhanced the ability of the medical staff to swiftly meet the needs of those with significant health issues. In addition, the mental health professionals provided staff support for all the responders and enhanced their stress management and overall functioning. These outcomes validate the need for health, behavioral health, and emergency response workers to train together, work together, and develop relationships so that integrated response is possible.

Challenges to interdisciplinary collaboration prior to events continue to exist. It is often up to DBH professionals to devise ways to be part of the emergency response community, such as leveraging personal relationships to gain access to preparedness projects.

> Comments From the Field:
> *I would often show up at planning meetings or exercises without a formal invitation or role. I would call this 'crashing the party'*
> **Rachel Kaul, Disaster Behavioral Health Professional.**

Though there may be informal mechanisms for getting to the table, most emergency planners would agree that early and intentional inclusion in the process for every necessary capability of response is preferred. In order for behavioral health to be seen as an essential function, it must be included throughout the emergency cycle as a matter of course and not just because of certain relationships or individual planner preferences.

This chapter will explore ways to achieve behavioral health and EM integration and strategies to address challenges. It will argue that, for real progress to occur, emergency managers must better understand what behavioral health clinicians do for a living so they can appreciate what this brings to health and medical preparedness, response, and recovery. To make this happen, DBH practitioners need to function within the EM context and communicate using like terms and concepts. Approaches to consider for enhanced collaboration and effective public health and medical response will be discussed.

THE FORMATION OF THE FIELD

The events of 9/11 created an undisputed need for large-scale psychological support across the country (Kaul & Welzant, 2005; Ozbay, Auf der Heyde, Reissman, & Sharma, 2013). In response, increased efforts to adopt an EM framework in relation to meeting the behavioral health needs of disaster survivors and responders began to emerge (Uhernik, 2008). One of the most significant developments occurred in 2003 when the Substance Abuse and Mental Health Services Administration, (n.d.) provided a funding opportunity (mentioned previously) accompanied by a planning document (U.S. Department of Health and Human Services, 2003) for states to develop an all-hazards DBH plan. Most states applied and used the funding to hire someone to write this plan to enhance their ability to deliver a behavioral health response. Though unintended, a valuable consequence of this initiative was the creation of a community of interest of DBH planners and coordinators who worked within their states to elevate the inclusion of behavioral health into emergency planning and response.

The new role created through the grant required behavioral health professionals to engage with nontraditional partners, such as emergency managers and public health professionals. Additionally, DBH coordinators provided stress management and crisis counseling training to emergency partners and tried to help other behavioral health colleagues understand the consequences of trauma and disaster for their populations and for the public at large. They wrote plans that sought to describe how, ideally, the DBH function fit into emergency planning, response, and recovery.

THE IDEAL OF INTEGRATION

Efforts toward effective integration described in most disaster plans created during that time recommended incorporating activities specific to behavioral health during all phases of response and broad collaboration with stakeholders. Even today, most state plans emphasize that behavioral health practitioners participate on emergency planning committees at the state and local levels and contribute to exercise development and play. Many of these plans describe how to create a response capacity by forming voluntary or funded teams to be available to provide crisis services during emergency events. Creating referral and resource linkages through coalition building or memorandums of understanding to the existing behavioral health and social service system are essential to support recovery. Fig. 2.1 provides an overview of potential DBH activities by phase of disaster that, if consistently utilized, can enhance any overall response.

Even after states created their all-hazards plans, challenges to implementation persisted. After the Substance Abuse and Mental Health Services Administration funding expired, many of the DBH coordinators and planners

Disaster behavioral health action timeline

Preparedness
- Engage in emergency managment and DBH training
- Identify stakeholder with whom to partner (e.g. human sevice providers, tribes, public health, emergency managment, law enforcement, healthcare providers)
- Provide training to stakeholders on DBH
- Establish crisis response capability by forming and training teams
- Establish and convene or participate in healthcare and behavioral healthcare coalitions
- Work with HPP* and PHEP** to insure inclusion of behavioral health in preparedness grant activities
- Provide relevant injects for excercises and participate
- Develop resoure lists and service provider lists for referrals

Response
- Staff emergency operations center
- Provide just-in-time training on DBH and on responder self care
- Activate DBH team members
- Develop clear mission assignment related to DBH based on assessment of need
- Provide DBH services in shelters, healthcare facilities and in the community as needed; evaluate services
- Identify or develop resources targeted for specific population; ensure language access and cultural competence
- Refer to existing service providers
- identify and apply for funding to support beahavioral health needs of community and affected populations

Recovery
- Implement grant funded services
- Develop and provide recovery specific training for stakeholders and for affected populations
- Engage in recovery mission assignments, coordination, technical assistance, and leveraging of steady-state assets/programs and services
- Build referral networks for behavioral health and human services
- Identify and document lessons learned
- Evaluate services throughout

FIGURE 2.1 Disaster behavioral health action timeline. This figure illustrates behavioral health action steps that can enhance overall disaster response.

struggled to maintained their positions or garner support for activities. However, some were successful in not only maintaining what had been established but were also able to expand their position with additional staff by leveraging other preparedness funding available through public health, mental health authorities, and EM agencies. The state coordinators continued to rely on their connections with each other by interacting at meetings, conferences, and training opportunities. There was general agreement that it was important to continue to learn from one another and strategize ways to further improve the preparedness and response capabilities of behavioral health in relation to emergencies, even as discrete funding for this languished. New driving factors had to be identified that could be leveraged to enhance understanding of the role behavioral health plays for individuals and communities in all phases of disasters. Among these drivers were evolving health policy documents and guidance for overall preparedness and healthcare system surge plans.

EMERGING FORCES DRIVING POLICY

It is important to consider the role policy plays in establishing practice standards and procedures. For emergency managers, policy guidance provides a roadmap to enhancing preparedness and response and establishes national priorities to promote at state and local levels. This encourages consistency and efficacy of disaster response. For DBH professionals, policy provides the legitimacy of their inclusion in nontraditional areas, such as emergency preparedness and disaster response and recovery. Policy opens doors to integration.

The primary plans that guide national and state emergency related activities are accomplished through interagency workgroup processes that include a wide array of federal partners, such as the Department of Homeland Security, various elements of US Department of Health and Human Services (HHS), and the Department of Transportation. Input from all areas of the federal government is essential. Vastly diverse agency priorities often compete for inclusion. Significant preparedness documents, such as the 2008 National Response Framework, mention mental or behavioral health as part of the overall public health and medical response but say very little about defined capabilities and activities (National Response Framework, 2008). However, following Hurricane Katrina, many in the DBH field realized that inclusion in national policy initiatives would be an important strategy to elevate behavioral health's place in disaster planning and response. Essential policy directives such as the 2006 Post-Katrina Emergency Management Reform Act (PKEMRA) (U.S. Department of Homeland Security, n.d.), the 2007 Homeland Security Presidential Directive (HSPD-21) (White House, 2007), the National Disaster Recovery Framework (2011), and the National Health Security Strategy's Implementation Plan (NHSS-IP, 2015–18) (Public Health Emergency, n.d.) were developed in response to the gaps and shortcomings realized during the Katrina response. All emphasize the importance of mitigating the mental health consequences of disasters to facilitate effective response but give no measures or examples on how this is to be done.

For emergency managers tasked with implementing preparedness and response policy, the lack of practical strategies and defined activities hindered efforts to accomplishing more broadly understood mitigation of health and behavioral health concerns. In addition, although behavioral health is defined as an integral part of Emergency Support Function #8 as part of Public Health and Medical Services, this capability is not routinely activated or included in medical response. It is often up to emergency managers to determine whether or not to request or include this capability in response.

For readers unfamiliar with the emergency support function (ESF) structure, these are a set of 15 standardized special activities frequently used to provide federal support in disasters (Federal Emergency Management Agency, 2008).

To address the ongoing challenge of defining the role of behavioral health in disasters and emergences, HSPD-21 called for the establishment of a Federal Advisory Committee for Disaster Mental Health. Established under the National Biodefense Science Board (NBSB), the Disaster Mental Health Subcommittee's recommendations included strategies to improve integration with EM and led to the development of the US Department of Health and Human Services Disaster Behavioral Health Concept of Operations (CONOPS) (Pfefferbaum, Flynn, et al., 2012). This CONOPS were the first national document to describe the conceptual framework and coordination for federal-level behavioral health preparedness, response, and recovery for disasters and public health emergencies. In the service of harmonizing a variety of federal efforts, its language is intentionally consistent with the National Preparedness Goal (NPG), the National Response Framework (NRF), and the NDRF. It also supports the goals and objectives of the National Health Security Strategy (U.S. Department of Health and Human Services, 2014).

The improvement in the understanding of the need to consider behavioral health has begun to emerge in other policy guides as well. An important example as a force for broader emergency preparedness standards is the HHS Centers for Disease Control (CDC) and Prevention's *Public Health Preparedness Capabilities* (2011). Released in 2011, it provides a fairly recent example of a critical planning document that describes the capabilities required to set national standards for state and local planning and inform cooperative agreements as well as grant activities. It emphasizes community resilience as one of five cross-cutting domains in the capabilities and integrates behavioral health capabilities as important to this throughout the document (Center for Disease Control (CDC), 2011). Resilience as a focus for emergency preparedness and response activities will be further discussed in this chapter.

Even with the inclusion of behavioral health in these EM policy initiatives over time, there is still a lack of concrete benchmarks and measures related to mental and behavioral health in the actual funding programs directed toward health preparedness. It seems that measures tied to funding are necessary to provide sustainable opportunities for behavioral health to be included in emergency preparedness activities. The HHS Assistant Secretary for Preparedness and Response's (ASPR) Hospital Preparedness Program (Public Health Emergency, 2016) had, at one time, emphasized mental health training and services as important elements to include in overall hospital preparedness effort. Due to an increasing number of mandated priorities linked to this funding, this specific benchmark was not included in the 2008 revision of the grant guidance. Many state DBH coordinators report this resulted in a decrease in their access to training inclusion in exercises and planning events. The CDC's Public Health Emergency Preparedness Program updated its guidance to awardees in 2015 and specifically requires coordination with behavioral health, a significant change from previous years. However, concrete indicators or strategies are left up to the awardees and there is little evidence that widespread effective coordination is occurring across the country (Center for Disease Control (CDC), 2015).

THE FOCUS ON RESILIENCE

In recent years there has been an increasing emphasis on resilience as a focus for preparedness, response, and recovery initiatives (Wulff, Donato, & Lurie, 2015). From a health and medical perspective, resilience relates to the behaviors people can learn or develop to withstand, adapt to, and recover from stress and adversity (Chandra, Acosta, Stern, Uscher-Pines, & Williams, 2011). Resilience is grounded in strong behavioral health core elements such as coping skills and social support. The growing interest in resilience has led to a greater general appreciation for the importance of incorporating psychological and emotional well-being into broader goals related to resilience building.

This is not a new concept. Behavioral health has been considered a critical element to overall human health, adaptability, and coping with adversity for decades (Plough et al., 2013; U.S. Department of Health and Human Services, 1999). As previously discussed, over the past 10 years, there has been a push to incorporate mental and behavioral health into the EM context which has led to efforts at the state and local levels to develop all-hazards DBH plans, enhance cross-disciplinary training, and increase exercise participation (Hawley et al., 2007; Reissman et al., 2007; Robertson, Pfefferbaum, Codispoti, & Montgomery, 2007). Federal emergency and disaster partners have worked to incorporate language pertaining to behavioral health into preparedness guidance and plans. Despite many promising steps, the recent Institute of Medicine (IOM) report, "Healthy Resilient, and Sustainable Communities after Disasters," emphasizes evidence of an ongoing lack of behavioral health policy integration into other disaster-related foci (Institute of Medicine, 2015). It points to a need for a national policy on DBH, beyond what now exists, to achieve successful and consistent collaboration between the EM community and the behavioral health professionals who engage in disaster and emergency response and recovery (Institute of Medicine, 2015). However, realities in the current funding environment make such policy development and implementation an uphill battle. There is an ongoing tension between current behavioral health reimbursement practices that rely on diagnosis and treatment and the emerging interest in health promotion and prevention as a priority in the health and behavioral healthcare sectors.

Preparedness funding designed to target and address health and medical related concerns began to decline in the years following Hurricane Katrina and continues today (Bevington, 2014; Schnirring, 2013; Weems, 2010). In addition, funding for mental health services has seen deep declines in almost every state (National Alliance on Mental Illness, 2011). The consistency and degree to which states have been able to support DBH preparedness has suffered. Recent assessments, such as a formal evaluation conducted by the Council of State and Territorial Epidemiologists in 2013 (Gould, 2014) and an informal survey project conducted by the Division of At Risk Individuals, Behavioral Health, and Community Resilience in 2014, conclude that states are struggling to develop or maintain adequate DBH planning capability. The numbers of original DBH coordinators who remain in the role have declined and newcomers face difficulty connecting with one another across state lines (Moskowitz & Klatt, Personal interview, 2014). As a result, ongoing improvement to integration of behavioral health into public health- and EM-based disaster planning varies greatly from state to state. There are limited numbers of experienced and trained DBH professionals who can rely on lessons learned and established relationships within the disaster community to support their efforts.

These developing limitations in the DBH field should raise serious concerns for emergency managers and leaders. They could result in inadequate health and medical emergency planning that, as evidence suggests, may increase health risks to public health and well-being of vulnerable populations prior to and after disasters (Herrman, 2012; Oldham, 2013; Osofsky, Wells, & Weems, 2014). Ineffectively incorporating behavioral health into health assessments impacts resource allocation and intervention application. This also limits the ability of the public health emergency preparedness and response system (of which behavioral health is an integral part) to gather relevant pre and post disaster data, thus creating an evidence base from which to enhance health and medical response (Pfefferbaum, Flynn, et al., 2012). The ongoing disconnect between the EM community and behavioral health continues to create barriers to effective collaboration, weakening broader preparedness efforts.

CENTRAL CHALLENGES TO ADDRESS

The challenges go beyond funding. Many are rooted in the day-to-day issues affecting collaboration between behavioral health and EM. Differences in culture, language, and mission objectives for those in either profession present major obstacles that need to be addressed (Robertson et al., 2007). Whether at the local, state, or federal level, separate and uncoordinated operational approaches are evident in many responses.

The State of State Behavioral Health Systems

Across the nation, state behavioral health systems are fragmented, under-resourced, and over-taxed with trying to meet the needs of existing consumers of services. This makes introducing and implementing emergency planning and preparedness or expanding services to new clients seeking help following a disaster extremely challenging for those interested in DBH. Many state mental health authorities do not have the capability to engage in preparedness strategies such as continuity of operations planning, capacity assessments, and risk analysis, or in providing training to staff on crisis intervention approaches such as Psychological First Aid (PFA). Stakeholders in the day-to-day systems must understand their defined roles and activities in relation to disasters before an event occurs in order to achieve coordinated engagement during an actual response.

Standards and Plans

A lack of yet agreed-upon standards and practices for DBH, such as those laid out for EM activities in National Incident Management System (NIMS), impedes planning and policy development (Reissman et al., 2007). Popular DBH approaches, such as PFA and crisis counseling, actually have little evidence to support them (North & Pfefferbaum, 2013). In addition, there is a lack of agreement in the disaster mental health community on data collection measures, processes, and use for both service provision and program evaluation. Psychological triage or program evaluation models vary, making clear conclusions about community assessment and intervention efficacy difficult. This is troubling to many in the health and medical response community and makes arguing a case for DBH inclusion in planning and funding problematic. In addition, while overall preparedness funding has increased in relation to physical consequences of disasters, mental and behavioral health preparedness remains unfunded and overlooked (Hawley et al., 2007; North & Pfefferbaum, 2013). Practitioners must focus on developing and instituting programs and approaches that can be evaluated and replicated to demonstrate value for the healthcare and emergency response communities.

Culture

Differences in professional culture are perhaps the largest obstacle to overcome in terms of integration of behavioral health into EM. Research indicates there are specific cultural and personality factors that exist in emergency response professions (Kronenberg et al., 2008; Paoline, 2003). People who gravitate toward these professions tend to form close-knit communities who trust each other more than those in other types of professions.

They frequently work in teams and adhere to documented protocols, procedures, or operational guidance. They rely on extensive training and planning together to excel. It may be the case that changing or adapting to new methods and approaches may require some convincing for people typically in these professions. Historically, there has also been significant stigma within the EM community associated with behavioral health needs (Rutkow, Gable, & Links, 2011). The need to recognize and address emotional impacts of emergency response is often downplayed in planning and response phases within EM. Approaches that rely on peer support strategies rather than on professional mental health intervention have historically been preferred by members of emergency service populations (Brown, 2003; Everly, Eyler, & Flannery, 2002).

Behavioral health practitioners typically display different characteristics in terms of personality and work style than EM and response professionals. The nature of the work requires great ability to relate to feelings and emotions as well as to empathize with distress. Clinicians often work independently and enjoy a great deal of autonomy in deciding practice methods. Confidentiality is a central tenant of the profession which often leads practitioners to keep their own counsel rather than seek input from others. Even as part of interdisciplinary medical care teams, the behavioral health role is usually distinct and often operates with little direction or oversight, unlike what is typically provided within a formal command structure. Most therapeutic approaches are eclectic, and there is not necessarily any single approach to addressing the needs of someone in crisis or exhibiting distress. In fact some research indicates that the success of interventions relies more on the qualities of the person conducting the intervention and less on the specific element of any approach (Wampold, 2001). In other words, there are few standard operating procedures for behavioral health like the ones emergency managers develop to which they can refer. There is not common language or a list of terms that can be used to bridge the differences between the two fields. With little understanding of what behavioral health activities and with issues around stigma contributing to a reluctance to consider the psychological elements of disasters, emergency managers are challenged to prioritize incorporating behavioral health into their day-to-day tasks and response work.

FUTURE DIRECTIONS

Even with promising developments in policy and practice, the disconnections between EM and DBH persist today. In order to continue to make progress, the behavioral health community may benefit by adopting EM strategies. Establishing standard operating procedures and mission assignments that align with other types of medical response capabilities would allow for greater understanding of what behavioral health professionals do during a response. The Health and Human Services Disaster Behavioral Health Concept of Operations promotes the inclusion of a Behavioral Health Liaison Officer (Behavioral Health LNO) in an Incident Command Structure. The Behavioral Health LNO is intended to responsible for ensuring coordination of DBH efforts for the federal response in collaboration with existing state, local, and voluntary organizational efforts. Some states include a behavioral health role in their all-hazards plans that describe their incident command structure. Including behavioral health professionals into structures familiar to EM and codifying their role promotes better understanding of the emergency culture for behavioral health practitioners and creates an appreciation for as well as a validation of behavioral health efforts for emergency managers.

Further emphasis is necessary to stress the lack of preparedness funding specifically aimed at including behavioral health. It is desperately needed if advances in practice and policy are going to be made. As federal grant applications and guidance are updated and revised attention should be given to include measures and examples of how behavioral health is best integrated into preparedness initiatives should be included.

Promoting a broader research agenda that can validate the contribution behavioral health makes to community resilience and to disaster recovery is of high importance. The limited evidence available indicates that early access to behavioral health can help disaster survivors and responders avoid the development of adverse and chronic psychological symptoms (Dieltjens, Moonens, Van Praet, De Buck, & Vandekerckhove, 2014; North & Pfefferbaum, 2013). A lack of rigorous evaluation of the benefits of specific approaches to disaster mental health has made it difficult to make the case for behavioral health integration to emergency managers and leaders in the response community. An evidence base that truly establishes best practices and identifies outcomes will help overcome doubt and the disconnect between the two fields that continues to exist.

In order to achieve true integration between behavioral health and EM, efforts toward greater understanding of culture, priorities, and practice must be prioritized for both disciplines. DBH practitioners could benefit from true immersion in the EM context and environment. Language and mission approach could be better aligned in relation to one another. Practice methods that establish a science base and standards could be developed and tested. Evidence could then be used to establish the need for funding and resources that would enhance DBH preparedness, and thus response and recovery. In this way, the psychological and emotional needs of disaster survivors and responders could be consistently and more effectively addressed, communities could more rapidly recover from adverse events, and be more prepared for future ones.

REFERENCES

Andrew, S. A., & Kendra, J. M. (2012). An adaptive governance approach to disaster-related behavioural health services. *Disasters, 36*(3), 514–532. Available from http://dx.doi.org/10.1111/j.1467-7717.2011.01262.x.

Bevington, F. (2014). *Have public health funding cuts impacted response capabilities?* Retrieved July 12, 2016, from < http://www.emergencymgmt.com/health/Have-Public-Health-Funding-Cuts-Impacted-Response-Capabilities.html >.

Brown, A. (2003). *Finally, a stress program designed for first responders.* Retrieved July 12, 2016, from < http://ehstoday.com/training/ehs_imp_12402 >.

Center for Disease Control. (2011). *Public health preparedness capabilities: National standards for state and local planning.* Retrieved July 12, 2016, from < http://www.cdc.gov/phpr/capabilities/DSLR_capabilities_July.pdf >.

Center for Disease Control. (2015). *PHEP B45 continuation guidance.* Retrieved July 28, 2016, from < https://www.cdc.gov/phpr/documents/hpp-phep-bp4-continuation-guidance-508-v4.pdf >.

Chandra, A., Acosta, J., Stern, S., Uscher-Pines, L., & Williams, M. V. (2011). *Building community resilience to disasters: A way forward to enhance national health security.* Santa Monica, CA: Rand Corporation.

Dieltjens, T., Moonens, I., Van Praet, K., De Buck, E., & Vandekerckhove, P. (2014). A systematic literature search on psychological first aid: Lack of evidence to develop guidelines. *PLoS One, 9*(12), e114714.

Everly, G. S., Eyler, V. A., & Flannery, R. B. (2002). Critical incident stress management: A statistical review of the literature. *Psychiatric Quarterly, 73*(2), 171–182.

Federal Emergency Management Agency. (2008) *Emergency support function annexes: Introduction.* Retrieved July 22, 2016, from < http://www.fema.gov/media-library-data/20130726-1825-25045-0604/emergency_support_function_annexes_introduction_2008_.pdf >.

Gould, D.W. (2014). Disaster mental health surveillance at state health agencies: Results from a 2013 CSTE assessment. *Paper presented at the 2014 CSTE Annual Conference, Nashville,TN.*

Hawley, S. R., Hawley, G. C., Ablah, E., Romain, T. S., Molgaard, C. A., & Orr, S. A. (2007). Mental health emergency preparedness: The need for training and coordination at the state level. *Prehospital and Disaster Medicine, 22*(3), 199–204. Available from http://dx.doi.org/10.1017/S1049023X00004659.

Herrman, H. (2012). Promoting mental health and resilience after a disaster. *Journal of Experimental & Clinical Medicine, 4*(2), 82–87.

Institute of Medicine (2015). *Healthy, resilient, and sustainable communities after disasters: Struggles, opportunities, and planning for recovery.* Washington, DC: The National Academies Press.

Kaul, R. E., & Welzant, V. (2005). Disaster mental health: A discussion of best practices as applied after the Pentagon attack. In A. R. Roberts (Ed.), *Crisis Intervention Handbook* (pp. 200–220). Oxford, NY: Oxford University Press.

Kronenberg, M., Osofsky, H. J., Osofsky, J. D., Many, M., Hardy, M., & Arey, J. (2008). First responder culture: Implications for mental health professionals providing services following a natural disaster. *Psychiatric Annals, 38*(2).

Moskowitz, S. & Klatt, D. (2014). Personal interview.

National Alliance on Mental Illness. (2011). *State mental health cuts: The continuing crisis.*

National Biodefense Science Board. (2010). *Integration of mental and behavioral health in federal disaster preparedness, response, and recovery: Assessment and recommendations.*

National Disaster Recovery Framework. (2011). Washington, DC: U.S. Department of Homeland Security.
National Response Framework. (2008). Washington, DC: U.S. Department of Homeland Security.
North, C. S., & Pfefferbaum, B. (2013). Mental health response to community disasters: A systematic review. *JAMA*, *310*(5), 507−518. Available from http://dx.doi.org/10.1001/jama.2013.107799.
Oldham, R. L. (2013). Mental health aspects of disasters. *Southern Medical Journal*, *106*(1), 115−119. Available from http://dx.doi.org/10.1097/SMJ.0b013e31827cd091Osofsky.
Osofsky, J. D., Wells, J. H., & Weems, C. (2014). Integrated care: Meeting mental health needs after the gulf oil spill. *Psychiatric Services*, *65*(3), 280−283. Available from http://dx.doi.org/10.1176/appi.ps.201300470.
Ozbay, F., Auf der Heyde, T. A. D., Reissman, D., & Sharma, V. (2013). The enduring mental health impact of the September 11th terrorist attacks: Challenges and lessons learned. *Psychiatric Clinics of North America*, *36*(3), 417−429. Available from http://dx.doi.org/10.1016/j.psc.2013.05.011.
Paoline, E. A. (2003). Taking stock: Towards a richer understanding of police culture. *Journal of Criminal Justice*, *31*(3), 199−214. Available from http://dx.doi.org/10.1016/S0047-2352(03)00002-3.
Pfefferbaum, B., Flynn, B. W., Schonfeld, D., Brown, L. M., Jacobs, G. A., Dodgen, D., ... Lindley, D. (2012). The integration of mental and behavioral health into disaster preparedness, response, and recovery. *Disaster Medicine and Public Health Preparedness*, *6*(1), 60−66. Available from http://dx.doi.org/10.1001/dmp.2012.1.
Pfefferbaum, B., Schonfeld, D., Flynn, B. W., Norwood, A. E., Dodgen, D., Kaul, R. E., ... Ruzek, J. I. (2012). The H1N1 crisis: A case study of the integration of mental and behavioral health in public health crises. *Disaster Medicine and Public Health Preparedness*, *6*(1), 67−71. Available from http://dx.doi.org/10.1001/dmp.2012.2.
Plough, A., Fielding, J. E., Chandra, A., Williams, M., Eisenman, D., Wells, K. B., ... Magaña, A. (2013). Building community disaster resilience: Perspectives from a large urban county department of public health. *American Journal of Public Health*, *103*(7), 1190−1197. Available from http://dx.doi.org/10.2105/AJPH.2013.301268.
Public Health Emergency. (n.d.) *National health security strategy and implementation plan*. Retrieved July 22, 2016, from <http://www.phe.gov/Preparedness/planning/authority/nhss/Pages/strategy.aspx>.
Public Health Emergency. (2016). *Hospital preparedness program (HPP)*. Retrieved July 22, 2016, from <http://www.phe.gov/Preparedness/planning/hpp/Pages/default.aspx>.
Reissman, D. B., Reissman, S. G., & Flynn, B. W. (2007). Integrating medical, public health, and mental health assets into a national response strategy. In B. Bongar, L. M. Brown, & L. E. P. G. Zimbardo (Eds.), *Psychology of Terrorism* (pp. 434−451). New York, NY: Oxford University Press.
Robertson, M., Pfefferbaum, B., Codispoti, C. R., & Montgomery, J. M. (2007). Integrating authorities and disciplines into the preparedness-planning process: A study of mental health, public health, and emergency management. *American Journal of Disaster Medicine*, *2*(3), 133−142.
Rutkow, L., Gable, L., & Links, J. M. (2011). Protecting the mental health of first responders: Legal and ethical considerations. *The Journal of Law, Medicine & Ethics*, *39*(1), 56−59. Available from http://dx.doi.org/10.1111/j.1748-720X.2011.00567.x.

Ruzek, J. I., Young, B. H., Cordova, M. J., & Flynn, B. W. (2004). Integration of disaster mental health services with emergency medicine. *Prehospital and Disaster Medicine, 19*(1), 46–53.

Schnirring, L. (2013). *Federal funds for disaster preparedness decline. CIDRAP.* Retrieved July 12, 2016, from <http://www.cidrap.umn.edu/news-perspective/2013/07/federal-funds-disaster-preparedness-decline>.

Substance Abuse and Mental Health Services Administration. (n.d.) *Crisis counseling assistance and training program (CCP).* Retrieved October 30, 2015, from <http://www.samhsa.gov/dtac/ccp>.

Uhernik, J. A. (2008). The counselor and the disaster response team: An emerging role. In G. R. Walz, J. C. Bleuer, & R. K. Yep (Eds.), *Compelling Counseling Interventions: Celebrating VISTAS' Fifth Anniversary* (pp. 313–321). Ann Arbor, MI: Counseling Outfitters.

U.S. Department of Health and Human Services (1999). *Mental health: A report of the Surgeon General.* Rockville, MD: HHS.

U.S. Department of Health and Human Services. (2003). *Mental health all-hazards disaster planning guidance.* See <http://mentalhealth.samhsa.gov/publications/allpubs/SMA03-3829/default.asp> (last checked 19 May 2008).

U.S. Department of Health and Human Services. (2014). *HHS disaster behavioral health concept of operations: U.S. Department of Health and Human Services.* Retrieved July 12, 2016, from <http://www.phe.gov/Preparedness/planning/abc/Documents/dbh-conops-2014.pdf>.

U.S. Department of Homeland Security. (n.d.) *Post-Katrina emergency management reform act.* Retrieved July 22, 2016, from <https://www.dhs.gov/keywords/post-katrina-emergency-management-reform-act>.

Wampold, B. E. (2001). What should be validated? The psychotherapist. In J. C. Norcross, L. E. Beutler, & R. F. Levant (Eds.), *Evidence-based practices in mental health: Debate and dialogue on the fundamental questions* (pp. 200–208). Washington, DC: American Psychological Association.

Weems, C. F. (2010). Hurricane Katrina and the need for changes in the federal funding of disaster mental health. *American Journal of Disaster Medicine, 5*(1), 57.

White House (2007). *Homeland security presidential directive/HSPD-21.* Washington, DC: George W. Bush.

Wulff, K., Donato, D., & Lurie, N. (2015). What is health resilience and how can we build it? *Annual Review of Public Health, 36,* 361–374.

An Emergency Management Perspective

Ronald Sherman

The words "hope for the best and plan for the worst" or some variation, hang on the walls of many EM offices. When emergency managers are not involved in response or recovery activities they are planning, exercising, and revising strategies based on exercise and real-life results. Then, where does DBH fit into all those plans? The National Disaster Recovery Framework (2011) prescribes the use of 15 ESFs that includes ESF-8, Public Health, and Medical Services (U.S. Department of Health and Human Services, 2009).

Many, if not most, emergency managers (including this writer) assume that ESF-8 covers everything related to health, including DBH. Indeed, it should. However, experience shows that it is not included on a routine basis. At the federal level, we tend to think of ESF-8 when there may be a need for widespread vector control or mass immunization after a disaster event. When DBH enters the picture at the federal level it is usually in the form of the Crisis Counseling Program (CCP), which is covered in Chapter 4, Why Is Integrating Emergency Management Essential to Disaster Behavioral Health? Challenges and Opportunities.

> I was a member of Federal Emergency Management Agency's Federal Coordinating Officer (FCO) cadre and served many times in this capacity during my 29-year EM career. Being in the FCO cadre and having been involved in over 250 Presidentially declared disasters, including some of our nation's worst, I may be in a unique position to talk about how DBH came to be of value to me.
>
> One of the first disasters I ever responded to on was a tornado that destroyed a small, rural town in Wisconsin in the late 1970s. Over 10% of the town's population was killed. The damage assessment team and I arrived the morning after the storm and began walking the tornado's path through the town. On my left, a woman, wearing a torn nightgown and barefoot, walked across our path. She was holding a dead cat by the tail and paid no attention to our team as she swung the cat's body onto a debris pile. There were tears streaming down her face as we asked her where she lived. She pointed across what might have been a street to a concrete slab that had been swept nearly clean by the twister. She then walked away in a daze as we stood there speechless.
>
> Later that morning we ran into the pastor from one of the destroyed churches and told him about the woman. He told us he had already driven to several surrounding towns (his truck had somehow survived the tornado) and had arranged for clerics from many faiths to come to his town to provide pastoral
>
> *(Continued)*

> **(Continued)**
>
> services. I had no clue what "pastoral services" were and asked the pastor to explain. He quickly described to me what I thought sounded very much like grief counseling. I thought I understood.
>
> A few days later a Disaster Assistance Center opened in a school bus storage facility outside of town and the center manager had a "Pastoral Services" table set up with the available cleric sitting there. Almost no one stopped at that table due to two main issues: The privacy issue had not been addressed nor had we taken into account that most people had never heard of pastoral services. Sadly, we had not considered asking mental health professionals to help us to help others. This experience was a learning moment I carry with me to this day.

MEETING THE NEEDS OF DISASTER SURVIVORS

Many of my coworkers, including local, Voluntary Organizations Active in Disasters (VOAD), state, and federal colleagues, thought (and some still think) that the psychological response to a disaster is something that a person, community, or country should and can "work through" by themselves. I believe this attitude leads to providing less adequate services to disaster survivors because it does not encourage the inclusion of DBH in the overall recovery. It also explains the reason that some EMs may not work to include DBH in their plans, exercises, or response operations.

Since Hurricane Katrina, I have seen this attitude become much less prevalent in the EM world. Our collective Katrina experience, in building and trying to manage many travel trailer parks in three states, exposed the need for what has become known as "wrap-around services." These services ranged from garbage pickup, mail delivery and laundry facilities to counseling (financial, spiritual, mental health, etc.), childcare, and recreation facilities. In the aftermath of Katrina, meeting emerging DBH needs was a huge challenge and we had many gaps in services as a result of our not seeking DBH assistance or there being a lack of DBH resources. The following are offered as first hand observations, rather than empirical findings:

- In the parks where we were able to successfully tie in social services agencies with DBH capability, there were far fewer incidents of drug and alcohol abuse, violence, domestic, and otherwise.
- Where DBH services were present, survivors seemed more capable of making decisions about their plans for housing, schools, and so forth.
- In those same parks, residents showed a higher level of resiliency—the ability to adapt to significant life changes, rebound, and move ahead with their lives.

- In parks where DBH was not integrated into wrap-around services, we saw troubling levels of substance abuse and physical violence and a marked lack of forward movement by the residents. These were the last of the temporary parks to close as their residents either would not or could not make decisions on the housing options offered to them.

So, where do DBH and EM meet? We need to meet in hallways, parking lots, cafeterias, planning sessions, and exercises—all prior to an incident. One reason for a lack of integration is the simple lack of knowledge by EMs about the nature and scope of what DBH can bring to the cause.

The lead writer of this chapter describes an encounter with a coworker, *from her own agency*, who asks, "What do you do?" Many emergency managers would probably ask the same question. Many EMs just do not know what to ask about the DBH aspect of Public Health. *We do not know what we do not know.* As EMs, our main responsibility is to link resources with needs. As emergency managers, we need to be aware of DBH as a vital component of successful disaster recovery. We need to be aware of the need to ask ESF-8 to obtain and provide DBH support. There are two points to be made here.

1. In our planning we need to list DBH as a needed resource and actively work to include it as another tool in our disaster toolbox.
2. Public health partners need to inform emergency managers of the full spectrum of services they can provide and push to include DBH in the planning process.

MEETING THE NEEDS OF DISASTER WORKERS

Another critical aspect of integrating DBH into ongoing disaster operations involves providing support to all disaster workers, at all levels, and in all disciplines. After Hurricane Andrew in 1992, there was a growing awareness of high stress levels among many disaster workers. These workers with high stress levels were often involved in the recovery process, especially those whose job was to deal directly, by phone, or in person, with survivors, and heard their compelling, upsetting, and often gut wrenching stories of loss.

There was a recognition that stress in response personnel, including leadership was contributing to compromised decision making, productivity, and staff turnover. In response, Federal Emergency Management Agency funded a Stress Management Cadre that was comprised of mental health practitioners and managed by the Substance Abuse and Mental Health Services Administration. Their efforts were directed toward all the disaster workers in the Joint Field Office (JFO) and any satellite offices. Their presence and availability were made widely known to all staff and use of their services was openly encouraged by management. As FCO, I would very publicly visit the Stress Management office hoping to show there was no stigma in visiting the "quiet room."

The services included one-on one talking sessions, group sessions, and even, if space allowed, providing a quiet room with a cot so an employee could simply remove herself or himself from a stressful situation. The stress management counselors, as they were called, would also wander through the office checking in on people and observing behavior. They would provide senior leadership with periodic updates on the overall psychological state of the staff.

Not all staff welcomed the counselors. Some saw their visits as intrusive, annoying, and stressful. Counselors had wide discretion in how they did their jobs. Some counselors led group sessions involving activities that were difficult to relate to stress management. Employee and management complaints about the counselors from very vocal individuals, coupled with a new administration that held a very different view of staff support, led to the demise of the program. In my opinion, this was the wrong move. I always told my staff, "If you don't take care of yourself, you cannot take care of the survivors." The stress counselors helped people take care of themselves. In retrospect, Substance Abuse and Mental Health Services Administration could have focused more effort on defining the appropriate range of counselor activities and their supervision. Federal Emergency Management Agency could have used DBH expertise to work with its own leadership to promote the importance of managing stress and ease the Stress Management Cadre's integration into the organizational culture.

Another approach to staff support was to provide an on-site First Aid Station staffed by nurses from the Public Health Service or the Veteran's Administration. They would check blood pressures, administer over-the-counter medications, provide first aid, and monitor any trends of stress-related health issues. While not delivering direct DBH services, the nurses could at least provide management with an assessment of staff well-being. This model of staff support has been readily accepted is routinely activated in large, long-term Joint Field Office operations. I believe its acceptability is rooted in the name. What is there not to like about First Aid?

OPPORTUNITIES FOR PROMOTING INTEGRATION

This section will explore, from a very local and personal perspective, areas where, when, and how emergency managers and DBH practitioners can meet and complement each other's efforts.

After a career at the national and federal level with Federal Emergency Management Agency, I am now involved with EM at the local level in a small Chicago suburb as Operations Chief for our Community Emergency Response Team (CERT) and as a Medical Reserve Corps (MRC) Unit Leader. Here is just one thing I recently experienced and learned.

As a thought experiment, imagine that as a local emergency manager you have just performed the annual review of your Emergency Operations Plan (EOP). Is ESF-8, Public Health and Medical (or something similar) included? If yes, does that section mention DBH? If your plan is like ours used to be, the answer is, "no."

We recently completed a review and updated our EOP. One of the glaring omissions was DBH. We were then driven to establish a relationship with our local hospital consortium whose members all offer standard mental health services. The lead member hospital does have a DBH Team and their points of contact are now in our Emergency Resource List. Additionally, the DBH contact is now on our notification list when our village activates the EOC. Our CERT and MRC members now attend the hospital consortium's meetings and their lead representative attends our planning meetings. As a result of this new relationship, we also met and have the local senior services agency and their crisis counselors as part of our volunteer team.

If your locality has a CERT or a MRC, you may be able to use them as a conduit to access DBH services if your public health entity is unable to do that for you. In fact, some of their members may be disaster behavioral specialists. If you do not have one of these groups, start one.

Visit either https://www.ready.gov/citizen-corps-partner-programs or https://www.medicalreservecorps.gov/ for reference and additional information.

We have developed the following checklist that may be used by both professions as they review plans, conduct exercises, or actually respond.

Activity Area	EM Indicators of DBH Integration
Planning	✓ ESF-8 included in EOP
	✓ ESF-8 includes DBH
	✓ DBH elements include appropriate health partners (e.g., hospitals, clinics, etc.)
	✓ DBH elements include appropriate other partners (e.g., schools, workplaces, private providers, etc.)
	✓ Community has CERT/MRC
	✓ CERT/MRC has DBH capability/resources
Exercises	✓ Public Health players include DBH
	✓ DBH included as exercise evaluators
EOC participation	✓ DBH has a separate seat next to ESF-8
Responder safety and health	✓ Public Health with DBH capability is on-site with responders
Public information guidance	✓ Public Health/DBH personnel are in the review chain for all outgoing public information, regardless of the medium used. See Chapter 11, Risk and Crisis Communications, for some excellent thinking on this topic.

As an emergency manager, one has to balance many critical and competing requirements. We are trained to ask:

1. "What are the needs out there?"
2. "Do they include search and rescue, water, food, shelter, debris removal?"

The challenge is deciding what the critical, current priority needs are, as well as finding, obtaining, and deploying the resources required to meet those needs. At the same time, we must deal with the many voices pushing their need as paramount. Responder and staff safety will always take first place in any needs list. However, where do DBH, stress management, and needs assessment fit in? These are all tough questions to answer, and we, in all professions, should remember that we rarely know what demands are on the plates of our partners. Additionally, our partners most likely do not know what demands are being placed on us. This is why we must meet in hallways, parking lots, cafeterias, planning sessions, and exercises—all prior to an incident. You cannot exchange business cards in the middle of a response.

REFERENCES

National Disaster Recovery Framework (2011). *National disaster recovery framework*. Washington, D.C.: US Department of Homeland Security.

U.S. Department of Health and Human Services. (2009). *Emergency support function #8: Public health and medical services annex*. Retrieved from < https://www.fema.gov/media-library-data/20130726-1825-25045-8027/emergency_support_function_8_public_health___medical_services_annex_2008.pdf >.

Making Integration Work

Rachel E. Kaul

There are examples of behavioral health personnel working in concert with EM and there are strategies that have proven effective to improve these efforts. Perhaps the best-known example is Federal Emergency Management Agency's Crisis Counseling Assistance and Training Program (CCP). This program, established through the Robert T. Stafford Disaster Relief and Emergency Assistance Act (1988), provides states with the opportunity to secure funding to address the psychological needs of disaster-affected communities. It exemplifies a collaborative approach between EM and behavioral health in that it is funded by Federal Emergency Management Agency, administered by Substance Abuse and Mental Health Services Administration, and implemented by state mental health entities. The mission of the CCP is to assist individuals and communities in recovering from the effects of natural- and human-caused disasters through the provision of community-based outreach and psycho-educational services. Federal staff at Federal Emergency Management Agency and Substance Abuse and Mental Health Services Administration essentially role model an integrated approach by working closely with each other to provide coordinated technical assistance and guidance to grantees to ensure program success and adherence to the grant's service parameters (Substance Abuse and Mental Health Services Administration, n.d.).

Work that resulted from the Substance Abuse and Mental Health Services Administration All-Hazards Preparedness grant referenced earlier provides more examples of strategies to enhance the behavioral health role within EM. Many of the original mental health coordinators found that staffing state emergency operation centers during disaster responses, as well as sitting side by side with planners, logisticians, and responders, created relationships and provided them with allies and proponents. Working long hours in the same environment as the other emergency response professionals created a common understanding of the vast needs created by a disaster and allowed DBH professionals to informally educate others about the psychological needs of the community and of the responders. The coordinators were not seen as outsiders but as part of the overall response, trusted, and respected.

An essential activity for the coordinators, which remains so today, is participation in exercises and tabletops put on by EM, hospitals, and public health agencies. In the beginning, the scenarios rarely included behavioral health injects. It was up to the coordinators to develop relationships and advocate for their inclusion. This proved to be extremely effective as professionals across disciplines were pushed to consider how people in varying degrees of distress could adversely impact response conditions. This helped them realize

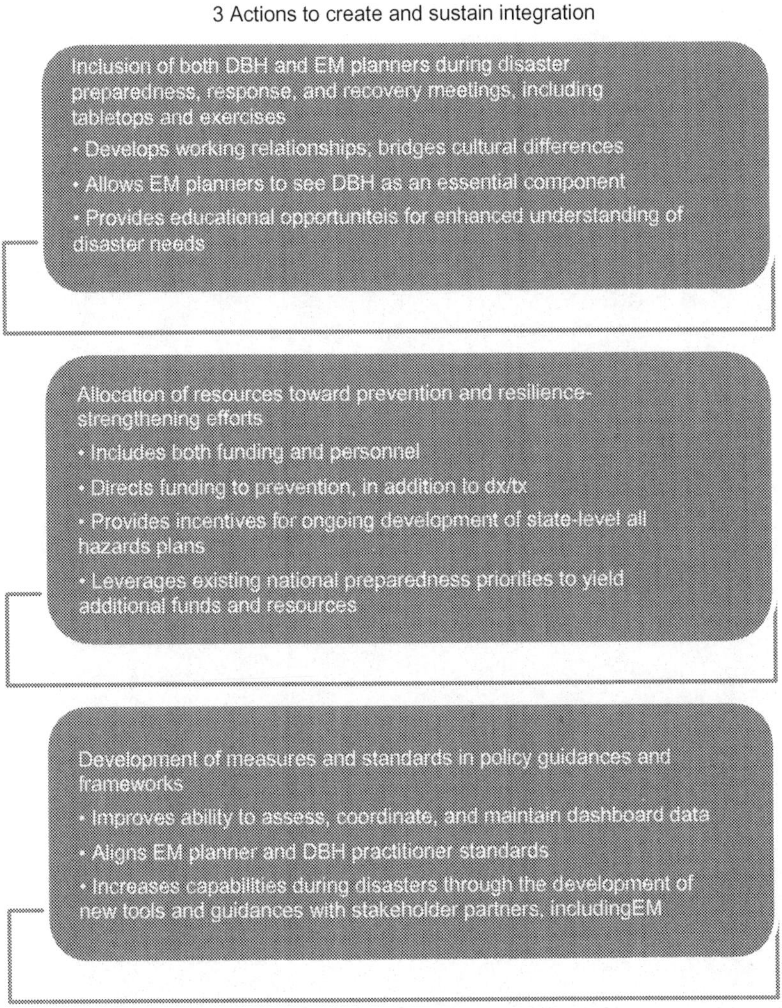

3 Actions to create and sustain integration

- Inclusion of both DBH and EM planners during disaster preparedness, response, and recovery meetings, including tabletops and exercises
  - Develops working relationships; bridges cultural differences
  - Allows EM planners to see DBH as an essential component
  - Provides educational opportuniteis for enhanced understanding of disaster needs

- Allocation of resources toward prevention and resilience-strengthening efforts
  - Includes both funding and personnel
  - Directs funding to prevention, in addition to dx/tx
  - Provides incentives for ongoing development of state-level all hazards plans
  - Leverages existing national preparedness priorities to yield additional funds and resources

- Development of measures and standards in policy guidances and frameworks
  - Improves ability to assess, coordinate, and maintain dashboard data
  - Aligns EM planner and DBH practitioner standards
  - Increases capabilities during disasters through the development of new tools and guidances with stakeholder partners, includingEM

FIGURE 2.2 Actions to create and sustain integration. This figure highlights three main recommendations to further support integration between EM and DBH.

how important having response personnel who were trained and in mitigating distress involved in order to make operations run more smoothly.

Leveraging existing national preparedness priorities, such as whole community planning and coalition, development as a preparedness strategy is another way to promote behavioral health inclusion. Federal Emergency Management Agency's emphasis on whole community planning encourages engaging human service and nontraditional stakeholders in preparedness efforts. Additional opportunities for resources and funds can be achieved by determining where these efforts are executed and how one can contribute.

Public health preparedness efforts of recent years have emphasized the formation of healthcare coalitions as a means to increase the capacity and ability to respond to emergency events. The inclusion of behavioral health entities in these coalitions has been encouraged. Planning tools and guidance developed by HHS provide examples of partners and strategies for building coalitions that include behavioral health personnel (U.S. Department of Health and Human Services, n.d.). The benefit of this approach is that it allows community-based programs to consider ways to leverage resources to alleviate negative community impacts of emergencies or to address needs when everyday services are taxed by a disaster.

These kinds of strategies are effective because they address some of the central challenges to integration: They allow cultural differences to be bridged by building relationships and trust, provide EM with a better understanding of the value of having behavioral health professionals on the team to maximize response and recovery efforts, and build broader stakeholder communities that allow different services and resources to be leveraged prior to, during, and after events (Fig. 2.2).

REFERENCES

Substance Abuse and Mental Health Services Administration. (n.d.) *Crisis counseling assistance and training program (CCP)*. Retrieved October 30, 2015, from <http://www.samhsa.gov/dtac/ccp>.

U.S. Department of Health and Human Services. (n.d.). *Disaster behavioral health coalition guidance*. Retrieved July 12, 2016, from <http://www.phe.gov/Preparedness/planning/abc/Documents/dbh_coalition_guidance.pdf>.

Chapter 3

Why Is Integrating Disaster Behavior Health Essential to Emergency Management? Challenges and Opportunities

Albert Ashwood[1], Steven Moskowitz[2], Brian W. Flynn[3], and Ronald Sherman[4]

[1]Oklahoma Department of Emergency Management, Oklahoma City, OK, United States,
[2]New York State Office of Mental Health, Albany, NY, United States, [3]Uniformed Services University of the Health Sciences, Bethesda, MD, United States, [4]Independent Consultant, FEMA Federal Coordinating Officer (Retired), United States

Through an Emergency Management Lens

Albert Ashwood

AN EMERGING PROFESSION

I have yet to ever meet someone who said they grew up wanting to be an emergency manager.

There is a simple reason for this. There is no agreed-upon definition of what an emergency manager is or what the responsibilities of the job entail. Some view the position as a planning job, consisting of the cubicle hours necessary to fill countless three-ring binders with reams of paper essential to document their many unread response and recovery plans.

Others feel that the position requires the physical prowess necessary to achieve the acts of heroism an emergency manager must display during times of disaster. Visions of Tommy Lee Jones rappelling out of a helicopter to rescue a victim perched on their flooded house come to mind.

Unlike the disaster behavioral health (DBH) professional, emergency managers usually have very few, if any, capital letters following their name on their business card. Doctors of emergency management are few and far between, and, quite frankly, I would be leery of anyone who promoted such a title.

The truth is that emergency management (EM) is an emerging profession. It does not have the long history of fire service, law enforcement, or emergency medical services. The Federal Emergency Management Agency (FEMA) was just recently created in 1979. The Robert T. Stafford Act, the federal law that provides disaster assistance, was passed by Congress in 1988. In terms of occupational history, EM is a comparatively recent addition.

When I began working for the Oklahoma Civil Defense Agency in 1988 our mission was very simple—prepare for nuclear attack. We spent 90 percent of our time writing Population Protection Plans, designed to relocate population in "target areas" to "host areas" in a short period of time. Numerous four-inch binders were filled with detailed steps on how people in metropolitan areas of Oklahoma City and Tulsa would be rapidly moved to the smaller communities of Stillwater and Bartlesville, respectively.

Of course, there were a couple of problems with this mission:

1. It is impossible to move that many people in the short amount of time between verified threat and impact. Readers can review any previous coastal hurricane evacuation with a week of lead-time to confirm this point.
2. As director of the Department of Emergency Management, it might be the case that I would need to take a sick day even if the Governor were to call this morning and tell me to get to work because there are nuclear bombs on the way. If I, as Director, did that, can law enforcement personnel be on the interstate highway directing traffic?

Regardless of what our primary mission was in those days, natural disasters were responded to on an "as requested" basis. The entire state could be leveled by a tornado or be three feet under water, but we would not leave our desk or even pick up the phone unless a local jurisdiction relayed to us that the situation was beyond their capability to respond and state assistance was requested. Names of places and people like Hugo (hurricane), Loma Prieta (California earthquake), Andrew (hurricane), Northridge (California earthquake), Katrina (hurricane), and Sandy (hurricane) have changed the way of doing business.

There are other differences between DBH professions and EM. Unlike the professional, whose background is assumed to include many years of education and experience, the typical emergency manager is someone who has been trained as a first responder. Many times the emergency manager is someone who has retired as a firefighter, law enforcement officer, or Emergency Medical Technician (EMT).

With this very limited history, it is obvious that the profession can be considered little more than a "work in progress."

Who are Emergency Managers and What Do They Do?

The Federal Emergency Management Agency (FEMA) training doctrine will say that the emergency manager will perform tasks associated within any of the four phases of EM: preparedness, response, recovery, or mitigation. It also indicates that each of these phases completes a continuous cycle from event to event. Nowhere in this doctrine does it say anything about the background, education, or experience of the emergency manager. It also says nothing about responding to the scene, performing search and rescue missions, enforcing the law, or providing emergency medical services.

My view of the role of the emergency manager can best be described by two simple words: *support* and *coordination*. If emergency managers concentrate on supporting others and coordinating resources necessary to supplement that support, then emergency managers can realistically be an asset to our first responders and, hopefully, make their jobs easier.

Of course, this support and coordination mission is irrelevant unless there is an assumed "customer" included. Therefore, despite significant differences, EM and DBH share the role, like all government, of serving the customer.

Who Is the Customer?

Every state emergency manager, including myself, has many customers: the FEMA, local jurisdictions, first responders, disaster victims, our staff, the general public, and so forth. I have always considered my primary customer to be at the local rather than the federal level.

The local level customer can be defined differently depending on which of the four overlapping phases of EM are being implemented. These are summarized in Table 3.1. During the response phase, the local customer might be the city, county, or tribal government requesting state resources to lessen the effects of the emergency incident. The local customer could also be individuals seeking information on where to go and what to do during their own personal crisis.

During the recovery phase customers are usually local governments seeking federal and state funds to help cover their emergency costs and long-term infrastructure damages. They can also be the individual disaster victim seeking financial assistance for immediate housing, asset replacement, and lost revenue needs.

In the mitigation phase, customers consist of local governments seeking funds necessary to implement actions to lessen the effects of future events. Customers can also be individuals seeking assistance to lessen their own potential crisis through the purchase of resources, such as storm shelters.

In the preparedness phase, customers encompass the entire population, as well as all levels of government and the private sector. Everyone needs to be prepared.

TABLE 3.1 Primary Customers for Emergency Managers in Various Event Phases

Event Phase	Examples of Primary Customers
Preparedness	Entire population
	All levels of government
	Private sector
	All service providers
Response	Victims/survivors
	All governmental levels
	All service providers
Recovery	Victims/survivors
	Local governments
	All service providers
Mitigation	Victims/survivors
	Local governments
	All service providers

In each of the phases, it is important to remember that others providing services, like the DBH professional, are also our customers. Emergency managers are also customers of DBH professionals. With this customer service model in mind, it makes perfect sense that the services we offer are *best provided when we take the time to learn the needs of our customers*. The private sector has known this simple truth and profited greatly from it for many years. Government at all levels is a service industry.

Customers of EM have the same general profile as those identified in the private sector. They have needs, passions, and feelings regarding every service provided and they are not afraid to share them with anyone who will, or they hope will, listen. When the customers are disaster victims, their negative feelings are often shared via every imaginable type of media or with other disaster victims, resulting in the development of an adversarial relationship between the customer and the service provider.

Impassioned customers often convey their discontent as a taxpayer who has been wronged by the bureaucratic machine. Government providers will respond to the accusations diplomatically in public, and, under their breath, refer to their antagonists as undeserving ingrates. In the many disasters I have worked over the years, I have found neither description to be accurate. This is an important area in which EM and DBH can jointly monitor, assess, and integrate efforts to promote understanding and reduce stereotyping, as well as promote effective communication and understanding.

Enter the need for a DBH professional—for both the customer and the service provider. Emergency managers, like first responders, usually shy away from victims' needs, when these needs are other than clothing, food, or shelter. At the same time, we are typically reluctant to ever view anyone who is providing the service as a potential victim, especially ourselves. Again, since many emergency managers come from first responder backgrounds, there is very little interest or desire to delve into the needs for what is sometimes seen as "touchy, feely" assistance, regardless of who needs it. In my experience, the need to integrate the services of both professions is obvious, yet rarely acknowledged.

Perhaps the best way to illustrate the need for this integrated effort is by sharing my own story of being both the victim and the service provider. Being able to see from both views has allowed me to realize how essential it is to address, in an integrated manner, the needs of customers and provided services following any natural or manmade disaster.

LOOKING AT CLOUDS FROM BOTH SIDES: A FIRST PERSON ACCOUNT FROM AN EMERGENCY MANAGER

Sunday, April 16, 1995

The Easter service at church was perfect. The minister discussed the women in the story of resurrection and the significance of God choosing women to perform practically every essential element. This was especially gratifying because my wife Cindy and I were accompanied by her sister Susie, a kindred spirit who led her own personal crusade to defend all underdogs, right all social wrongs, and, above all, promote the importance of women and their equality in every aspect of life on this earth.

I too have always believed women to be equal, if not superior. However, I often thought Susie would try out new arguments on me when I might unintentionally say something she could attack as general male thinking. I knew that the blessing of the minister's message would set the stage for a wonderful, uneventful Easter dinner we would be hosting after church.

Susie, an attorney for the Department of Housing and Urban Development (HUD) in Oklahoma City, had invited a new coworker to dinner, George, who had recently moved to the Oklahoma City office from California. He made a good impression on us by bringing gifts to our five-year-old son Donald and one-year-old daughter Rachel.

This was a different Easter for us. A few days earlier Cindy and I had been watching *The Ten Commandments* on television and began discussing Judaism and, particularly, the Passover Seder. After realizing how little we knew, Cindy called Susie, who we knew had a close Jewish friend. Never one to pass up a teaching opportunity, Susie began our Easter afternoon by preparing a traditional Seder, complete with matzo, bitter herbs, and, of course, wine, all while retelling the story of Exodus from Egypt.

Following this solemn and educational offering, we enjoyed our American Easter feast of ham, baked beans, and potato salad, before adjourning to the back yard for the children's Easter egg hunt. It turned out to be such a wonderful day. None of us would have ever believed that it would be our last day together.

Wednesday, April 19, 1995

I clearly remember thinking, during my 40 minute commute to work, how beautifully blue the sky was. The sun was shining and spring had officially sprung.

My day would consist mostly of office work and reviewing FEMA reimbursement documents for previously declared disasters. As the state's Recovery Manager, most of my time was spent with infrastructure project proposals, project closeouts, and educating local governments on documentation requirements. There was not much to my career that would fall under the umbrella of "exciting" or "sexy."

My office was small and stacked with reams of paper. It was adjacent to the state operations area or "bullpen" as we called it. I was shuffling papers, seated at my desk, when the first notice came. "Albert, you better come out here! There has been an explosion," said our duty officer. "They say it's the federal courthouse!"

All local television stations had interrupted their programing to announce that an unexplained explosion had indeed occurred in downtown Oklahoma City. One station had a news helicopter in the air, streaming live video of a nine-story structure with half of its entirety missing in a cloud of billowing smoke. The reporters said it was the federal courthouse, but they were wrong. I knew they were a block too far south. It was the Alfred P. Murrah Federal Building, home to numerous federal agencies, including HUD. It was where Susie worked. In fact, the video showed in great detail where her office was located, or more accurately, where it used to be. I immediately rushed back to my office, grabbed the phone and frantically attempted to call her, but the lines were all busy. Looking back, I see how ridiculous this idea was. The helicopter clearly showed her office gone. Why I thought phones would still be working is beyond me. My director entered the bullpen and told me to head to the scene with a couple of hand-held radios for backup communications.

While traveling my three-mile journey downtown, I thought of Susie, prayed for her safety, and wondered how such a terrible accident could have happened. Surely, it was some sort of freak natural gas explosion. As I drew closer to the building, the plume of black smoke loomed larger, yet the sky around it was still that beautiful blue I had noticed earlier.

I parked my truck two blocks away and ran to the site of the disaster. Approaching from the south, I engaged victims who were wandering aimlessly with cuts and abrasions. Private business workers from downtown tried to assist them. On the upper floors of the building, I observed two women leaning out of what used to be a window, screaming for help with their evacuation. I thought, "Where are the professionals? The firefighters, police officers, EMTs? Why aren't they here?"

As I made my way around the building, I found them. The north side of the Murrah Building consisted of what used to be an exterior wall on three sides, with an enormous pile of rubble in between. Medical triage was being performed at the corner of the structure and first responders were coordinating all rescue missions. I was amazed by the number of civilians in the debris pile. They were feverishly trying to locate coworkers who had been standing next to them 20 minutes earlier.

My job was simple. Locate the Incident Commander (IC) and offer any and all resources the state owned. I could see the Command Post at the other corner of the building and I worked my way through the debris to get there. I kept my eyes scanning the scene for Susie. Her blonde hair would be easily recognizable. Then, I realized there was no color, only shades of gray provided by the falling ash.

As I approached the Command Post, I heard response personnel discussing the need to cordon off the scene and evacuate civilians still in harm's way. I also heard the discussion of the bomb and where it was located.

The bomb? It was the first time I even considered this was not some tragic accident. Terrorism? In Oklahoma City? It made no sense. This was not Beirut. It was not even New York City or Los Angeles. This was Oklahoma City—the heartland, the buckle of the Bible Belt. Why would foreign terrorists come here?

"Move north!" came the scream. "We've located another device!" In an instant, individuals emerged from the pile and formed a mob, sprinting north to safety. I too moved north, with the Command Post. In a matter of minutes, the only personnel left at the scene were the dedicated firefighters, police officers, and EMTs performing their duties, with disregard to their personal safety.

Fortunately, the threat was based on some confiscated ordnance located in one of the federal law enforcement offices. In retrospect, this warning allowed us all to establish a defined perimeter at the scene with accountability of each first responder present. It also allowed us to begin to compile a list of those who were missing.

We had cell phones in 1995, but they were few in number. Most of them, like mine, consisted of a phone, similar to the one connected to your landline, complete with cord and battery. However, cell phones were useless due to the high volume of users. I was able to communicate with my office through my hand-held radio. After relaying my initial situation update, the department's executive secretary told me that my wife had called and wanted to know if I had located Susie. My heart sank. I had hoped the two had already connected. After all, Susie was constantly late for work or perhaps she was not even scheduled to be in the office that day. I asked the executive secretary to call Cindy and have her go to a local hospital that was receiving disaster victims. By going there, she would be connected with other hospitals and churches, serving as reception areas.

The remainder of the day included a tornado warning, the arrival of FEMA Urban Search and Rescue (USAR) Teams, countless responders from every governmental agency, and a never-ending army of reporters. Everyone was accounted for except for the presumed victims, which included Susie.

Sometime late in the evening, I was able to call my wife. She had joined her parents at a friend's house. "Have you heard anything?" Cindy begged. "No, but there are still on-going search and rescue missions," I said. I crumbled a bit as we talked. I had been in what was left of the building. She had merely watched the television. I knew in my heart that if we had not heard anything, the odds of finding Susie alive were not good. Before we ended our call she told me how proud she was of me and the job I was performing. It made me feel ashamed. I felt like I was not only giving her and her parents false hope, but also not doing all I could to locate her sister in that pile. It was my opportunity to provide the good news to my family, but I had nothing to offer. I have never felt so depressed.

The last survivor, a young lady who was freed after having her leg amputated by a surgeon using a pocket knife, was pulled from the debris pile on the night of the 19th. Then the spring rains came.

Saturday, April 22, 1995

Since the previous Wednesday, much had transpired. Search and rescue, or more accurately, search and recover missions continued at the site. At the same time, field offices for the Federal Bureau of Investigation (FBI) and FEMA were established. I relocated with FEMA and initiated my duties as State Coordinating Officer (SCO) for the Presidential Declaration.

The Oklahoma Restaurant Association adjourned from their annual conference and immediately established an impromptu hostel for rescue workers at Oklahoma City's local convention center. The facility set a standard I have yet to see surpassed, complete with gourmet meals (not military meals ready to eat or MREs), secluded sleeping areas, fresh apparel, and even massage stations. Members of the USAR teams from across the country were impressed beyond belief. Perhaps most impressive was the operation performed by the State Medical Examiner's (ME) Office in conjunction with the State Funeral Directors' Association. They joined forces to establish "The Compassion Center" at a Christian church located three miles north of the disaster site. The center was designed to house all family members of those missing. The ME's operations officer provided regular briefings on the recovery and identification process. Clergy and professional counselors were available at all times to meet the ongoing needs of those experiencing the agonizing wait.

The process was simple. When a loved one was recovered and identified, the family contact was phoned at their provided number and asked to come to the Center at an agreed-upon time. The family would be greeted at an isolated door, away from the crowd still waiting, and led upstairs to a private room. The Compassion Team usually consisted of one or two funeral directors, professional counselors, representatives from the American Red Cross and the Salvation Army, and clergy of the family's faith. The lead spokesperson would convey the bad news and offer any details regarding the location of the body. Prayers were offered and the family was allowed all the time necessary to share their feelings and attempt to figure out their next step. I honestly do not believe a more caring, compassionate process could have been developed if the parties were given a year to plan. I am so proud of my fellow Oklahomans and truly amazed by their efforts.

I remember walking from the site to the field office that Saturday morning in the hardest, coldest rain I feel I have ever experienced. My US Army-issued poncho seemed to do little more than keep the water out of my eyes. The phone rang and I hurried under a nearby tree to answer it. "Honey, we got the call," Cindy said. "They want us to be at the Compassion Center at four. Can you make it?" "Of course," I said. "I'll meet you outside."

Over the next few hours I felt nauseous, waiting for 4 p.m. to arrive. I continued to do my job. The fact is, it was the one thing that kept me from thinking about everything else that was going on. To this day, people tell me how amazed they were that I could continue to do my job with all that was happening in my personal life. I tell them all the same thing, "Work was easy. Going home was hard."

We met at the church at the assigned time. Cindy and her parents looked beaten. We all felt helpless. The process was carried out as described. We shared prayers and hugs, and slowly adjourned to face the tasks we each had in front of us. For me, it was back to work. For my family, it was additional notification to family and friends, the obituary, and arrangements for the memorial service. As I said, work was much easier. As I walked back to my truck, I felt the pain of loss, but I also felt relief that the search was over. I looked back at the Compassion Center, knowing there were still so many waiting.

Back in the truck, I immediately called my brother to give him the news. His family had been watching our children at our home 40 miles away. For some reason, in the middle of telling him the news, I burst into tears. I am not even sure why. It was not that I received news I was not expecting. For some reason, this simple act of talking to my brother brought on emotions like I have never experienced.

The next evening, I took a break from work. Cindy and I drove home to tell our son about the bad man who destroyed the building where his Aunt

Susie worked and that she would always be with him, but she would not be able to visit him anymore. He actually took the news better than we had hoped. I think we entered a long discussion on how Batman could catch the bad man and justice could be served.

May Through June 1995

The remains of the building were imploded at the end of May. There were three victims still in the pile. The firefighters felt sure they knew where they were located, but the engineers nixed any recovery effort due to the continued instability of what was left of the Murrah Building. Following the implosion, the firefighters' theory on location proved correct and the last three families were notified.

During this period, memorial services were held, the President visited, funds were raised, and the Compassion Center evolved into Project Heartland, an institution that provided free mental health services to any and all victims of the disaster for as long as was necessary. For the responders and administrators, such as myself, the hot topic was Post Traumatic Stress Disorder (PTSD). Professional counselors provided by the Federal Public Health Service would take a seat at your workstation once or twice a week to ask you how you were feeling and try to engage in a more in-depth conversation. I hated these inquiries. I do believe my blood pressure rose each time one of these "touchy-feely" types approached. Work and structure ruled my world. The busier I could be and the more hours I could work was just fine with me.

I attended Susie's memorial service. I also retrieved her personal belongings from the police department, her autopsy report from the ME's office, and, with her father, her undamaged Honda Accord parked in the building's underground garage. It seems she was on her way to the ninth floor for computer training when the explosion occurred. As usual, she was running late. Many in her training survived. Her friend George and 166 others did not. He had been in this city for such a short time. I often think about the wonderful Easter Sunday we shared.

Editor's Note by Brian W. Flynn

When we received and reviewed Albert's contribution to this book, we were powerfully moved. His insights are invaluable, his credibility is unquestionable, and his first person account of simultaneously being an emergency manger and a victim is remarkable. His disclosure of the difficulty he had in writing the most personal portions of this has only heightened our admiration and gratitude.

His first person account has also prompted me to share an incident regarding Albert that might be instructive to readers. It illustrated a pitfall and a misstep made by me in my attempt to integrate DBH and EM.

(Continued)

> **(Continued)**
>
> I was in Oklahoma City very shortly after the bombing as part of the Federal Emergency Management Agency operation there. My job was to lead the Federal effort in attempting to assure that behavioral health (BH) needs were adequately and appropriately addressed. As my role rapidly evolved, I needed to meet with Albert, in his role as SCO, to discuss and get his support for BH operations. To that end, I called his office and asked for an appointment that day to see him. I was told by his assistant that he was not available. Never being one to be easily deterred, I went to his office in person and repeated my request to his assistant. I got the same response. I got even more assertive. After all, this meeting was important. Finally, she gave me a look of professional but unquestionable firmness (and annoyance) and said, "Mr. Ashwood cannot meet with you today because he is attending his sister-in-law's memorial service."
>
> I cannot remember ever being so embarrassed and regretful in my professional career than I was on that moment. My insensitivity and distorted sense of self-importance was stunning. However, I learned several important lessons that day that have guided my work ever since. They are:
> - As important as DBH or any other specific issue/program is, it may and should not be the top priority at all times.
> - In the wake of disasters, especially the immediate response phase, we never know the personal impact of the event on our colleagues. This is especially true for those of us who are coming to a locality from far away.
> - Those of us in BH need to take great care in practicing what we preach.
> - One cannot get it right every time, but one sure better try to.

Lessons Learned

The Oklahoma City Bombing was a watershed in my life, both personally and professionally. Every significant event in my life is now categorized as occurring before or after the bombing.

I have learned that I do not like being a victim. I do not like sharing my feelings and I prefer to live a private life with my family. I work hard to keep the bombing out of my mind but hate the fact that others forget. I cringe every time I hear a disaster survivor say, "God was watching over me." They have no idea how that sounds to the families of the fatalities. Each year I despise the month of April and beg for May 1st to arrive, yet I refuse to ever miss an anniversary service. Above all, I have struggled to share my story in this writing. However, with all this said, I must admit that I believe the bombing has made me a better emergency manager.

Through my experience, I realize words like *closure* are offered by the unaffected. There is no such thing. Watershed events are never forgotten. They are simply kept in a place where life is allowed to continue. I also realize these events do not have to be of a magnitude similar to the bombing.

Loss is loss. It does not matter whether it is an act of terrorism, a tornado, or a traffic accident. How can I expect my customers to feel or act any differently than I do?

I also realize that BH is a needed service in which I have no expertise. Luckily, there are subject matter experts in this field. DBH professionals have been integrated in every event this department has had a part in since that April day in 1995. The well-being of both victims and service providers are incorporated in all planning efforts. We have also worked hard to train all response and recovery staff that words matter and sensitivities exist when discussing disaster situations. We have worked hard to assure that phrases like, "We're lucky there were so few fatalities," or, "There was a great deal of damage, but we've had worse tornadoes," will not be uttered by service providers in the state.

These training practices and DBH integration should be mandated throughout the EM profession at all levels of government. Surely, this will occur as these professions continue to evolve. We must all realize that true recovery encompasses more than a check from the government. After all, is this not simply just good customer service?

As a fire chief friend likes to say, "Our work day is somebody's worst day."

Through a Disaster Behavioral Health Lens

Steven Moskowitz

One of my favorites reads in my youth was a classic science fiction novel by Robert Heinlein, entitled *Stranger in a Strange Land*. It tells the story of Valentine Michael Smith, a human who comes to Earth in early adulthood after being born on the planet Mars and raised by Martians.

This would pretty well sum up my entry as a social worker and former not-for-profit administrator entering the world of EM by way of my new job within our state mental health agency. While not from Mars, my education and professional background was vastly different from what I encountered entering the world of EM. Concepts like the "strengths perspective of personal interaction" and "cognitive behavioral interventions" had guided the world I left, while this new land of EM spoke in a language called "ICS" and possessed a laser-like focus on broad planning strategies, processes, and procedures within a highly defined structure.

It may not have been as significant a transition as Valentine Michael Smith encountered in trying to fit into earth culture after being raised on Mars, but the difference in basic philosophies and priorities I found upon entering the world of EM was meaningful. That is not to say that I did not find aspects of this new environment somewhat familiar or others even comfortable. There was a sense of competent efficiency I could relate to in this highly organized environment where comprehensive planning was not only valued, but seen as essential. My previous work in the management of human services had given me a deep respect for the kind of commitment to the processes of planning I found when I attended my first meetings of our Human Services Group as part of Emergency Support Function (ESF) 6.

I was impressed as well by the respect our state emergency management organization was demonstrating toward the needs of the people impacted by the disasters as our group met to build plans for recovery centers, mass sheltering, housing, and feeding. As the lead of the sub-group on Mental Health,[1,2] I was included in all of these planning efforts—a level of respect

---

1. New York State uses the term mental health to represent both the mental health and substance abuse efforts represented by the term *behavioral health* found in some other states and at the federal level.
2. While this section diverges from use of DBH throughout the book, it should not detract readers from the content. It reflects the author's real-world work environment and the editors' desire to respect that context.

and inclusion for the role of mental health that I had not always found outside of my own domain of social work.

Over time and with training, I became familiar with the basics of Incident Command Structure and how the National Response Framework guided our planning efforts on the state level. Our Human Services Group was well integrated into the emergency management organizational structure and as a result, I had the opportunity to interface with the full-time EM staff and other support agencies across the full spectrum of activities found in a large state's emergency management organization.

With that exposure however, I had a growing awareness of how mental health assistance and perspectives were narrowly defined as a specific element within our emergency management agency's recovery operations. Disaster mental health (DMH) assistance was always included as a key support element at shelters and assistance centers following a large-scale event. However, the broader perspective and implications of the emotional and psychological impact of disasters on those both directly and indirectly affected was not considered helpful in the broader scope of comprehensive planning beyond those ESF-6 recovery activities at that time.

This awareness was somewhat slow to evolve. After all, when your state faces three major natural disasters within a 13-month span, as New York experienced with Hurricane Irene, Tropical Storm Lee, and Super Storm Sandy, EM prioritized ensuring the safety and well-being of the population, restoring critical services, and supporting the community in the daunting process of recovery. These experiences, both in the field directing direct services and back in the Emergency Operations Center (EOC) (see chapter 10: Integration in the Emergency Operations Center (EOC) for more on EOC integration) directing recovery activities, provided me with opportunities to see that the perspectives and knowledge I brought from the world of mental health were working well along side of the world of EM. However, they were far from being integrated in a way that could benefit overall EM.

Within EM, mental health seemed to be a narrowly defined commodity; one of the resources to be quantified and, then deployed, if and when, the resource was requested. This is very much in keeping with the structured, hierarchical culture of EM in which the control perceived to be necessary to function effectively rests in part with the ability to readily access and utilize any needed resources as efficiently as possible. This is reflected in the Federal Emergency Management Agency federal resource typing system, which commodifies virtually every type of physical resource that may be needed during response, which includes BH personnel as a part of Public Health and Medical Teams.

What makes BH more than just a deployable resource, however, and why the integration of mental health more broadly into EM planning is critical to its continual evolution, lies in the nature of what disaster behavioral specialists know. By education and training, BH folks have insight into the impact on the people that are ultimately affected by a disaster event. This insight is based on several areas of knowledge. First, sufficient research has been done to provide BH professionals with a keen sense of the emotional and psychological impact large-scale traumatic events have on individuals and groups. This, in turn, informs the core training for DMH professionals.

For example, DBH professionals approach each event aware that its nature, be it human-caused or natural, will provide strong indicators as to what may be expected as typical reactions. Further, these reactions will vary widely in their expression from affecting emotions to altering typical behaviors to challenging a survivor's spiritual beliefs. This ability to anticipate how individuals may react to a given critical incident can dramatically improve EM planning.

This became apparent during our state's response to the Ebola virus outbreak in 2015, when EM worked, under great social and political pressure, to define procedures to protect the public from the disease. The most commonly developed plans called for the quarantine of individuals who had, or may have been exposed to, the virus and isolation of those infected. Our state plan called for a comprehensive screening program primarily at key international airports for all individuals returning from one of the three Ebola-affected countries. Individuals who had symptoms were moved immediately to isolation into one of the designated Ebola treatment facilities in the state. Those who may have had direct contact with people with the virus but were asymptomatic were placed in quarantine for 21 days.

As might be expected, the plans for the screening program were a product of extensive discussion and reflected consideration of not only of the immediate health issues, but also security concerns, travel logistics, and hospital staffing. However, what went unappreciated until after the first individuals were placed in quarantine and isolation, were the significant issues tied to the emotional and behavioral impact of such a program on the individuals and immediate families of those in isolation and quarantine, as well as of the healthcare workers engaged in the care and support of those directly affected. In the first instance, several individuals placed in isolation began expressing considerable stress as a result of the uncertainty of their health status and was exacerbated by the physical barriers of their isolation. Their isolation limited access to the common means of support usually provided by partners, family, and loved ones. When facing fears associated with acquiring a life-threatening disease and yet deprived of usually available emotional supports, isolation eventually led to severe stress reactions and considerable anxiety. In response to these needs, mental health assistance was requested and DMH counselors were made available for counseling via Skype connections.

The effect on the healthcare workers in the designated Ebola hospitals became the second set of BH issues experienced during the outbreak. While nurses and health aides are regularly confronted by life-threatening illness and physical trauma, some of the psychosocial aspects of infectious disease lead to a greater-than-usual level of stress and anxiety for these professionals.

In both instances, these issues had a significant impact as anxiety and stress reactions related to infectious disease have a particular quality because they represent anxiety related to uncertainty and the unknown. Uncertainty as to whether the individual was exposed or may be exposed to the pathogen, and the unknown nature of the ultimate health outcomes of such exposure. That infectious disease events generate fear that exceeds the reality of the situation is a result of a process referred to as the "social amplification of risk" where the occurrence of an adverse event in which the risk was either previously unknown or ignored represents the potential for significant consequences (Slovic, 2001).

Further adding to the challenging psychosocial environment was that expert knowledge (in this case, on causes and transmission of the virus) was compromised as widely conflicting information was provided by the media, medical community, and politicians. This generated an overall atmosphere of a lack of trust in otherwise trusted "official sources." Thus, the act of seeking objective information—often a tool that aids in coping—instead served to further raise anxiety about the future.

The inclusion of a BH perspective could have aided the front end of response planning in several ways, such as providing psychoeducation in the form of tips sheets with practical techniques for addressing the stress and anxiety of working in an environment of unknowable fear and social stigma. Such tip sheets were eventually created and distributed. If the tips sheets were to have been part of proactive planning, this may have served to diminish the negative consequences felt by so many in the course of the outbreak. A BH perspective could also have helped to frame crisis messaging, both to the public and to the healthcare workers implementing the isolation and quarantine plans. Hopefully, the lessons learned from the Ebola virus outbreak will inform response plans for the next inevitable encounter with responding to infectious disease.

WELLNESS IN CHAOS

Few engaged in EM and/or response would argue that terms such as "vicarious trauma," "coping strategies", and "self-care" are not the stuff of casual workplace conversation. Yet, the very nature of being an emergency manager or responder means that each and every person doing such work will invariably find themselves exposed to trauma on multiple levels, which will have an effect on their emotional and psychological reactions to those events (Palm, Polusny, & Follette, 2004). This is due to a number of factors, including the impact that even indirect exposure to trauma has on those who choose a profession that, simply by its nature, will bring them into contact, either directly or indirectly, with circumstances that result in the suffering and or loss on the part of the people they are committed to serve.

Being responsible for the organizing and training of a statewide cadre of DMH responders, I became increasingly familiar with the psychological impact of disaster on those exposed to traumatic events as I coordinated several large-scale event responses. Interacting directly with the counselors I assigned to recovery centers and shelter operations who became first-hand witnesses to grief and loss in their work, I gained an acute awareness of the emotional challenges they experienced. Often, they described how emotionally challenging it was to simply drive up to and enter a recovery center. It required them to pass through neighborhoods of flooded homes where damaged furniture and belongings were piled at the side of the road-damaged beyond repair. The keepsakes and heirlooms gathered over lifetimes reduced to rubbish spoke silent stories of loss and heartbreak for the survivors of the storms.

Time spent in the Emergency Operations Center (EOC) following several major events, where operations ran for days and weeks, broadened our perspective to include the men and women who staffed the various elements of the EOC operation—the Planning and Logistics Branches, the Individual and Public Assistance teams, the functional branch representatives, even the Public Information staff. Many of these folks had begun working the event days before the storm made landfall. Days into the response, the stress brought on by the intensity of the effort to find a means to counter the damage and destruction wrought by the storm was becoming palpable. Gone was the easy jocularity of the days before zero hour and the easy banter of professionals who were confident that that they possessed the skills and knowledge necessary to do their best when crunch time came. I came to realize that the staff in the Operations Center, just like the DMH counselors working in the field, were exposed to many of the unique stressors that are particular to disaster response.

Disasters occur with little or no warning and unfold in ways that cannot be too accurately predicted. By definition, the damage they cause is widespread, affecting large groups of individuals or whole communities. While DMH responders and emergency managers are not placed in situations where they are personally at risk of injury, disasters still place biological demands on them as they lead to repeated activation of the physiological stress response (Halpern, Tramontin, & Vermeulen, 2010). This is especially true of emergency managers who, unlike DMH responders—deployed only on occasion and then returned to their "day jobs"—emergency managers respond regularly to many types of potentially stress inducing events and thus remain constantly prepared for action in rapidly changing circumstances. This reinforces the hyperarousal, hypervigilance, and threat-scanning elements of the stress response.

As taught to DMH responders, the impact of these stressors can compromise one's ability to think clearly, make sound judgments, or set priorities—all of which can undermining one's effectiveness to do their job. These challenges are not limited to the job site either. It may manifest in home life as well, such as having difficulty sleeping, a sudden loss of appetite, or even adverse changes in mood leading to being more prone to conflicts with coworkers, friends, and loved ones. DMH responders are taught to recognize these stressors, to acknowledge their existence as well as their effect. This insight represents a first step in planning how best to manage stressors and avoid compromising job performance and/or personal life.

Since such stress may be considered an occupational hazard, DMH training emphasizes, what is broadly termed, "self-care" as an essential component of preparing to go into the field. Good self-care integrates numerous elements which focus on both mind and body, such as maintaining social supports both at home and at work, engaging in activities that are pleasurable (whether physical sport or quiet reading), and continuing faith practices if relevant.

All emergency managers and disaster responders can benefit from identifying the symptoms of stress and learning how to engage in self-care activities. While disaster-related stress affects all who engage in the work, self-care can decrease the negative outcomes of such exposure. Understanding self-care concepts and skills, such as identifying and seeking support, scheduling self-care activities, and monitoring your level of burnout, are effective ways to break a cycle of stress. These strategies could be easily generalized to fit into the basic training for all staff who work in an EOC. Self-care training, with its discussions of self-awareness, self-regulation, or the need to make a plan to share one's stress indicators with a coworker are admittedly foreign to the traditional culture of EM. It will, nevertheless, lead to greater effectiveness—an easily recognized shared goal of both DMH and EM.

The concepts and skills of the field of mental health may seem just as alien and initially unwelcome to professionals in the world of EM as the earthbound characters found Michael Valentine Smith, but science fiction has existed for generations. Ultimately it is just another way to convey meaningful values and inspire a new way to consider much of what we already know.

It is not by coincidence that topics of mental health are more and more prevalent at EM Planning Division meetings and professional emergency preparedness seminars. It is due to the maturing of both the fields of EM and mental health. As each begins to recognize that at its core, the effort to assist individual and communities adversely affected by disaster will benefit greatly from the hard-earned wisdom and knowledge of the other.

REFERENCES

Halpern, J., Tramontin, M., & Vermeulen, K. (2010). *Disaster mental health: Essential principles and practices*. New Paltz, New York: Institute for Disaster Mental Health, State University of New York.

Palm, K. M., Polusny, M. A., & Follette, V. M. (2004). Vicarious traumatization: Potential hazards and interventions for disaster and trauma workers. *Prehospital and Disaster Medicine*, 19(1), 73–78.

Slovic, P. (2001). *The perception of risk*. London: Earthscan Publications Ltd.

Making Integration Work

Brian W. Flynn and Ronald Sherman

The authors of the prior sections of this chapter have described why integration of EM and DBH is important. They have explained why integration is not just desirable but mandatory if meeting the needs to survivors and those who serve them is to be accomplished. So, how can this be accomplished? Since this chapter has EM as the lead, let us explore several steps that emergency mangers can take to bring the content of this chapter to life.

IDENTIFY BEHAVIORAL HEALTH RESOURCES

Depending on the nature of the location and level at which disaster preparedness is taking place, the availability of BH resources will vary considerably. These variations are the result of several factors, including the nature and type of BH professionals in the area (e.g., psychiatrists, psychologists, clinical social workers, etc.), the level of interest and training in DBH activities, and a wide array of organizational, administrative, and financial factors. The latter include the ability of these BH workers to take time away from existing service requirements, the ability to absorb financial loss from existing income sources and streams (e.g., individual or organization loss of patient care reimbursement), and options available to continue services to pre-event patients and clients.

Part of this identification process involves the potential need for, identification of, and availability of BH sub-specialists. Examples might include specialists in child and family issues, substance abuse, and geriatrics to name a few. It is also important to recognize that BH professionals are employed in many ways that may make them more or less available for disaster work. Examples include private business and industry, schools, health care systems, government entities, and individual and group practices.

Implicit in the above discussion is the recognition, described in other portions of this book, that very few BH professionals have the specialized training to optimize their contributions in all phases of disasters. Assessing the appropriateness of the training and experience of these professional is critical. Emergency managers might consider linking with a known and trusted DBH professional to help, and even lead, this vetting process. As noted in this chapter and others, the integration of a skilled DBH professional is invaluable to emergency mangers. At the same time, well-meaning and well-intentioned DBH professionals can actually create additional challenges.

EXPLORE SCOPE AND LIMITS OF PRACTICE AND AVAILABILITY

Even when available and appropriately trained BH professionals are identified and able to provide services, there are important factors to consider regarding what they do, when, and for how long. As noted above, emergency mangers are advised to enlist a known and trusted BH specialist to help in considering these issues. Key questions include:

- What are the applicable licensing/certification factors involved? How do these affect not only the ability of DBH professionals to serve, but the scope and nature of what they can do under applicable laws and regulations? See Chapter 12, Navigating External Factors: Legal, Ethical and Political Issues for more detail on this topic.
- Through what organizations and processes can DBH resources be activated?
- How long will these resources be available?
- How will replacements and transitions be handled?
- How will information flow between EM at various levels and DBH leaders?
- How will feedback in areas such as unmet or changing needs be provided and obtained?

ORIENTATION TO THE WORLD OF EMERGENCY MANAGEMENT

The world of EM is foreign to most BH professionals. Few are exposed to the nature, culture, and language of EM. For that reason, it is important that BH professionals working in disasters be exposed to, and trained, in a wide variety of elements of the emergency manager's world. This includes an awareness of the Incident Command System, a strong relationship with the local Public Health entity, and at least the contact information for the local EM person or agency.

Both emergency managers and DBH professionals often come to their jobs in disaster with other backgrounds and, too often, insufficient training to perform optimally in their newly acquired occupational roles. To the extent possible, early training should include at least introduction to the similarities and differences between their primary training and what is allowed and required to function optimally in disaster situations.

ASSURE BH FOCUS ON BOTH SURVIVORS AND WORKERS

Emergency mangers are acutely aware in all their activities, policies, and actions that they have many core responsibilities. This is far from a singular task. Among these EM responsibilities are the support for survivors and the support for response personnel.

At the risk of overgeneralizing, DBH professionals typically focus on the needs of primary victims and survivors. They may need increased awareness that their services are also needed by responders and other workers also. Similar to how the provision of various types of support to disaster survivors is different then traditional BH work, appropriately supporting emergency managers and other disaster responders is different than servicing survivors. Albert Ashwood's compelling account of his experiences at the start of this chapter could not have conveyed this point more powerfully.

As part of the training and orientation for DBH professionals, emergency managers play an important part in helping make these important distinctions. They can also help DBH workers understand the most efficacious ways of relating to and interacting with emergency managers.

As Mr. Ashwood's account so clearly demonstrates, disasters activate professional roles as well as our very personal and often idiosyncratic reactions and responses. Both professions need to understand and accept this reality. Through acknowledging this dual process, leaders can be aware of how professional roles and personal responses impact each other. This will allow them to become aware of how this process enhances and/or impairs the junction of occupational roles and human response.

For their part, DBH professionals will need to be able to support and educate EM and other workers in a number of areas such as self-care, referral, consultation, and communication as a BH intervention. The latter is described in more detail in Chapter 11, Risk and Crisis Communications.

ESTABLISH AND SUPPORT A POSITIVE EM CULTURE

It is often said that the best way to reduce adverse BH consequences in disaster work is to effectively and appropriately manage the response. In practice, this requires meeting three criteria:

- Existence of sound preparation.
- Availability of scalable, appropriate, and adequate resources.
- An EM organizational and operations culture that places this as a core value, focuses on shared values and responsibilities and is positive and supportive of all workers.

As a result, these three requirements encourage acknowledging and addressing individual and collective psychosocial reactions of personnel.

INCLUDE DBH PROFESSIONALS IN EXERCISES AND DRILLS

Only through drills and exercises can plans, including plans for integration, become viable. Without these elements, plans are likely to remain as ineffective shelf documents. DBH workers should be incorporated into all drills and exercises. Emergency mangers should consider using DBH in the design and evaluation of exercises, in addition to full participation of DBH professionals as players in drills and exercises.

FOCUS ON SHARED NEEDS AND CHALLENGES

Emergency managers have specialized needs that link nicely with the knowledge and skills of many BH professionals. Strategically, it makes sense to identify these natural matches. It is the low-hanging fruit of more complex integration. As examples, communications challenges as well as handling survivor desires and expectations come easily to mind.

As noted by Mr. Ashwood, there is almost a guaranteed friction between what victims want and need and what disaster response and recovery personnel are able to provide. While seldom being able to actually address most of the practical recovery needs of disaster survivors, DBH can play a critical role in helping all parties understand the dynamics behind this type of conflict and provide consultation regarding how to retain productive communications during times of tension.

With regard to communications, words matter. What we say and how we say it matters in every domain. Chapter 11 contains extensive information in this regard. Even well-meaning words can sometimes cause more pain. All disaster workers, and especial leaders in all disaster-related professions, should be trained in and continuously sensitized to how language impacts others, promotes shared understanding, and can unfortunately create more pain. Emergency managers and DBH professionals have shared experience and expertise in this area that can and should form a very productive alliance.

MONITOR, EVALUATE, AND REVISE THE NATURE AND SUCCESS OF INTEGRATION

No attempts at integration between EM and DBH will get it all correct, all of the time, in the first attempts. For this reason, early discussions between the two professions, in addition to the factors stated in this section above, should explore how to monitor, evaluate, and based on these processes, adapt integration strategies and approaches.

Many have said that, because of the diverse needs in disaster preparedness, response, and recovery with which BH professionals can assist, there is some role that most every DBH professional can play. Not everyone needs to provide, or is necessarily good at, direct services to survivors or workers. Many DBH professionals have extensive training and experience in monitoring and evaluating programs and processes. These types of skills can be very helpful in assuring that appropriate changes are made over time and that these changes are based on objective information.

Similarly, many emergency managers are skilled in monitoring and evaluating programs and processes and in implementing immediate changes. The progression of debris removal, power restoration, opening of critical transportation routes, and restoring the fresh water supply are usually highest on the priority list. Changes in priorities or methods can be made very quickly. DBH professionals need to understand that their EM counterpart does not have the luxury of waiting for the results of a long-term study or survey before making decisions.

Chapter 4

Why Is Integrating Emergency Management Essential to Disaster Behavioral Health? Challenges and Opportunities

Anthony H. Speier[1] and Ronald Sherman[2]
[1]*Louisiana State University Health Sciences Center, New Orleans, LA, United States,*
[2]*Independent Consultant, FEMA Federal Coordinating Officer (Retired), United States*

Through a Disaster Behavioral Health Lens

Anthony H. Speier

The term "integration" has increasingly become synonymous with transformational change strategies in health care and emergency management (EM). In the arena of health care, the Patient Protection and Affordable Care Act (Public Law 111−148) has dramatically changed the landscape for both public sector and private sector health care. Based on a population-based health and wellness model, integrated care emphasizes a team approach involving health professionals of various disciplines working collectively with the patient as an active partner in the direction and delivery of their care. The most notable characteristics are the emphasis on collaboration, shared responsibilities, and problem-solving approaches for both immediate patient health care and long-term patient well-being (Johnson, Sanders, & Strange, 2014; Kohl, Brach, Harris, & Parchman, 2013; Kuramoto, 2014).

In a similar fashion, the practice of EM is being transformed through evolving compliance strategies associated with operationalizing Presidential Policy Directive 8 (PPD-8) (U.S., DHS, 2011) (http://www.fema.gov/pdf/prepared/npg.pdf). Issued by President Obama in 2011, it directs the federal government to develop a National Preparedness Goal and a National Preparedness

System for keeping the nation safe. Of special importance is the emphasis on an integrated "all-of-nation" approach to preparedness which specifies the shared responsibility at all levels of government, public and private sectors of communities, and individual citizens. The basic approach is an emphasis on building resilient communities through mastering the core capabilities outlined in the national preparedness system. The National Preparedness Goal and related federal documents stress the importance of resilient communities and the necessity of a wholly *integrated* approach across the various levels of government and society. Detailed content is readily accessible on this process as an updated second edition of the National Preparedness Goal and Strategies (2015). (https://www.dhs.gov/presidential-policy-directive-8-national-preparedness; http://www.fema.gov/media-library/assets/documents/25959) http://www.fema.gov/core-capabilities Retrieved July 11, 2016.

From a readiness-for-change and systems perspective, EM and DBH are experiencing broad change initiatives within their respective disciplines. These dynamic changes foster an organizational environment conducive for sustainable change to emerge and become integrated into the larger system initiatives (Chang et al., 2012; Cutter et al., 2013; Medina, 2015). The readiness for organizational change coupled with a philosophical orientation, which emphasizes risk reduction models, makes the integration of EM into DBH a change process for which the time has come.

While it is heartening that the culture supporting organizational change is present in both disciplines, the hard work necessary for operationalizing integration of EM into DBH is still in its infancy. Historically, EM and DBH, have essentially functioned as parallel activities that infrequently share a common mission. EM is the comprehensive process associated with preparing, planning, responding, and recovering from major disaster incidents. EM occurs at the community, county, state, regional, and national levels. DBH is the practice of addressing the psychological impact of such incidents on people (survivors and responders) and their communities. As this broad process of building national preparedness evolves, the relationship between EM and disaster behavioral health (DBH) is being encouraged by policy makers as a positive strategy with significant potential for advancing improved public health and enhanced national preparedness.

In the pages which follow, the following will be addressed: (1) the origins and evolution of EM and DBH as unique disciplines within the disaster response and recovery environment; (2) how the disaster management cycle of prevention, preparedness, response, and recovery can be strengthened; and (3) the approaches available for integrating EM into DBH.

THE CULTURE OF EMERGENCY MANAGEMENT

Modern EM has evolved from the Civil Defense (CD) activities of the mid-twentieth century Cold War era, which focused primarily on preparations for potential nuclear bomb attacks. Adults, who were school-age children in the 1950s, may recall the emergency preparedness drills in schools across the nation that required getting under your desk and waiting until told to return to your seat. Your job as a child participant was to learn the drill, known as "Duck and Drop," and to assure your class was flawless in its response (Heath, Ryan, Dean, & Bingham, 2007). Both the drill and the instructions were clear and easy to follow. Fortunately, the nation was never tested by an actual nuclear incident, so there is no direct evidence that lives would have been saved by using this response strategy. However, we know well that existing science certainly suggests otherwise. During the 1950s, the culture of emergency preparedness was decidedly quasi-military, with the central government firmly managing the preparedness agenda (Palmer & Sells, 1965).

The response culture within federal EM regarding natural disasters did not formally develop as a comprehensive program on a large scale until the middle of 1970s. Prior to the Disaster Relief Act of 1974 (U. S. Government Printing Office, 1974), major disaster incidents were handled at the local and state level with federal assistance and provided on an ad-hoc basis through legislation specific to the incident. Response and coordination issues continued to challenge the effectiveness of federal response initiatives through the 1980s. Further reorganization and the appointment of trained emergency managers turned Federal Emergency Management Agency into a recognized global leader in EM—specifically regarding preparation, mitigation, and response to natural and human-caused incidents. Post-9/11 reorganization of national security and the creation of the Department of Homeland Security led to additional changes in Federal Emergency Management Agency's organizational structure and functional role in disasters. Criticism of the functionality and response capabilities of the federal government in response to Hurricane Katrina called for additional reform of the agency (Drabek, 2007; U. S. Government Printing Office, 2006). Additional policy and regulatory changes have occurred continuously throughout the years following major incidents.

With respect to domestic disasters, emergency managers operate within a command and control environment, specifically the National Incident Command Structure. This is the operational conduit for information-gathering and decision-making regarding the mobilization of resources in response to an event. Preparation and mitigation activities are also addressed in a tightly managed organizational model. The current mission statement for Federal Emergency Management Agency is to support our citizens and first responders to ensure that as a nation we work together to build, sustain, and improve our capability to prepare for, protect against, respond to, recover from, and mitigate all hazards (https://www.fema.gov/media-library/assets/videos/80684).

Over the last 40 years, the roles of EM agencies at the local, state, and federal levels of government have changed both in priority and operational policies. Today, emergency managers at all levels of government engage in training activities. This systematic training process assures the development of expertise within the five mission areas of prevention, protection, mitigation, response, and recovery, which are specifically identified in the Nation's National Preparedness Goal. (U.S., DHS, 2011).

The dynamic relationships between stakeholders and responders in any large scale disaster have increased awareness among disaster responders and emergency managers of additional interpersonal characteristics, skill sets, and situational factors that can influence the overall success of any response initiative (O'Sullivan, Kuziemsky, Toal-Sullivan, & Corneil, 2013; Paturas, Smith, Albanese, & Waite, 2016). Emergent issues which require further planning include the interpersonal dynamics of the individual stakeholders, the role of critical thinking, decision-making styles, interagency communications, agency relationships, and personal flexibility. Situational factors as reported by Glick (2014) include the complexity of the disaster incident, its severity, and a variety of situational factors associated with the uniqueness of the event. Emergency managers at all levels of government must respond to the threat of disasters by identifying vulnerabilities in communities to certain threats, such as flooding, earthquakes, and so forth. When a disaster incident occurs, emergency managers must also recognize a host of factors when shaping their response strategies. This includes real-time assessment of impact, damages, and the level of threat to survivors and the community's infrastructure. In addition to assessment and initiating various response actions, emergency managers must assure the accurate recording of disaster-related damages and the mobilization of resources necessary to assist survivors and their communities return to their pre-disaster status.

THE CULTURE OF DISASTER BEHAVIORAL HEALTH

The discipline of DBH is relatively new as a specialty area within the professional behavioral health community. DBH's origins come from the traditions of community mental health, substance abuse, public health, and human services. Morris (2011) provides a comprehensive review of the disaster crisis counseling model, tracing early response models from more generic crisis intervention programs of the early 1970s. Project Outreach, created in 1972, was the first example of a NIMH funded Crisis Counseling Program (CCP) designed as a human services model for crisis counseling services implemented in response to a major disaster incident. Project Outreach employed 50 trained workers under the supervision of a licensed mental health counselor, providing brief supportive counseling and outreach services to survivors of the flooding associated with Hurricane Agnes which inundated the Scranton/Wilkes-Barre Pennsylvania metropolitan area. During the same year, two additional major flood incidents occurred in West Virginia (Buffalo Creek Dam failure), and in South Dakota (Rapid Creek). In all three incidents, communities experienced loss of housing and infrastructure, and over 480 persons were killed. The psychological impact was so substantial that it prompted Congress to include "Crisis Counseling" as a federal program within the 1974 Disaster Relief Act. The CCP is authorized under Sec. 416. Crisis Counseling Assistance and Training (42 U.S.C. 5183) of the Robert T. Stafford Disaster Relief and Emergency Assistance Act which states:

> *"The President is authorized to provide professional counseling services, including financial assistance to State or local agencies or private mental health organizations to provide such services or training of disaster workers, to victims of major disasters in order to relieve mental health problems caused or aggravated by such major disaster or its aftermath." (p. 47)*

Today, the Crisis Counseling Assistance and Training Program is administered by the Federal Emergency Management Agency (FEMA) through the Substance Abuse and Mental Health Services Administration (SAMHSA). The State Mental Health Authority (SMA) applies for CCP grants in compliance with program guidance provided by SAMSHA to the SMA. The CCP provides short-term crisis counseling to survivors and responders, as well as education and training to other relief agencies involved in disaster recovery operations. The CCP is basically a parallel program that is not directly involved with EM recovery activities, except on a limited situation-specific basis. Examples of collaborative endeavors include training of various Federal Emergency Management Agency staff regarding the program services and how counseling or other CCP services can be accessed. Major activities of the CCP include having a presence in the Disaster Recovery Centers where survivors go to access various disaster relief programs, providing assistance in shelters, and providing door-to-door outreach to survivors in areas affected by the disaster incident. The CCP is not a clinical treatment program. The CCP is a short-term supportive counseling and coping resource designed to assist persons during the response and recovery phases of the disaster.

From a mental health historical perspective, the CCP was a novel approach based on human services, psycho-educational model which assumes most people are mentally healthy before the disaster incident and, with supportive counseling, will recover and move forward. As an intervention model, CCPs rely on three basic strategies which include:

1. Outreaching to disaster survivors
2. Identification of persons who are experiencing symptoms of acute stress disorders, or persistent psychological or psychiatric disorders
3. Implementation of educational and training strategies for providers and survivors

For more information, it is possible to find more information online about the Crisis Counseling Assistance & Training Program (https://www.fema.gov/recovery-directorate/crisis-counseling-assistance-training-program).

In contrast to the CCP model, the original Community Mental Health Center model (CMHC) was based on a medically-driven approach consistent with reducing and managing the psychiatric symptoms of persons with psychiatric disorders. CMHC staff roles did not accommodate easily to the needs of disaster survivors. Early CCP evaluation reports generally noted the apparent disconnect (including poor communication and cooperation) between CMHCs and the disaster-related CCP. Much of what providers experienced was not necessarily reluctance to respond, but a lack of synchrony between the CMHCs organizational design and the human services model of the CCP (Baisden & Quarntelli, 1981; Heffron, 1977; Rochefort, 1984).

Today, many of these organizational and funding obstacles, especially prevalent during the 20th century, have significantly less impact on the CMHC's ability to effectively deliver CCP program services. CMHCs in many communities have undergone several organizational transformations, such as introducing new funding streams, workforce development initiatives, and innovative program models, which include a range of community-based as well as clinic-based services (Parks, 2015). New funding and program models are positioning former CMHC programs into more flexible and comprehensive community program models, which are consistent with current population health service delivery models (Findley, Pottick, & Giordano, 2015; Hoge, Wolf, Migdole, Cannata, & Gregory, 2016; Peek, 2013).

In fact, in many communities CMHCs are now the primary contractor for CCP programs following a disaster. However, with respect to access and delivery of clinical services to survivors, significant barriers remain. The Stafford Act, as currently interpreted, assumes that sufficient pre-disaster community resources are available to address the increased surge in demand for short- and long-term clinical treatment services. Currently, formal clinical treatment models are not routinely funded following major natural disasters (Madrid et al., 2015; Norris & Bellamy, 2009; Pandya et al., 2010; Scheeringa, Cobham, & McDermott, 2014).

A BRIEF REVIEW OF DISASTER BEHAVIORAL HEALTH LITERATURE (1977–2016)

In the mid-twentieth century, opinions regarding the extent of mental health treatment needed following a disaster incident varied among academics and practitioners. Views among practitioners varied from believing extensive intervention was needed whereas others believing only minimal interventions were necessary (Morris, 2011). Still, another perspective was to actively address the disorganized features of the disaster response process as a means for reducing the stress many disaster survivors experienced. Logue, Hansen, & Struening (1979) conducted one of the first quantitative inquiries into the extended recovery period experienced by some survivors of natural and technological disasters, such as Three Mile Island and Love Canal. The results from the early study of the Hurricane Agnes flooding in Pennsylvania indicate that numerous stressors are associated with prolonged recovery. These include both physical and psychological stressors. Flood survivors reported a recovery period from emotional distress from one to 24 months, with 54% reporting a recovery period of 18 months and 39% of flood survivors reporting physical distress for more than six months. Reports on the difficulties associated with implementing the CCP programs included a number of organizational and preparedness issues among local providers (Heffron, 1977; Logue et al., 1979). These early reports emphasized the need for improvements in four arenas: (1) participating in disaster preparedness activities; (2) building strong interorganizational relationships; (3) developing policies specific to disaster communications and response/recovery conditions; and (4) addressing how outside assistance can be effectively managed during and after an incident.

From a contemporary perspective, researchers and program planners have continued to refine and document their methods for investigating the complexity of human responses to disasters. Numerous articles have been published regarding the characteristics of disaster incidents, the people most and least vulnerable to exposure, human stress reactions in the short- and long-term, individual coping strategies, and the relative health status (or resilience) of communities (North & Pfefferbaum, 2013, Norris et al., 2002). This body of published literature provides planners and emergency managers access to empirical evidence across many natural and technological disasters. As an aid to planning, this information can foster the integration of DBH (information about the impact and recovery of survivors) with EM (procedural elements).

For example, the diversity and complexity of human reactions described in the literature suggests the relationships among an individual's psychological status, level of exposure to the disaster, environmental conditions, having a prior mental disorder, possible prior trauma disorder/exposure, and other risk vulnerabilities for individuals and communities are often associated with prolonged recovery trajectories in both natural and human-caused technological events (Masten & Obradovic, 2008; Palinkas, 2012; Picou, Marshall, & Gill, 2004). Additional factors which can influence an individual's recovery include the following (Goldman & Galea, 2014; Lowe, Tracy, Cerda, Norris, & Galea, 2013; Norris et al., 2002):

1. One's perception of loss is as devastating as one's actual loss;
2. Elevated and persistent stress coupled with a sense of a loss of control regarding one's personal and collective future;
3. Pre-existing mental disorders;
4. Prior trauma exposure including PTSD;
5. Female gender;
6. Younger age;
7. Middle adulthood;
8. A history of chronic stressors;
9. Minority status;
10. Low socio economic status;
11. Low levels of social support.

Over the last decade, researchers have also investigated factors that may influence positive recovery outcomes for individuals and communities. Studies show that the vast majority of persons who experience disasters demonstrate the ability to "bounce back" from the paralyzing emotions encountered and move forward with constructive decision-making (Bonanno, 2004; Bonanno, Brewin, Kaniasty, & LaGreca, 2010; Bonanno, Galea, Bucciarelli, & Vlahov, 2007; Norris, Stevens, Pfefferbaum, Wyche, & Pfefferbaum, 2007). Interest in resilience as a critical component for recovery has generated evidence that multiple factors influence its emergence at the community as well as the individual level. Additional factors associated with recovery include the availability of social support within the family and community and the perceived availability of that social support (Norris, Tracy, & Galea, 2009). The role of social support and related concepts have produced broader and multi-faceted perspectives on our understanding of how people manage and resolve stressors under a range of contexts and circumstances (Chang, 2010; Weil, Lee, & Shihadeh, 2012).

INTEGRATION THROUGH THE ASSIMILATION OF TWO CULTURES

In its simplest form, when large-scale incidents occur, it is the role of government (at the local, state, and federal levels) to respond by helping the affected people recover and communities to return to pre-incident status. The difficulty for responders is that people react differently to disasters based on a number of external and internal factors. Some of us face different challenges and experience more extreme exposure to disaster-related trauma. Some circumstances are more challenging than others making the recovery process more difficult and prolonged. However, the research evidence clearly indicates that the vast majority of persons exposed to disasters recover without formal behavioral health treatment (Bonanno et al., 2010; Masten & Obradovic, 2008; Norris et al., 2009).

EM and DBH have very distinct cultural traditions, orientation, and perspectives regarding their roles and responsibilities within the disaster cycle. As such, the demands of any situation or response activity may be interpreted, and a response solution built on, entirely different information and analysis of the same situation. Hence, solution building is often driven by the different disaster response roles of either DBH or EM. For example, as an emergency manager, the primary mission is to move communities toward a return to social and economic viability, which often initially involves actions necessary to keep people safe and domiciled in secure and sanitary conditions. From a DBH perspective, people experience better recovery outcomes when they have a choice in developing their recovery strategies, and when social networks and communities stay intact. As such, issues of preference and choice are not always considered a priority for temporary housing plans. Emergency managers are acutely aware of the necessity of working within their defined protocols, which will also involve other local, state, and federal entities.

SYSTEMIC CONSIDERATIONS

People who have survived a large-scale disaster incident will often recall many of the negatives associated with their individual response and recovery experience. In contrast, participants in an "After Action Review," where responders assess the incident, much time is spent recalling and analyzing what went well as well as what systems issues need to be improved. Both perspectives (survivor and responder) are equally valid and are driven by personal experience. However, neither can be considered a comprehensive perspective. In order for sustainable and systemic change to occur, it is necessary to begin addressing our performance in disaster incidents within the context of the disaster cycle phases of preparedness, prevention, mitigation, response, and recovery. By utilizing the frameworks for all phases of a disaster incident, we create a multifaceted and dynamic context for taking into account multiple perspectives from a widespread and diverse group of survivors and responders. This is the optimal context for initiating sustainable integration of EM and DBH.

The systemic demands of responding to disasters increases in complexity as the event unfolds—the response contingencies escalate and task demands on responders become more intense. For example, during the response phase of a disaster, emergency managers spend much of their time reacting to unfolding and often unanticipated mini-events which are embedded in the macro-level event. Examples often involve replacing failed generators at shelters, getting emergency food supplies such as "Meals Ready to Eat" (MREs) to a hospital suddenly without power, providing crisis interventions to persons with symptoms of mental illness exacerbated by their current exposure to traumatic material, securing additional water craft for expanded search and rescue missions, and addressing flooded cemeteries with the difficult issues of recovery of gravesites and re-interment of loved ones. Within the "chaos" of the incident-response environment, emergency managers are confronted with high stress levels. Prior exposure to traumatic events, chronic stressors, and personal characteristics of emergency managers and responders increase the likelihood of high stress reactions (Benedek, Fullerton, & Ursano, 2007; Gordon & Lariviere, 2014; Schutt & Marotta, 2011). DBH staff assigned to operations centers mitigates these effects by providing a range of stress management activities and techniques which are non-threatening and help responders maintain response functionality.

REFERENCES

Baisden, B., & Quarntelli, E. (1981). The delivery of mental health services in community disasters: An outline of research findings. *Journal of Community Psychology, 9*, 195−203.

Benedek, D., Fullerton, C., & Ursano, R. (2007). First responders: Mental health consequences of natural and human-made disasters for public health and public safety workers. *Annual Review of Public Health, 28*, 55−68.

Bonanno, G. (2004). Loss, trauma, and human resilience. *American Psychologist, 59,* 20−28. Available from http://dx.doi.org/10.1037/0003-066x59.120.

Bonanno, G., Brewin, C., Kaniasty, K., & LaGreca, A. (2010). Weighing the cost of disasters: Consequences, risks, and resilience in individuals, families and communities. *Psychological Science in the Public Interest, 11,* 1−49.

Bonanno, G., Galea, A., Bucciarelli, D., & Vlahov, D. (2007). Psychological resilience after a disaster: New York City in the aftermath of the September 11th terrorist attack. *Psychological Science, 17,* 181−186.

Chang, E., Rose, D., Yano, E., Wells, K., Metzger, M., Post, E., ... Rubenstein, L. (2012). Determinates of readiness for primary care-mental health integration (PC-MHI) in the VA health care system. *Journal of General Internal Medicine, 28,* 353−362. Available from http://dx.doi.org/10.1007/s11606-012-2217-z.

Chang, K. (2010). Community cohesion after a natural disaster: Insight from a Carlisle flood. *Disasters, 34,* 289−302.

Cutter, S., Ahearn, J., Amadei, B., Crawford, P., Eide, E., Galloway, G., ... Zoback, M. (2013). Disaster resilience: A national imperative. *Environment, 55,* 25−29.

Drabek, T. (2007). Emergency management and homeland security curricula: Contexts, culture, constraints. *Paper presented.* Alberta, CA: Annual Meeting of the Wstern Social Sciences Association.

Findley, P., Pottick, K., & Giordano, S. (2015). Educating graduate social work students in disaster response: A real time case study. *Clinical Social Work Journal,* 1−9. Available from http://dx.doi.org/10.1007/s10615-015-0533-6.

Glick, J. A. (2014). Decision making by leaders in high consequence disasters: A study of decision making by U.S. federal coordinating officers. *Dissertation Abstracts International Section A, 74.*

Goldman, E., & Galea, S. (2014). Mental health consequences of disasters. *Annual Review of Public Health, 35,* 169−183.

Gordon, H., & Lariviere, M. (2014). Physical and psychological determinants of injury in Ontario firefighters. *Occupational Medicine, 64,* 583−588.

Heath, M., Ryan, K., Dean, B., & Bingham, R. (2007). History of school safety and psychological first aid for children. *Brief treatment and crisis Intervention, 7,* 206−223. Available from http://dx.doi.org/10.1093/brief-treatment/mhm011.

Heffron, E. (1977). Project outreach: Crisis intervention following natural disaster. *Journal of Community Psychology, 5,* 103−111.

Hoge, M., Wolf, J., Migdole, S., Cannata, E., & Gregory, F. (2016). Workforce development and mental health transformation: A state perspective. *Community Mental Health Journal, 53,* 323−331.

Johnson, T., Sanders, D., & Strange, J. (2014). The affordable care act for behavioral health consumers and families. *Journal of Social Work in Disability & Rehabiliation, 13,* 110−121. Available from http://dx.doi.org/10.1080/1536710X.2013.870517.

Kohl, H., Brach, C., Harris, L., & Parchman, M. (2013). A proposed health literate care model would constitute a systems approach to improving patient's engagement in care. *Health Affairs, 32,* 357−367. Available from http://dx.doi.org/10.1377/hlthaff.2012.1205.

Kuramoto, F. (2014). The affordable care act and integrated care. *Journal of Social Work in Disability & Rehabilitation, 13,* 44−86. Available from http://dx.doi.org/10.1080/1536710X.2013.870515.

Logue, J., Hansen, H., & Struening, E. (1979). Emotional and psychical distress following hurricane Agnes in Wyoming Valley Pennsylvania. *Public Health Reports, 94,* 495−502.

Lowe, S., Tracy, M., Cerda, M., Norris, F., & Galea, S. (2013). Immediate and longer-term stressors and the mental health of hurricane Ike survivors. *Journal of Traumatic Stress, 26*, 753–761.

Madrid, P., Sinclair, H., Bankston, A., Overholt, S., Brito, A., Domnitz, R., & Grant, R. (2015). Building integrated mental health and medical programs post-disaster: Connecting children and families to a medical home. Prehospital and disaster. *Medicine, 23*, 314–321.

Masten, A., & Obradovic, J. (2008). Disaster prevention and recovery: Lessons from research on resilience in human development. *Ecology & Society, 13, 9*.

Medina, A. (2015). Promoting a culture of disaster preparedness. *Journal of Business Continuity & Emergency Planning, 9*, 281–290.

Morris, A. (2011). Psychic aftershocks: Crisis counseling and disaster relief policy. *History of Psychology, 14*, 264–286. Available from http://dx.doi.org/10.1037/a0024169.

Norris, F., & Bellamy, N. (2009). Evaluation of a national effort to reach hurricane Katrina survivors and evacuees: The crisis counseling assistance and training program. *Administration and Policy in Mental Health and Mental Health Services Research, 36*, 165–175.

Norris, F., Stevens, S., Pfefferbaum, B., Wyche, K., & Pfefferbaum, R. (2007). Community resilience as a metaphor, theory, set of capacities, and strategy for post disaster readiness. *American Journal of Community Psychology, 41*, 127–150. Available from http://dx.doi.org/10.1007/s10464-007-9156-6.

Norris, F., Tracy, M., & Galea, S. (2009). Looking for resilience: Understanding the longitudinal trajectories of response to stress. *Social Science & Medicine, 68*, 2190–2198.

Norris, F. H., Friedman, M. J., Watson, P. J., Byrne, C. M., Diaz, E., & Kaniasty, K. (2002). 60,000 disaster victims speak: Part I. An empirical review of the empirical literature, 1981–2001. *Psychiatry, 65*(3), 207–239.

North, C. S., & Pfefferbaum, B. (2013). Mental health response to community disasters: A systematic review. *Journal of the American Medical Aaaociation, 310*(5), 507–518. Available from http://dx.doi.org/10.10001/jama.2013.107799.

O'Sullivan, T., Kuziemsky, C., Toal-Sullivan, D., & Corneil, W. (2013). Unraveling the complexities of disaster management: A framework for critical social infrastructure to promote population health and resilience. *Social Science and Medicine, 93*, 238–246.

Palinkas, L. (2012). A conceptual framework for understanding the mental health impacts of oil spills: Lessons learned from the Exxon Valdez oil spill. *Psychiatry, 75*, 203–206.

Palmer, G., & Sells, S. (1965). Behavioral factors in disaster situations. *Journal of Social Psychology, 66*, 65–71.

Pandya, A., Katz, C., Smith, R., Ng, A., Tafoya, M., Holmes, A., & North, C. (2010). Services provided by volunteer psychiatrists after 9/11 at the New York City family assistance center: September 12-November, 20, 2001. *Journal of Psychiatric Practice, 16*, 193–199. Available from http://dx.doi.org/10.1097/01.pra.0000375717.77831.83.

Parks, J. (2015). Integrated care: Working at the interface of primary care and behavioral healthIn L. E. Raney (Ed.), (pp. 193–216)). Arlington, VA: American Psychiatric Publishing, Inc, 2015. xvii, pp 276.

Paturas, J., Smith, S., Albanese, J., & Waite, G. (2016). Inter-organisational response to disaster. *Journal of Business Continuity & Emergency Planning, 9*, 346–358.

Peek, C. (2013). *The National Integration Academy Council. Lexicon for behavioral health and primary care integration: Concepts and definitions developed by expert consensus. AHRQ Publication No.13-IP001-EF*. Rockville, MD: Agency for Healthcare Research and Quality.

Picou, J., Marshall, B., & Gill, A. (2004). Disaster, litigation and the corrosive community. *Social Forces, 82*, 1492–1522.

Rochefort, D. (1984). Origins of the "third psychiatric revolution": The community mental health centers act of 1963. *Journal of Health Politics, Policy & Law, 9*, 1–30.

Scheeringa, M., Cobham, V., & McDermott, B. (2014). Policy and administrative issues for large-scale clinical interventions following disasters. *Journal of Child Adolescent Psychopharmacology, 24*, 39–46. Available from http://dx.doi.org/10.1089/cap.2013.0067.

Schutt, J., & Marotta, S. (2011). Personal and environmental predictors of posttraumatic stress in emergency management professionals. *Psychological Trauma: Theory, Research, Practice, and Policy, 3*, 8–15.

U.S. Department of Homeland Security, National Preparedness Goal. (2011) Retrieved from <http://www.fema.gov/pdf/prepared/npg.pdf>. Accessed on November 8, 2016.

U.S. Government Printing Office. (2006) A FAILURE OF INITIATIVE. Final Report of the Select Bipartisan Committee to Investigate the Preparation for and Response to Hurricane Katrina, February 15, 2006. Committee of the Whole House on the State of the Union. <www.gpo.gov>. WASHINGTON 2006.

U.S. Government Printing Office. (1974). *Disaster Relief Act of 1974, Public Law 93-288*. Retrieved form <https://www.gpo.gov/fdsys/pkg/STATUTE-88/pdf/STATUTE-88-Pg143-2.pdf>. Accessed on November 8, 2016.

Weil, F., Lee, R., & Shihadeh, S. (2012). The burdens of social capital: How socially-involved people dealt with stress after hurricane Katrina. *Social Science Research, 41*, 110–119, 110-112-119.

WEB LINKS

<https://www.dhs.gov/presidential-policy-directive-8-national-preparedness; http://www.fema.gov/media-library/assets/documents/25959>. Retrieved on June 6, 2016.

<http://www.fema.gov/core-capabilities>. Retrieved on July 11, 2016. <https://www.fema.gov/media-library/assets/videos/80684>. Retrieved on July 12, 2017. <https://www.fema.gov/recovery-directorate/crisis-counseling-assistance-training-program>. Retrieved on June 16, 2016.

Through an Emergency Management Lens

Ronald Sherman

WHY DOES DISASTER BEHAVIORAL HEALTH NEED EMERGENCY MANAGEMENT?

This may sound a bit harsh or absolute, but the primary answer to that question is that EM "owns" most disaster preparation, response, and recovery activities. All actions in every phase are becoming increasingly formalized. Just as in professional sports, nobody plays without not only understanding who owns the team, but also agreeing to the owner's contract. Basically, in order for DBH personnel to effectively do their jobs, they must understand the structure and agree to play by EM rules and within EM structures. Those in all specialty fields coming into an EM structure or collaborating with emergency managers need to know this, and DBH is no exception.

While this may be a new playing field for many behavioral health practitioners, it is the reality of today's EM world. We, the emergency managers, are the ones who can open the door for you, show you the ropes, and get you a seat at the table.

DBH needs EM *information* to appropriately target its resources, understand changing situations, evolving priorities and changes in operational direction. EM has access to the damage assessment information, especially the locations and demographic profiles of affected populations that can be used to start the DBH needs assessment. While there are limitations on what information EM professionals can share about individuals affected by the disaster, we are still the best source. Please see Chapter 12, Navigating External Factors: Legal, Ethical, and Political Issues for a detailed description of the possibilities, limitations, and challenges of sharing of personal information.

DBH needs *access* to EM in order to describe their contributions, obtain sanction, and be recognized as a legitimate partner in the emergency community. Emergency managers need to know what the field of DBH is doing and it is in the best interest of DBH for EMs to know what is being done, where needs are most acute, and what new needs are emerging. That will happen only if you attend the regular planning meetings and provide input for situation reports.

Let us explore a basic example. Depending on the incident, special credentials (badges) may be needed to access a disaster area or an Emergency Operations Center. EM can facilitate getting those credentials for DBH personnel. If DBH personnel cannot gain access, they cannot attend the meetings and verbally report on behavioral health related activities. If they cannot get in, and are not part of the team, you will not be asked for input that could be included in the EM reports. The DBH information might include a request for resources or support from another agency or group that DBH professionals need to do their job.

If DBH personnel need *support*, in any form, they will not receive it from emergency managers, unless the emergency managers know what is required to address DBH needs. Frankly speaking, emergency mangers control almost all types of resources in disasters. Emergency managers can provide support to DBH personnel in many areas including:

- Public Affairs—Emergency managers can offer support in getting out the message DBH needs to disseminate.
- Logistics—Emergency managers may be able to offer workspace and IT/communications assistance.
- Situation Updates—DBH personnel need to know what is going on; either through regular briefings or situation reports.
- Planning—Emergency managers can support DBH personnel by sharing information on future actions and by incorporating DBH into the overall plan.

> Three reasons DBH needs EM on its side:
> Information
> Access
> Support

The public health group (typically Emergency Support Function 8), including the DBH component, need to ensure that their requirements and activities are coordinated as well as integrated with other initiatives. These initiatives can include basic public health measures like boiling guidance for drinking water, evacuations, mobile home park construction for temporary housing, power restoration, town meetings, and any number of other event specific initiatives.

While all of these examples may not seem related to DBH, they are. Each activity has some element of assuring that messages are heard and understood, and that those responsible for these seemingly unrelated activities understand the psychological and emotional impacts of the event.

In these examples, DBH specialists can offer support that can not only help other domains, but can also perform critical DBH missions. Remember, DBH is far more than counseling. A prime example of this is the role of DBH in assisting in the crafting of public messages by helping to create realistic expectations and address fears and rumors. Please see Chapter 11, Risk and Crisis Communications, for insight and ideas on how DBH specialists can assist emergency managers in the realm of monitoring trends and disseminating information.

HOW CAN INTEGRATION BE FACILITATED?

Now we have addressed *why* DBH needs to integrate with EM, we must consider factors that help or hinder DBH in becoming integrated with EM. First, we will examine several factors that can facilitate integration.

Within the EM community, there is an *increasing appreciation* for the need for, and the value of, DBH. When I first started working in the field of disaster assistance, as it was known in the 1970s, our awareness of any DBH issues related to disasters was quite limited. In the years that followed, EM became increasingly aware during our work with survivors that DBH factors were making our jobs more difficult. Survivors could not give information accurately because they were so stressed. Some survivors lashed out at responders out of fear and frustration. Applicant stress was causing errors in applications being filed for disaster assistance and emotionally overwhelming the unprepared application takers. There was hostility toward damage assessment personnel who were visiting survivors' homes to verify losses. Disaster workers quit their jobs because of the high levels of stress. There was a general inability, among all involved, to manage the stress.

Since that time, two important trends have developed. First, as emergency managers, our awareness of the mental health component of disasters has dramatically increased. We are also starting to get comfortable with using the words "disaster behavioral health." Second, as emergency managers' awareness has increased, so has willingness to include more partners, especially BDH professionals, in the preparedness, response, and recovery processes as active participants and advisors. There was a very practical reason for this shift; we realized that stress among survivors and disaster workers were compromising our ability to do our job.

More emergency managers are *now trained* to understand and appreciate how the inclusion of other professions and disciplines in all of our work can yield benefits for the disaster survivors and disaster workers. It is no longer unusual to see a faith-based organization as part of the disaster preparedness and response organization. Twenty-five years ago that would not have been the case. The same change is also slowly occurring in regards to the DBH profession. Sometimes it is even driven from a surprising source— EM.

I recently attended a meeting of our local hospital consortium. This was a quarterly emergency planning gathering. During the all-day session, two major questions emerged:
1. Who should be involved in emergency planning and exercises?
2. What other partners are we missing and how do we find them?

The host hospital has a robust psychiatric division with both in-patient and out-patient services, and is also a teaching hospital. When I asked about their

(Continued)

> **(Continued)**
>
> DBH capabilities, I was met with blank stares. The first response was something along the lines of, "Do you mean disaster *mental* health?" My response was, "Yes, and much more." A robust discussion followed and DBH will be the focus of the next quarterly meeting. All eight member hospitals promised to bring a representative who could address the issue.

While there are not any new laws that require either EM or DBH to explore integrating their activities, there are *emerging practices* within the EM community that show more openness to engaging with partners who have not traditionally been at the emergency planning, response, and recovery tables. An example is the inclusion of faith-based organizations as mentioned above. In order to facilitate integration, DBH personnel needs to educate emergency managers and help to expand EM's traditional notion and understanding of what the world of public health truly encompasses.

FACTORS HINDERING INTEGRATION

Now, consider a few factors that can hinder integration.

Some emergency managers react to hearing the words "disaster behavioral health" with an immediate leap to thinking it solely means dealing with mental illness issues. For some, there is likely a stigma attached to mental illness. Some will say it is not their job to deal with that type of problem. DBH specialists need to break through that wall in ways that are non-threatening to the emergency managers. Or, perhaps not break through the wall, but find a creative way around it. The behavioral health profession is very accustomed to facing and combatting stigma (http://store.samhsa.gov/shin/content/SMA06-4176/SMA06-4176.pdf). It is worth reminding the DBH community that stigma is alive and still an issue in the EM community.

Past negative experiences with an individual DBH practitioner can profoundly and negatively influence an emergency manager's attitude toward the entire field and lead them to exclude DBH from being part of the team. The bad experience could have come from the rare experience of dealing with someone who showed up with a pushy, know-it-all attitude, or someone who chose to act solely on their-own without informing EM of their activities. This is a tough challenge for all to overcome. It can be overcome by DBH professionals who can demonstrate an understanding of where they fit into the EM structure and convincingly explain how they can support all aspects of the organization. In the end, it will be new and positive experiences that will overshadow old, negative ones.

Identifying who is a credentialed and verified responder, who can be allowed access to a disaster area/site, or be allowed to join the response team can be quite difficult. How is an emergency manager to know that a purported DBH professional really is just that? Without help, there is no way for an emergency manager to know who is legitimate and appropriately trained. In this writer's experience it is not unusual for supposed "experts" to suddenly appear at disaster sites, usually when we are grasping at straws for solutions. At that point, emergency managers may accept any perceived help we can get which can lead to the experiences described in prior paragraph. This is a great example of how a solid DBH-EM relationship developed *prior to* an incident can eliminate, or at least reduce, this concern.

Most emergency managers do not know a lot about the public health structure (e.g., state and local), let alone the DBH structure, or even *its existence*. They may know about federal entities like the Substance Abuse and Mental Health Services Administration (SAMHSA), but that will do little to help them understand the social services delivery systems (often already broken) in their local area and state. Emergency managers are used to dealing with people from other public entities and government agencies. If a state or local Behavioral Health Authority uses contracted private individuals or companies to provide social services, hopefully include the DBH component, the average EM may balk at having a private sector person involved. This can be solved by addressing two challenges covered earlier:

1. Having an established relationship between EM and DBH.
2. Ensuring proper and acceptable credentialing and badging.

STRATEGIES TO HELP DBH GAIN AND SUSTAIN INTEGRATION

Culture and History

Understanding the culture, mission, and structure of EM will help enable DBH to create and pave paths to integration. A short history of one emergency manager's culture and history may be useful as an example, especially since emergency managers' career paths can be quite varied.

> When I entered the field of EM, it was with a very small federal agency named the Federal Disaster Assistance Administration. The Federal Emergency Management Agency (FEMA) did not yet exist. The job announcement listed two requirements: the ability to adapt and respond to rapidly changing requirements and to travel nationwide and to all United States territories and protectorates on short notice. Who could turn down such an offer?
>
> "Emergency management" was not even a phrase or concept then. We were called *disaster assistance specialists*. There were 10 full-time employees in the Chicago Regional Office of the then Federal Disaster Assistance Administration,
>
> (*Continued*)

> **(Continued)**
>
> which was one of ten nationwide. I was the only one who had any previous exposure to the mental health world, and it was for 18 months as a psychiatric intern at a state mental health hospital where the patient population consisted almost entirely of physically violent males. Most of the other staff came from military, legislative, or public works backgrounds. Our training as disaster specialists today would be categorized as on-the-job or just-in-time training. My experience is not unique. Many of us found ourselves in EM through prior experiences that one might think would never lead to this career. This stands in stark contrast to the current state of the EM profession, *and it is now a profession*. See Chapter 8, Expanding the Tent: How Training and Education Partnerships With other Professions Can Enhance Both EM and BH, for some excellent thoughts on this topic.

Although great strides have been made to making EM a profession there is still variations in how the role is fulfilled. Depending on location and funding, the position or responsibilities of an emergency manager is variable. It may be a part-time job held by someone who is also the Township Road Commissioner and a full-time farmer in the township. He or she is a staff of one.

On the other end of the spectrum, in a large city or county, the position may be titled Director of Homeland Security and Emergency Management. Depending on locale and funding, the job may be one of the many hats the person wears or it may be their sole endeavor. The job may be a full-time profession, consist of a once-a-month meeting with the county board, or just a once-a-year review of an Emergency Operations Plan.

In the field of behavioral health, in general, there is a more structured, academic road to certification and licensure that might be unfamiliar to many EM professionals. Many of them may not know the difference between a social worker and a psychiatrist. They may not have even heard about DBH. Emergency managers may ask, "What is DBH and can you help? We have to get a generator over to the hospital, bring in a portable control tower to re-open the airport and open ice and water distribution stations. What do you do and where do you fit in?" These types of questions present a challenge for both professions. It also provides the opportunity to open a discussion about how the two can complement, and even enhance the performance of both. As stated throughout this book, these discussions are far more productive when they occur before a disaster happens. Nobody, especially emergency managers, has any time for these discussions in the midst of a response. Frankly, if an emergency manager is faced with someone trying to start this dialog when an active response is underway, the reaction is likely to be frustration and hostility. This will make future discussion even more complex and difficult.

Mission

This chapter's title asks why DBH would want to integrate its capabilities with those of EM. The answer is easy. Both professions share a common mission to help disaster survivors respond to, recover from, and move on from life-changing incidents. If DBH is not part of the EM package, then it is not possible to deliver the full spectrum of services that survivors require. This means that, collectively, we are not doing our jobs.

A sense of shared mission can also help DBH and EM through times when integration is difficult. In the military, the sense of shared mission and unit cohesiveness has long been acknowledged to produce better results. As DBH and EM work increasingly closer together, a reminder of our shared mission will help through the most trying circumstances.

How Can DBH Become Indispensable to EM?

As discussed throughout this section, there are many advantages to both DBH and EM when integration occurs. As noted in the prior paragraph, strain and challenges will inevitably occur. One of the ways that DBH can insure that integration persists is to establish and nurture a relationship where EMs believe they need DBH to accomplish their work, and vice versa. As noted earlier, it is true that neither can optimally do their jobs without the other. Following are a few examples of what EM will always need from DBH:

- Messaging guidance: Starting with evacuation notices, DBH integration with all public information and messaging activities, as described in Chapter 11, can provide EM with anticipatory guidance for survivors and workers.
- Needs assessment: EM needs, and DBH can provide, assessments of existing and changing survivors' needs and mood, identification of geographic areas where needs are expanding or contracting, and explanations of human needs assessment techniques.
- Worker monitoring: DBH can alert EM to signs of workforce stress, identify stress differentials by role and function/work location, point out the impact of personnel and operational policies on reducing or exacerbating stress, and teach how to identify and intervene when decision-making ability is impaired.

DBH professionals can make themselves indispensable to EM professionals by first recognizing that emergency managers are professionals, albeit at different levels of experience and engagement. Then, they can further enforce collaboration by providing at least the three critical services described above. Emergency managers can encourage integration by doing what they do best: identifying needs and resources, locating the gaps, and recognizing DBH as a resource that can fill many gaps.

How Does All This Look at the Various Parts of the Disaster Cycle?

A very simplified view of the disaster cycle includes three phases: preparedness, response, and recovery. These phases are notional at best and can overlap. However, for the purpose of this section, the phases do provide a framework for describing what DBH practitioners can do to further integration.

In the *preparedness* phase, DBH can get to know the culture and history of EM and, most importantly, the EM organizational structure, in order to find out where DBH can fit in. It may be necessary for a DBH person to literally invite themselves to the party. During this phase is the time to make sure there are established relationships with any local public health entity or private services provider. This is also the time to learn about credentialing and badging requirements and to establish a process through which DBH authorities can assist in the vetting process.

A small part of EM preparedness is training in classroom-type settings. More importantly, this includes participation in plan review meetings, exercises, drills, and simulations. These provide great opportunities for DBH professionals to showcase how they can support and enhance EM efforts as well as receive valuable training. Several factors need to be observed during emergency response exercises. First, the levels involved need to be noted. Depending on the event, there may be local, state, and federal agencies involved. Second, the types of organizations engaged in the efforts can range vastly. This can include school districts, social services groups, and faith-based and other volunteer organizations. Third, the type of event is also important to note. It is necessary to think about scenarios such as an active shooter exercise at a local elementary school, a hazardous materials drill near a senior citizen's residential facility, or a simulated flood that destroys thousands of homes. Who would be involved in the response? How does DBH become part of that planning and preparedness? These preparedness activities offer many opportunities for DBH integration.

In the *response* phase, DBH professionals can expect to be part of a seemingly hectic situation, especially if they are in an Emergency Operations Center. This is the time when it can seem there is no role for DBH. Nothing could be further from the truth. A response may last a few hours, a few days, or span several weeks depending on the disaster event. During this phase, DBH specialists can contribute by providing worker and survivor monitoring, assessments, and recommendations. It is also the time for DBH to show flexibility in supporting EM as the situation evolves and needs and resources change.

The *recovery* phase presents a different array of opportunities for DBH to support EM. This is the time when ongoing needs assessment of survivors, who may be making crucial and life-changing decisions about their recovery process, is important information to provide to EM. Again, assistance in crafting public messaging can go a long way toward explaining to survivors how long it will take to build a mobile home park, how much longer they can expect to be in a hotel room with no kitchen, or why flood insurance payments may be smaller than expected.

As the recovery phase comes to an end it leads into the preparedness phase, completing the cycle. This is the time for program evaluation for all involved in delivering disaster services. DBH professionals should be fully involved in after-action meetings and help document areas for future improvement. The areas for improvement then become part of the preparedness and planning effort for the next event. There will be another.

Hopefully, in this section, readers have gained an increased appreciation for why DBH needs integration with EM. In addition, specific steps that can be taken to enhance integration have been provided. In my several decades as an emergency manger, I have seen the relationship between EM and DBH change dramatically. It continues to evolve. For an increasing number of emergency managers like myself, DBH has evolved from a question mark to an exclamation point. Behavioral health is an essential part of all EM considerations at every phase.

Making Integration Work

Anthony H. Speier

As discussed earlier in this chapter, the authorization in the Stafford Act is, in effect, the national policy statement regarding the legitimacy of DBH services. Under the Stafford Act, funding has been available to provide DBH services for over the past 40 years. However, this legislated legitimation has not led to an integrated emergency response culture. Why is this? The reasons are perhaps similar to the divide between physical health and mental health. For many decades, mental health has been conceptualized as separate from general health issues of the body. Over the last decade, there has been significant effort at reintegrating the care and management of physical and mental health (Mollica et al., 2004; Osofsky, Osofsky, Wells, & Weems, 2014; Shim & Rust, 2013), in particular addressing the person as a whole and the impact physical disorders have on an individual's mental health and vice versa.

Federal Emergency Management Agency's whole community framework for prevention, mitigation, response, and recovery provide the beginnings for advancing the integration of DBH and EM. Embracing the whole person mind/body perspective opens the door for considering the behavioral aspects of EM as a logistics issue. Emergency managers have significant expertise compiling essential data to inform decision-making and its consequences. Incorporating the DBH into ESF #8 related EM decisions will improve the outcomes of decisions, increase the efficiency of operations, and reduce the likelihood of unintended consequences.

The fundamental strategy for initiating change involves introducing a series of psychological and social decision-making considerations prior to moving forward with action at all stages of the disaster cycle. This approach requires replacing the planning principle of a "one-size fits all," hierarchical decision-making model with a collaborative decision-making model based on the principles of community-based participatory planning (O'Sullivan et al., 2013; Viswanathan et al., 2004). Recent work with the whole community approach has also documented the value of reaching out to diverse communities and the benefits of encouraging inclusion within the larger community (Chandra et al., 2013; Plough et al., 2013; Sobelson, Wiginton, Harp, & Bronson, 2015). The box below provides a few reminders of behavioral actions and attitudes that can build trust, demonstrate the power of transparent of community partnerships, and empower disaster survivors as partners in their recovery.

Reminders for Disaster Recovery via an Integrated EM and DBH approach
- What we can predict we can communicate.
- Sharing information dissolves ambiguity.

(Continued)

(Continued)

- We can mitigate the impact of our psychological and social vulnerabilities.
- We can anticipate human responses to certain situations.
- We can strengthen our whole community through engagement in *All Hazards* preparedness.

EMPOWERMENT AT WORK: AN EXAMPLE OF STATE SYSTEM INTEGRATION

The Louisiana Office of Homeland Security and the Department of Health and Hospitals (DHH) have enjoyed a close and dynamic relationship throughout a wide range of historic and routine disaster incidents. The state-level ESF #8 is tasked through Department of Health and Hospitals and works closely with other state agencies in managing all public health disaster events. The Office of Behavioral Health leads DBH preparedness, response, and recovery operations. Over the years of building collaborative working relationships, a large measure of trust and respect has been nurtured and continues to be reflected in the broad areas of integration enjoyed by these agencies. Since Hurricane Gustav, the state's ESF 8 Command Center maintains a DBH resource desk and serves as staff advisors on DBH to the incident commander. DBH staff provides stress management teams to emergency managers and first responders. At local and state levels, DBH professionals provide support to EM staff and leadership. As personnel change, the potential for losing certain aspects of the integrated response is a concern. To mitigate this institutional loss, DBH administrative staff continue to update DBH planning and response procedures and exercise formats to include DBH responsibilities. DBH roles are now an expectation within the activities tasked through ESF #8 and codified into many of the planning response and recovery operations at the state and local level (Speier & Prats, 2015).

THE PROMISE OF AN INTEGRATED DBH AND EM

It is well-established common knowledge in the professional literature that early intervention mitigates general psychological distress and may reduce the later onset of post-traumatic stress disorder and other related mental disorders. Based on this evidence, the National Biodefense Science Board (NBSB) commissioned the Disaster Mental Health Subcommittee to study and make recommendations regarding DBH and EM integration. The National Biodefense Science Board unanimously put forth the subcommittee recommendations for the integration of DBH into mainstream EM procedures (ESF#8) across the disaster cycle (Pfefferbaum et al., 2010).

The National Biodefense Science Board recommendations focus on integration activities at the federal level and recognized organizational, financial, and other administrative impediments at the state and local levels. Reflecting on the lack of progress since 2010 suggests that the motivation for meaningful integration at the national level has not reached the critical threshold of acceptance, which must clearly proceed implementing an integrated model. Another approach toward integration is starting at the local level. A basic principle of disaster response and planning is that all disasters are local incidents, meaning that the people and communities directly affected are the most invested in their recovery. At the local and state levels, sustainable integration can be achieved if we embrace the importance of behavioral health as a crucial public health initiative. By investing in local health initiatives, communities can begin to advocate for the integration of DBH into their *All Hazards* planning and response models.

In this chapter, we have demonstrated that the pre-conditions for organizational change require stakeholders within communities to learn to work together. It is through collective initiatives, such as emergency preparedness drills, that we learn about each other's organizations and job responsibilities. Getting to know each other builds trust and facilitates positive communication as well as problem solving. My experience in Louisiana taught us that working, training, and planning together required an investment of time and a long-term commitment to become a member of the team. Limited integration is possible independently on multiple levels. However, without strategic initiatives in place at the federal level, which formalizes disaster roles and funding throughout the disaster cycle, integration initiatives will continue to have only minimal success. Integration of systems is successful when the people involved see value for all involved. Simply stated, if DBH staff and emergency managers learn how to utilize the strategies that have been successful for both entities in carrying out their mission, then both parties will be able to move forward with integration. Our core responsibility is to do better at our jobs and improve recovery outcomes for survivors and their communities. There is an ethical obligation to move forward with an integrated model of EM and DBH.

REFERENCES

Chandra, A., Williams, M., Plough, A., Stayton, A., Wells, K., Horta, M., & Tang, J. (2013). Getting actionable about community resilience: The Los Angeles county community disaster resilience project. *American Journal of Public Health, 103*, 1181–1189.

Mollica, R., Cardeza, B., Osofsky, H., Raphael, B., Ager, A., & Salama, P. (2004). *Lancet, 364*, 2058–2067.

Osofsky, H., Osofsky, J., Wells, J., & Weems, C. (2014). Meeting the mental health needs after the gulf oil spill. *Psychiatric Services, 65*, 280–283. Available from http://dx.doi.org/10.1176/appi.ps201300470.

O'Sullivan, T., Kuziemsky, C., Toal-Sullivan, D., & Corneil, W. (2013). Unraveling the complexities of disaster management: A framework for critical social infrastructure to promote population health and resilience. *Social Science and Medicine, 93*, 238–246.

Pfefferbaum, B., Flynn, B., Schonfield, D., Brown, L., Jacobs, G., Dodgen, D., ... Lindley, D. (2010). The integration of mental and behavioral health into disaster preparedness, response, and recovery. *Disaster Medicine and Public Health Preparedness*, 6, 60–66.

Plough, A., Fielding, J., Chandra, A., Williams, M., Eisenman, D., Wells, K., ... Magana, A. (2013). Building community disaster resilience: Perspectives from a large urban county department of public health. *American Journal of Public Health*, 103, 1190–1197.

Shim, R., & Rust, G. (2013). Primary care, behavioral health, and public health: Partners in reducing mental health stigma. *American Journal of Public Health*, 103, 774–776. Available from http://dx.doi.org/10.2105/AJPH.2013.301214.

Sobelson, R., Wiginton, C., Harp, V., & Bronson, B. (2015). A whole community approach to emergency management: Strategies and best practices of seven community programs. *Journal of Emergency Management*, 13, 349–357. Available from http://dx.doi.org/10.5055/jem2015.0247.

Speier, A., & Prats, R. (2015). Post hurricane Katrina and the preparedness, response, recovery cycle: Integrating behavioral health into a state's disaster response capabilitiesIn D. Nameth, & J. Kuriansky (Eds.), *Ecopsychology: Advances from the intersection of psychology and environmental protection* (Vol. 2, pp. 15–34). Santa Barbara, CA: ABC-CLIO.

Viswanathan, M., Ammerman, A., Eng, E., Gartlehner, G., Lohr, K. N., Griffith, D., Whitener, L. (2004). Community-based participatory research: Assessing the evidence. Evidence Report/Technology Assessment No. 99 (Prepared by RTI–University of North Carolina. Evidence-based Practice Center under Contract No. 290-02-0016). AHRQ Publication 04-E022-Rockville, MD: Agency for Healthcare Research and Quality.

Section II

Key Areas of Integration

This section explores several specific factors critical to establishing and sustaining integration. While sharing many characteristics, every disaster is different, every community that experiences a disaster is different, and every disaster response is different. Chapters in this section explore how these variations impact the roles and functions of both emergency managers and behavioral health professionals, and what this means in terms of integrating their efforts.

These chapters also explore the creative ways in which training and educational opportunities might be integrated and expanded. They also review the benefits for disaster survivors that can result from the integrated efforts of the disaster behavioral health and emergency management professions.

Chapter 5

Integration in Disasters of Different Types, Severity, and Location

James M. Shultz[1], Marianne C. Jackson[2,3], Brian W. Flynn[4], and Ronald Sherman[5]

[1]*University of Miami Miller School of Medicine, Miami, FL, United States,* [2]*Federal Emergency Management Agency (Retired), New York, NY, United States,* [3]*NYC Emergency Management, New York City, NY, United States,* [4]*Uniformed Services University of the Health Sciences, Bethesda, MD, United States,* [5]*Independent Consultant, FEMA Federal Coordinating Officer (Retired), United States*

Through a Disaster Behavioral Health Lens

James M. Shultz

Emergency management (EM) and disaster behavioral health (DBH) integration is progressively evolving but varies markedly by level of disaster response (ranging from local to multinational). Prior to discussing multiple levels of response, it is important to explain how the complexity of disaster events underlies the need to achieve coordination among levels of response.

COMPLEX SYSTEMS THINKING FOR EM AND DBH INTEGRATION IN DISASTERS

The application of "complex systems thinking" provides useful insights that are particularly relevant to the theme of integrating EM and DBH across different levels for disasters and catastrophes (Cavallo, 2014; Cavallo & Ireland, 2014). It is evident that as disasters increase in scope, they require the activation of higher levels of incident management, bringing more response assets into play and involving more expansive jurisdictions.

The ability to enlarge the response network to match the demands of the disaster is a mainstay of EM worldwide.

As incidents scale up in magnitude and intensity they frequently acquire new properties and transform in complexity. It is important to discern that disasters (or catastrophes) are not simply large-scale emergencies. Conversely, crises and emergencies are not "mini-disasters." Whatever the metric that is used to measure the size or force of a disaster hazard, frequently a twofold increase in "dose" produces more than a "double batch" of disaster consequences.

Physicist Dirk Helbing, a renowned complexity scientist, captured this in a disarmingly simple phrase when he wrote, "disasters cause disasters" (Helbing, Ammoser, & Kühnert, 2005). Beyond the small-scale emergency level, the propagation of a hazard is not a linear process. One explanation for why disaster effects can mushroom unpredictably is that we live in a world of "globally networked risks" (Helbing, 2013). Human communities dwell within densely populated "risk landscapes" that influence the occurrence, types, and severities of disaster events (Shultz, Espinola, Rechkemmer, Cohen, & Espinel, 2016; Shultz, Galea, Espinel, & Reissman, 2016). The take-away for EM/DBH integration is that management of larger events is not just a matter of increasing proportions; the level of difficulty also amplifies rapidly (Helbing et al., 2015).

As a pragmatic example, a multidisciplinary author team worked together to elucidate the cascade of events that were set in motion during the deadly railway crash that occurred in Santiago de Compostela, Spain on the evening of July 24, 2013 (Shultz, Garcia-Vera, et al., 2016). Authors described the physics of train derailment, the patterns of physical injury sustained in a tumbling passenger carriage, and the expanding rings of psychosocial impact that subsumed injured crash survivors, bereaved loved ones, rescue personnel, citizen responders, local community residents, and the national population of Spain.

DIFFERENT LEVELS OF DISASTER RESPONSE

One of the defining dimensions of EM/DBH integration is the level of response necessary to manage the extreme event. Response levels range from local to international. The scale of the disaster is a primary determinant of the scope of the response that must be activated to manage the event.

The Event Dictates the Level of EM Response

As the disaster enlarges to affect a larger population or geographic area, event management necessarily expands to include EM offices for all disaster-affected communities and the overarching EM command structure for the region. However, there is more to matching the level of the EM response than just the spatial dimensions of the disaster.

As the severity of disaster impact increases, event management moves upward to higher administrative levels. One way to conceptualize the level of EM coordination is to consider the relationship between the demands posed by the disaster and the response capabilities of the disaster-affected community. Quarantelli (2005) has created a rank ordering of terms to describe critical incidents that trigger an urgent response, to which "crisis" has been added (Shultz, Espinola, Rechkemmer, Cohen, & Espinel, 2016; Shultz, Galea, Espinel, & Reissman, 2016):

- *Crisis* — Capacities exceed demands—with capacity to spare.
- *Emergency* — Capacities meet (and may somewhat exceed) demands.
- *Disaster* — Demands exceed capacities. Outside help is needed and may direct the event.
- *Catastrophe* — Demands overwhelm and obliterate capacities. Outside help takes over direction of the event by default.

A community can deal with a crisis or emergency using local resources. Indeed, most events are crises or emergencies that can be handled by dispatching a small number of trained and experienced responder units. The incident command for many of these events does not flow through the local emergency operations center but is handled, for example, by fire/rescue services at the station level or by law enforcement at the precinct level.

An important boundary is crossed when the event rises to the level of a disaster (or catastrophe). The concept that a disaster crosses a critical threshold is implied in the definition: *"a disaster is an encounter between forces of harm and a human population in harm's way, influenced by the ecological context, that creates demands exceeding the coping capacity of the affected community,"* (Shultz, Espinel, Flynn, Hoffman, & Cohen, 2007; Shultz, Espinel, Galea, & Reissman, 2007; Shultz, Espinola, Rechkemmer, Cohen, & Espinel, 2016; Shultz, Galea, Espinel, & Reissman, 2016). In contrast to crises or emergencies, disasters are population phenomena, affecting entire communities or larger geographic areas (Shultz, Espinola, Rechkemmer, Cohen, & Espinel, 2016; Shultz, Galea, Espinel, & Reissman, 2016). A disaster event is too big for a community to respond adequately utilizing only its own resources. Therefore, outside assistance must be requested, or assistance will be offered proactively, pending acceptance by the local authorities. The EM response typically shifts from the community jurisdiction, guided by local leadership, to an overarching incident command structure managed at a higher level.

Major disasters are locally rare but globally common, so disaster professionals should heed the guidance to "think locally, act globally" (Shultz & Cohen, 2015). Disasters often call upon EM professionals to deploy to events occurring outside their home territories. Actually, this is a fundamental precept of incident command management, creating mobile resources that can be directed where they are needed, guided by flexible, interchangeable command structures. Not unlike a set of nesting Russian dolls, incident command is scalable; as the event enlarges, the command system expands upward and outward until it reaches the level that can completely encompass the dimensions of the event and organize the underlying layers of response assets.

The Event Defines the Level of DBH Response

Most crises or emergencies do not warrant, and do not generate, a formal DBH response. This makes sense based on classic and much-cited DBH research. Norris et al. (2002a, 2002b) found that events characterized by: (1) few deaths or injuries, (2) limited destruction and property loss, (3) minimal disruption of social support systems, and (4) no malicious human intent produce minimal psychological impact for the affected population. Numerically, most critical incidents fulfill all four of these criteria, do not cause pronounced psychological effects for the population, and typically do not trigger a DBH response.

Local DBH Response for a Crisis or Emergency

In the instance where "DBH-related" services are provided in response to a crisis or an emergency, the goal is to match these services to the specific psychosocial needs generated by, or exacerbated by, the event. This may entail practical assistance (e.g., housing and basic needs provided by the Red Cross to a family that has been displaced following a house fire), bereavement support (e.g., grief counseling following the deaths of two community members in a tornado that has damaged several homes), or individualized psychosocial patient care available within the health system (e.g., psychological consultation or vocational rehabilitation for a school van crash survivor who has sustained an amputating injury).

Local DBH Response for a Disaster or Catastrophe

At the level of a natural or technological disaster or a human-perpetrated act of mass violence, it is appropriate to assess the need for a DBH response and to provide DBH intervention if warranted based on the results of the assessment. Due to the magnitude and demands placed on the affected community by a disaster or catastrophe, local resources are insufficient to mount a concerted, coordinated DBH response during the impact, response, and early recovery phases. Moreover, DBH, as well as EM personnel may be affected by the event. Those responders who are affected in the line of duty will have diminished capacity to help others and will, in fact, require prioritized psychosocial support for themselves. Taken together, it is clear that outside resources will undoubtedly be needed. DBH efforts are likely to be coordinated, or at least initiated, at higher levels based outside the local community even if local DBH teams and mental health resources are incorporated into the response. This will be discussed in the next section on state-and-regional DBH response.

An Example of State-and-Regional Level DBH Response for a Disaster or Catastrophe: State of Florida, USA

DBH response at a regional, subnational level—the level that subsumes one or more states, provinces, or territories—takes many forms. In the United States, progress toward achieving EM/DBH integration varies considerably by state. Several states have designed and operationalized a comprehensive DBH response capability. The State of Florida is at the forefront and will serve as a case example.

Within Emergency Support Function 8 (ESF-8), "Health and Medical Services," Florida's public health system is charged with deploying DBH services, specified as Core Mission #9 (http://www.floridadisaster.org/documents/CEMP/2014/2014%20Finalized%20ESFs/2014%20ESF%208%20Appendix_finalized.pdf). In turn, ESF-8 is one component of Florida's all-hazards Comprehensive Emergency Management Plan (CEMP, available at: http://www.floridadisaster.org/cemp.htm). DBH services are provided not only to disaster-affected citizens but also to disaster response personnel who experience distress, anxiety, grief, and loss during state- and federally-declared disasters in the State of Florida. In the State of Florida, DBH services include mental health, stress, and substance abuse considerations for survivors and responders, and also addresses the behavioral health care infrastructure, persons with preexisting serious behavioral health conditions, individual and community resilience, and risk communication and messaging (Florida Department of Health, 2015). The provision of DBH services for responders is a key part of ensuring responder "force protection."

A Standard Operating Guidelines document describes the coordination of DBH activities among the Florida Department of Health (FDOH), Florida Department of Children and Families (DCF), Florida Crisis Consortium (FCC) partners, American Red Cross, and Florida Crisis Response Team. The FCC is the FDOH-designated organization that provides behavioral health response personnel to support public health operations within the State Emergency Response Team (SERT) and respond to mission requests from Florida counties. The DBH concept of operations stipulates that DBH is a subfunction of ESF-8, coordinated by FDOH in partnership with the FCC to perform two types of activities: (1) monitoring the need for DBH services and (2) providing DBH resources as necessary for actual mission requests.

Monitoring needs for DBH services. As one aspect of EM/DBH integration, DBH services are directed using the incident command management structure. Specifically, the FCC Clinical Director is the Technical Specialist (TS) for

DBH within the ESF-8 Planning Section. Provision of DBH services and support is recommended based on the presence of multiple event-related "triggers" including high levels of traumatic injury, mass fatalities, disproportionate impact on children or vulnerable populations, severe damage and destruction, disruption of mental health services, prolonged event duration, and widespread loss of personal property. If one or more triggers are present, the Situation Unit is tasked with collecting indicator data to forecast the behavioral health effects on the community with special emphasis on disruption of mental health treatment infrastructure and impacts on vulnerable populations. DBH resources are only deployed if there is a reviewed and approved mission request from the disaster-affected county.

Deployment of DBH resources based on a mission request. The selection of resources assigned for the DBH mission request is determined on a hierarchical basis: (1) existing service provider from the area, (2) established response team, and (3) ad hoc response team identified through the FCC. Based on situational assessment, preidentified, drilled, and trained Florida DBH resources may be selected based on event-specific needs. During a Presidentially declared disaster, Florida DBH assets may be supplemented by federal mental health teams (MHTs). The FCC also maintains a pool of qualified DBH responders through contractual agreement with the FDOH. Importantly, DBH responders provide appropriate—nonclinical—interventions under the supervision of the licensed clinician. DBH teams do not provide mental health services; if such services are required, referrals are made by the team's clinician or the FCC Clinical Director.

Additionally, ESF-8 provides the following additional "DBH-related" services: Public information, clinical guidance, and transition to recovery support. Included among the options is a direct request from DCF to bring in Crisis Counseling Program (CCP) personnel and services to Florida.

Regional level response. At the regional level, the State of Florida has established Regional Disaster Behavioral Health Assessment Teams (RDBHATs) to conduct onsite DBH needs assessments in the early aftermath of disaster and DBH Response ("Strike") Teams to provide services in the field. All team personnel have been carefully vetted predisaster including background checks and "asset-typing" based on credentials and skills. This allows the formation of response teams composed of a specified complement of assets.

The mission of the DBH Response Teams is "to support counties in mitigating the adverse effects of disaster-related trauma by promoting and restoring psychological well-being and daily life functioning in impacted individuals, responders and communities" (Florida Department of Health, 2014).

The response teams work with the FCC in a manner that exemplifies EM/DBH integration by providing information to assist State and county

DBH planning and also enriching the understanding of behavioral health needs throughout all phases of a disaster. Teams are created and trained to be able to continuously improve community and county behavioral health preparedness, monitor behavioral health, and facilitate local partnerships to meet identified behavioral health needs following disaster impact, and provide guidance to meet identified behavioral health needs of disaster responders.

National DBH Response for a Disaster or Catastrophe: United States of America

Within the hierarchy of the executive branch of the US government, the DBH response, and this includes EM/DBH integration, is coordinated by the Division for At-Risk Individuals, Behavioral Health, and Community Resilience (ABC); Office of Policy and Planning (OPP); Office of the Assistant Secretary for Preparedness and Response (ASPR, 2014); based within the US Department of Health and Human Services (HHS). This federal level DBH structure is described in the *HHS Disaster Behavioral Health Concept of Operations (DBHCONOPS)*, 2014, available at: (http://www.phe.gov/Preparedness/planning/abc/Documents/dbh-conops-2014.pdf).

The DBH concept of operations' document defines DBH in this manner: "disaster behavioral health ... includes the interconnected psychological, emotional, cognitive, developmental, and social influences on behavior, mental health, and substance abuse, and the effect of these influences on preparedness, response, and recovery from disasters or traumatic events."

Aligned with the EM approach that focuses on the EM life cycle, DBH actions are classified by disaster phase into those focusing on preparedness, response, and recovery. DBH preparedness activities, including planning, training, and conducting exercises (up to and including full field simulations) aim to "mitigate the behavioral health impacts of disaster." DBH response activities emphasize risk communication, disaster responder force protection, and early supportive interventions for survivors [psychological first aid (PFA), crisis counseling] that can be delivered by trained paraprofessionals. DBH recovery activities focus on the evolving behavioral health needs for disaster survivors as they grapple with losses and hardships in the aftermath.

The DBH concept of operations document is comprehensive in scope and many of the operational elements that are described receive focused attention in other chapters within this volume. Pertinent to the current discussion, what distinguishes the national level of DBH response in the United States is the strong integration of DBH into the overarching public health response priorities. This acronym-rich document provides detail regarding the linkages among multiple public health entities (DBH is located within ASPR but coordinates with many other parts of HHS) as well as numerous preparedness and response partners both governmental and nongovernmental. The tie-in to EM is clearly evident; when the National Response Framework (FEMA, 2015) and National Disaster Medical System activate in response to a disaster declaration, the DBH activities are mobilized as part of ESF-8 (Health and Medical Services). DBH functions operate within well-devised incident command-and-control structures.

Another defining feature of the national response level is the description of the roles played by national DBH leadership. DBH leaders serve at the federal level as members of the Federal Disaster Behavioral Health Group and in a liaison capacity to the HHS Incident Response Coordination Team. Depending upon the nature of the incident, federal DBH resources and assets may be made available to disaster-affected jurisdictions; examples include the CCP, the Disaster Distress Hotline, and federal MHTs. Equally important, behavioral health force protection is implemented to safeguard HHS personnel who are responding to the disaster in a variety of capacities that frequently subject them to high stress and potential health and mental health risks.

As described in the previous section, much of the DBH response is orchestrated at the state or community level close to the disaster scene, so an important part of the federal DBH leadership role focuses on federal-to-state support and coordination of resources and personnel.

International DBH Response for a Disaster or Catastrophe Focusing on Contrasts to United States National Response

International DBH response diverges in important ways from the national level, particularly if the national comparison is the United States. Here are several distinguishing features.

1. *Terminology.* Although the United States preferentially uses "disaster behavioral health" as official jargon, DBH is a relatively unfamiliar expression throughout the rest of the world. Instead, "mental health and psychosocial support" is the phrasing that is recognized and used worldwide by the World Health Organization, United Nations agencies, Inter-Agency Standing Committee (IASC), and numerous organizations involved in disaster and humanitarian response. MHPSS is the "coin of the realm" for international communications on the theme of DBH.
2. *Types of disaster events.* In the United States, DBH services are usually activated for sudden-onset disasters that are circumscribed in time and space. Internationally, for similar discrete disaster events occurring elsewhere in the world, MHPSS services resemble the DBH response in the United States in important respects. The DBH/MHPSS response for

such events tends to occur in the early postimpact phases when exposures to hazards have ceased or greatly diminished and while the disaster survivors are dealing with losses and hardships in the aftermath. However, across many international settings, MHPSS is frequently applied to individuals who are affected by protracted humanitarian crises and complex emergencies. Often these events involve active armed conflict where exposures to the forces of harm are ongoing and continuous.

3. *Survivors' prior exposure to trauma.* In the United States, many disaster survivors have experienced relatively few traumatic life events or significant losses and life changes prior to their disaster experience. DBH services therefore are targeted for the disaster stressors associated with this rare, isolated, and distressing experience. Examples include DBH responses to "flood fighters" along the rising Red River in Fargo, North Dakota (Shultz et al., 2013), storm surge victims of Hurricane Sandy in New Jersey (Neria & Shultz, 2012), and surviving family members of the students and teachers killed in the rampage shooting in Newtown, Connecticut (Shultz, Muschert, Dingwall, & Cohen, 2013).

In other nations, MHPSS services are often used with individuals who have sustained a multiplicity of traumas and losses lifelong. Even when the MHPSS response is activated in response to a sudden-onset disaster event, the exposures associated with the immediate disaster are overlaid upon a lifetime of prior trauma. Examples include MHPSS support for survivors of the 2010 Haiti Earthquake (Shultz, Marcelin, Madanes, Espinel, & Neria, 2011) or the 2013−16 West Africa Ebola outbreak (Shultz, Baingana, & Neria, 2015).

4. *Settings for delivery of DBH services.* In the United States, DBH services may be provided in disaster recovery centers, shelters, public health clinics, neighborhood centers, other community venues, or even the homes of survivors. Only a subset of DBH service recipients have been displaced from their homes and some of these have already returned home but are grappling with losses and trauma exposures. The numbers of survivors specifically served by DBH services are typically dozens to hundreds.

International MHPSS responses, particularly during humanitarian crises, are often provided in camp settings where survivors are conflict-induced forced migrants, either refugees or internally displaced persons (IDPs). The most fluid and dynamic complex emergency calling for MHPSS resources during the 2014−16 time period is the war in Syria and Iraq. MHPSS can only be provided in rare instances within Syria but massive refugee flows have arrived in Jordan and Lebanon and more recently throughout Europe. The harrowing mass exodus of Syrian refugees to Europe has challenged the ability of receiving nations to accommodate the influx. MHPSS services lag behind the priorities of resettlement and provision of basic needs. Almost all refugees and IDPs have been traumatized and through forced migration, they have lost homes and all possessions as well as livelihoods and community connections. The numbers in need of MHPSS support are in the millions from this event alone.

5. *DBH providers.* In the United States, the DBH providers include a limited number of volunteer mental health professionals including American Red Cross disaster mental health personnel and trained paraprofessionals such as Medical Reserve Corps members. For international responses, MHPSS services are provided by humanitarian actors working under the aegis of the United Nations organizations that comprise the disaster response "cluster system" or for a broad array of disaster relief and response organizations such as Doctors Without Borders (MSF). Training is provided for laypersons to become MHPSS paraprofessionals, supervised by clinicians, and delivering nonclinical interventions in camp and community settings.

6. *Scientific basis for DBH interventions.* In the United States, most DBH providers are volunteers. Selection of interventions often depends upon the preferences of the available providers. In general, there is a lack of evidence to underpin many, if not most, individual and collective interventions aimed at disaster survivors who have not been diagnosed with clinically significant symptom elevations for common mental disorders (CMDs).

Developing an evidence base for low intensity early psychosocial support is extremely difficult methodologically. Nevertheless, the need for a scientific underpinning for early DBH interventions has been widely recognized (NBSB, 2008; Shultz & Forbes, 2013). Popular interventions, even those *informed* by evidence, remain largely untested. The federally funded Crisis Counseling Program, in use for decades, has never undergone evaluation regarding its efficacy. The various versions of PFA that have been promoted and widely used during responses in the past decade have not been assessed for efficacy. Many states have response teams certified in the application of critical incident stress management (CISM), including critical incident stress debriefing (CISD) that has been scientifically discredited as ineffective for preventing psychological consequences following disaster exposure. The National Organization for Victim Assistance (NOVA) also promotes DBH interventions that have not been evaluated.

Nationally within the United States, there is a lack of uniformity in terms of the choice of DBH interventions and differences in the types and training of DBH response team members. The intervention options generally lack scientific evidence; usually there is no systematic screening of survivors to determine possible symptom elevations for CMDs;

there is a general absence of scientific evidence for the intervention efficacy; referral systems for patients with severe symptoms or suicidal ideation, intent, or plan are lacking; and follow-up of survivors who have received DBH services is rarely performed.
7. *DBH/MHPSS response: Response sophistication and coordination turned upside down.* On the international front, global guidelines for MHPSS in emergency response were developed by consensus and released in 2007 by the Inter-Agency Standing Committee (IASC, 2007). In the past decade there has been an acceleration of both the evidence base for MHPSS assessment and intervention and the transformation of science into applied tools and interventions.

At the international level there is a very strong connection between global mental health (GMH) and MHPSS applications in emergency settings. There is a very favorable relationship to academia. Research and practice operate symbiotically. The multinationals that work continuously with vulnerable populations on a broad spectrum of sustainable development initiatives, and are onsite to support the populations during humanitarian emergencies and armed conflicts, are the same "players" that provide MHPSS services during disasters, pandemics, and other emergency situations.

So here is an irony; as this chapter goes to print, the author (a US citizen based at a US university) is actively working with the World Health Organization professionals and a large cadre of students based in the United States and European universities to conduct a rapid "desk review" for the nation of Ecuador that just sustained a 7.8 Mw earthquake on April 16, 2016. The protocol for the desk review has been modified directly from the battery of assessment tools prepared by the World Health Organization. This review will be used to coordinate MHPSS response using in-country resources, supplemented by international personnel from multiple agencies that are already operating in Ecuador, informed by culturally-appropriate guidance from seasoned GMH researchers from multiple universities. Earthquake survivors will be assessed using validated screening instruments, and for those with elevated symptoms, evidence-based and field-tested psychological interventions that will be applied by a variety of professionals from Inter-Agency Standing Committee-affiliated agencies, with appropriate training, supervision, and evaluation. The ultimate irony is that, if there was a similarly destructive earthquake occurring later this year inside the United States, it is unlikely that any of this coordination—especially with the World Health Organization, the United Nations, and an international network of researchers—would take place. The DBH response would depend upon the local makeup of the available volunteer teams. Any research would likely be conducted by local researchers from the affected area (research dollars flow to local universities following a disaster), many of whom do not routinely work in DBH/MHPSS.

CONCLUDING COMMENTS AND TAKE-HOME LESSONS FOR EM AND DBH PROFESSIONALS

First, internationally, MHPSS (DBH) in emergency settings has grown up, guided by the World Health Organization and supported by the United Nations cluster system. Science and practice are operating in close partnership to the benefit of disaster-affected populations. These quantum advances are available for use by EM and DBH professionals who are willing to consider the merits of international perspectives, best practices, and lessons learned.

Second, hierarchical, incident management organizational structures are not optimal for MHPSS (DBH). Dealing with the psychosocial needs of disaster-affected populations does not fit well with a regimented, top-down, command-and-control framework. EM personnel and DBH personnel need to find flexible structures that adapt to the longer time horizons and the nonlinear processes that operate when dealing with MHPSS (DBH) concerns. Effective response needs to be tailored to the cultural, community, family, and individual needs of disaster survivors.

Third, protocol-driven, one-size-fits-all approaches do not work for MHPSS (DBH). As one example, perhaps some of the appeal of the early debriefing approaches (after a tough call or deployment, everyone participates, everyone talks about the "worst part") was the systematic nature of this approach—just add a page to the incident command flip-book and dedicate an hour at the end of the shift. However, it is clearly known that debriefing is not psychologically healthy or helpful for all responders and debriefing is especially inappropriate for civilian survivors.

Fourth, researchers need to think like responders. And response personnel need to take on the "mind" of a researcher. The brain vs brawn dichotomy serves no one well, especially for DBH response. We need more practical and pragmatic researchers who are field-experienced. We need more curious and observant responders who have flexible minds to develop themselves and their teams into "learning organizations." Disasters keep morphing in complexity. Rigid systems are no match whether the "challenger" is Mother Nature or very clever, independently thinking humans who plot to cause harm.

Fifth, adding the international dimension to the consideration of response levels creates opportunities to learn from the best practices of other types of responder units facing a broader spectrum of disaster challenges. This includes observing how EM and MHPSS (the rest of the world's DBH) is integrated across the globe (Table 5.1, Fig. 5.1).

TABLE 5.1 Current Developments in Mental Health and Psychosocial Support (MHPSS*) for International Disasters and Humanitarian Emergencies

1. Global mental health (GMH) inform MHPSS

Global mental health (GMH) initiatives are ongoing and continuous regardless of disaster occurrence. The primary role of the World Health Organization (http://www.who.int) based in Geneva, Switzerland, is to direct and coordinate international health within the United Nations (UN) system. Mental health (http://www.who.int/mental_health/en/) is one of the World Health Organization focus areas and currently much of the activity is coordinated around the Mental Health Gap Action Program (mhGAP, http://www.who.int/mental_health/mhgap/en/).

The GMH focus is appropriate and increasingly valued. Globally, mental, neurological, and substance use disorders (MNS) account for 14% of the global burden of disease, affecting an estimated 450 million persons. Mental illnesses represent the leading cause of disability-adjusted life years (DALYs), accounting for 37% of DALYs from noncommunicable diseases (NCDs, http://www.nimh.nih.gov/about/director/2011/the-global-cost-of-mental-illness.shtml).

MHPSS during emergencies (http://www.who.int/mental_health/emergencies/en/) is one subset of the GMH activities provided by the World Health Organization, other United Nations organizations, and non-United Nations partners.

2. Mental health is included in the Sustainable Development Goals, 2015–30

The United Nations Sustainable Development Goals (SDGs) for 2015–30 provide internationally recognized guidance for worldwide humanitarian action. Based on international advocacy, rallying under the banner of "there is no health without mental health," the SDGs do include mental health as one of the "targets" subsumed under Goal 3: Health and well-being ("Ensure healthy lives and promote well-being for all"). A set of indicators to assess progress toward achieving improved mental health will be incorporated into the SDGs. This is an important advance; mental health was not included in the predecessor set of Millennium Development Goals (MDGs), 2000–15.

The World Health Organization (2015) publication, *Health in 2015: From MDGs to SDGs* provides a detailed description of Goal 3 (http://www.who.int/mediacentre/news/releases/2015/mdg-sdg-report/en/).

3. MHPSS and the United Nations "Cluster System" for humanitarian response in disasters

Regarding EM/MHPSS integration, when disaster strikes, the United Nations organizes a multicomponent response using the "cluster system." The work of the clusters is guided by the United Nations Office for the Coordination of Humanitarian Assistance (UNOCHA) and response actions are triggered by disaster impact. The World Health Organization is the lead agency for the Health Cluster and this includes mental health and MHPSS services.

The clusters (listed alphabetically) focus on:
- Agriculture: [Food and Agriculture Organization (FAO)]
- Camp Management: [Office of the High Commissioner for Human Rights (UNHCR), International Organization for Migration (IOM)]
- Early Recovery: [The United Nations Development Program (UNDP)]

(Continued)

TABLE 5.1 (Continued)

- Education: [The United Nations International Children's Emergency Fund (UNICEF), Save the Children]
- Emergency Shelter and Non-Food Items: [Office of the High Commissioner for Human Rights (UNHCR), International Federation of the Red Cross and Red Crescent (IFRC)]
- Emergency Telecommunications: [World Food Program (WFP)]
- Food Security: [World Food Program (WFP), Food and Agriculture Organization (FAO)]
- Health: [World Health Organization/Pan American Health Organization (WHO/PAHO)]
- Humanitarian and Emergency Relief Coordination: [Office for the Coordination of Humanitarian Assistance (OCHA)]
- Information Management: [Office for the Coordination of Humanitarian Assistance (OCHA)]
- Logistics: [World Food Program (WFP)]
- Nutrition: [The United Nations International Children's Emergency Fund (UNICEF)]
- Protection: [Office of the High Commissioner for Human Rights (UNHCR), with UNICEF for Child Protection and the United Nations Population Fund for gender-based violence]
- Water, Sanitation, and Hygiene (WASH): [The United Nations International Children's Emergency Fund (UNICEF)]

Disasters create demands across multiple dimensions and the cluster system is designed to meet these needs in a flexible manner, dividing tasks and responsibilities among agencies best suited to assess needs and deliver goods and services.

Regarding MHPSS, the actions of many clusters that provide practical assistance to disaster survivors also may have favorable mental health effects by diminishing stress and distress. It is well known that the key elements of early psychological intervention and support are the provision of safety, calming, connection, self-efficacy, and hope (Hobfoll et al., 2007) and the work of most clusters contributes to these elements.

4. Scientific and programmatic guidance for MHPSS services

Mental health experts within the World Health Organization have spent the past decade creating a comprehensive set of guidelines and tools for the provision of MHPSS. They work collaboratively with many GMH scientific experts and humanitarian response organizations.

When disaster strikes, the MHPSS response is guided by the "4Ws": *Who is Where, When, doing What (4Ws) in Mental Health and Psychosocial Support* (IASC, 2012). This guidance is highly pragmatic; the 4Ws come with Excel worksheets, activity codes, and mechanisms for monitoring activities and coordinating with the clusters.

One of the foundational documents, the *Inter-Agency Standing Committee Guidelines on Mental Health and Psychosocial Support in Emergency Settings* (IASC, 2007) was developed by the Inter-Agency Standing Committee, comprised of multiple United Nations agencies (generally those involved with the cluster system) and multiple outside (non-United Nations) entities including several international Red Cross societies (ICRC, IFRC) and InterAction, an umbrella organization for more than 100 response agencies. This publication introduces the Intervention Pyramid for MHPSS in Emergencies. The Pyramid has four tiers and the services provided by multiple "clusters" contribute to all tiers: (1) basic services and security, (2) community and

(Continued)

TABLE 5.1 (Continued)

family supports, (3) focused, nonspecialized supports, and (4) specialized services. The Health Cluster includes teams of GMH generalists and specialists who deliver the tier 3 and 4 support that is specific to MHPSS.

Especially relevant for tier 1, understanding that humanitarian emergencies create demands for practical needs—and that lack of survival needs also produces psychological distress—the World Health Organization has created the *The Humanitarian Emergency Settings Perceived Needs Scale (HESPER)*. Early assessment using the HESPER helps to organize the optimal levels of involvement of the relevant clusters that must be brought to bear for effective response.

The World Health Organization and Inter-Agency Standing Committee have designed a complete toolkit of assessment instruments. These instruments are subdivided into three categories of tools: (1) for coordination and advocacy (these include the 4Ws and HESPER), (2) for MHPSS through health services, and (3) for MHPSS through different sectors, including through community support.

The World Health Organization, in collaboration with The War Trauma Foundation and World Vision, has together developed a PFA field guide to help structure early intervention for disaster-affected populations. This is one of the interventions subsumed under tier 3, focused Although there is lack of evidence for the efficacy of PFA (Shultz & Forbes, 2013), the World Health Organization developers of this simple, well-illustrated PFA model (now available in more than a dozen languages), are taking steps to evaluate the intervention, beginning with qualitative analyses of actual field applications.

5. Evidence-based intervention guidance

The World Health Organization Health cluster interventions are based on scientific evidence (e.g., trauma-focused cognitive behavioral therapy, TF-CBT). As one of the steps in recent, the World Health Organization has commissioned extensive evidence reviews to assess the efficacy of psychological interventions when used with survivors in disasters and humanitarian emergencies.

Some of the GMH science is supported by Grand Challenges Canada, the US National Institute on Mental Health (NIMH), and Wellcome Trust. Numerous GMH pilot studies and a number of transition-to-scale studies have been conducted with disaster-affected populations demonstrating the efficacy and effectiveness of various interventions.

Guidelines for the systematic implementation of evidence-based treatments have been developed (e.g., the World Health Organization's *mhGAP Humanitarian Intervention Guide*). This guide covers all mental health, neurological, and substance abuse (MNS) conditions and disorders. In 2016, the World Health Organization will officially receive transfer of copyright from Columbia University to adapt interpersonal psychotherapy (IPT) for global dissemination. IPT and the briefer interpersonal counseling (IPC) version have demonstrated efficacy for reducing symptom elevations for major depression, posttraumatic stress disorder (PTSD), and anxiety disorders in both clinical and humanitarian emergency settings. An IPT/IPC expertise and training center is planned in Medellin, Colombia with the World Health Organization participation.

6. EM/MHPSS integration when providing GMH interventions

Matching response partners to disaster needs, including MHPSS, is complex. As one critical resource for organizing and monitoring the response, OCHA has established

(Continued)

TABLE 5.1 (Continued)

ReliefWeb (reliefweb.int) as a shared online portal for the humanitarian community. ReliefWeb that serves as a repository for current and historical information on disaster events and complex emergencies. Cross-referenced by disaster, country, and organization, this service compiles information in real time from disaster situation reports and response organization updates to provide a comprehensive look at the human impact of disasters.

Once responder organizations are in the field and MHPSS has been prioritized, personnel must be trained and supervised to deliver the psychosocial assistance and the evidence-based interventions. However, in many Low and Middle Income Countries (LAMICs), there is a lack of trained mental health providers. Fortunately, excellent science has been brought to bear on this almost-universal need for MHPSS providers. It is possible to rapidly train community health workers and paraprofessionals to serve as counselors to deliver the interventions such as IPC/IPT. This "task-shifting" approach has been found to be effective therapeutically and offsets the extreme shortages of mental health professionals in LAMICs. This success is now leading to the training of paraprofessional volunteers in highly developed nations also.

Another recently launched online resource for GMH initiatives generally, and MHPSS applications specifically, is the Mental Health Innovation Network (MHIN, http://mhinnovation.net). Based at the Center for Global Mental Health at the London School of Hygiene and Tropical Medicine, MHIN serves as a platform for promoting state-of-the-art GMH innovations and cutting-edge advances to the evidence base. MHIN also facilitates networking among researchers, policy-makers, humanitarian actors, and multinationals including the United Nations, the World Health Organization and the Inter-Agency Standing Committee.

Note: MHPSS* is the international terminology that is equivalent to DBH in the United States.

FIGURE 5.1 The cluster system for multisectoral coordinated response in disasters. *From Cluster Coordination, by © United Nations. Reprinted with the permission of the United Nations. Retrieved from http://www.unocha.org/what-we-do/coordination-tools/cluster-coordination.*

REFERENCES

Cavallo, A. (2014). Integrating disaster preparedness and resilience: A complex approach using system of systems. *Australian Journal of Emergency Management, 29,* 46–51.

Cavallo, A., & Ireland, V. (2014). Preparing for complex interdependent risks: A system of systems approach to building disaster resilience. *International Journal for Disaster Risk Reduction, 9,* 181–193. Available from http://dx.doi.org/10.1016/j.ijdrr.2014.05.001.

FEMA. (2015). *National Response Framework.* Available at: http://www.fema.gov/national-response-framework.

Florida Department of Health. (2014). *Fact sheet: Disaster behavioral health response teams.* Available at: http://www.floridahealth.gov/programs-and-services/emergency-preparedness-and-response/healthcare-system-preparedness/disaster-behavioral-health/#heading_2.

Florida Department of Health. (2015). *Disaster behavioral health standard operating guidelines.* Available at: http://www.floridahealth.gov/programs-and-services/emergency-preparedness-and-response/healthcare-system-preparedness/disaster-behavioral-health/#heading_2.

Helbing, D. (2013). Globally networked risks and how to respond. *Nature, 497,* 51–59. Available from http://dx.doi.org/10.1038/nature12047.

Helbing, D., Ammoser, H., & Kühnert, C. (2005). Disasters as extreme events and the importance of network interactions for disaster response management. In S. Albeverio, V. Jentsch, & H. Kantz (Eds.), *The unimaginable and unpredictable: Extreme events in nature and society* (pp. 319–348). Berlin: Springer.

Helbing, D., Brockmann, D., Chadefaux, T., Donnay, K., Blanke, U., Woolley-Meza, O., & Perc, M. (2015). Saving human lives: What complexity science and information systems can contribute. *Journal of Statistical Physics, 158,* 735–781.

Hobfoll, S. E., Watson, P., Bell, C. C., Bryant, R. A., Brymer, M. J., Friedman, M. J., ... Layne, C. M. (2007). Five essential elements of immediate and mid-term mass trauma intervention: Empirical evidence. *Psychiatry, 70*(4), 283–315.

Inter-Agency Standing Committee. (2007). *IASC Guidelines on Mental Health and Psychosocial Support in Emergency Settings.* Geneva, Switzerland. Available at: https://interagencystandingcommittee.org/system/files/legacy_files/guidelines_iasc_mental_health_psychosocial_june_2007.pdf.

Inter-Agency Standing Committee. (2012). *Who is Where, When, doing What (4Ws) in Mental Health and Psychosocial Support: Manual with Activity Codes.* IASC Reference Group for Mental Health and Psychosocial Support in Emergency Settings. Geneva, Switzerland. ISBN: 978-9953-0-2496-7. Available at: http://www.who.int/mental_health/publications/iasc_4ws.pdf.

National Biodefense Science Board. (2008). *Disaster Mental Health Recommendations Report of the Disaster Mental Health Subcommittee of the National Biodefense Science Board.* Available at: http://www.phe.gov/Preparedness/legal/boards/nprsb/Documents/nsbs-dmhreport-final.pdf.

Neria, Y., & Shultz, J. M. (2012). Mental health effects of Hurricane Sandy: Characteristics, potential aftermath, and response. *JAMA, 308*(24), 2571–2572. Available from http://dx.doi.org/10.1001/jama.2012.110700.

Norris, F., Friedman, M. J., Watson, P. J., Byrne, C., & Kaniasty, K. (2002a). 60,000 disaster victims speak: Part I. An empirical review of the empirical literature: 1981–2001. *Psychiatry, 65,* 207–239.

Norris, F., Friedman, M., & Watson, P. (2002b). 60 000 disaster victims speak, part 2: Summary and implications of the disaster mental health research. *Psychiatry, 65,* 204–260.

Office of the Assistant Secretary for Preparedness and Response (2014). *HHS Disaster Behavioral Health Concept of Operations (DBHCONOPS).* ASPR, U.S. Department of Health and Human Services (HHS). Available at: http://www.phe.gov/Preparedness/planning/abc/Documents/dbh-conops-2014.pdf.

Quarantelli, E. L. (2005). Catastrophes are different from disasters: Some implications for crisis planning and managing drawn from Katrina. The Social Science Research Council. *Understanding Katrina: Perspectives from the social sciences.* Retrieved from http://understandingkatrina.ssrc.org/Quarantelli/.

Shultz, J. M., Baingana, F., & Neria, Y. (2015). The 2014 Ebola outbreak and mental health: Current status and recommended response. *JAMA, 313*(6), 567−568. Available from http://dx.doi.org/10.1001/jama.2014.17934.

Shultz, J. M., & Cohen, M. A. (2015). Disaster Health Briefing: Disaster health maxim: think locally, act globally. *Disaster Health, 2*(3−4), 146−150. doi: http://dx.doi.org/10.1080/21665044.2014.1090274.

Shultz, J. M., Espinel, Z., Flynn, B. W., Hoffman, Y., & Cohen, R. E. (2007). *DEEP PREP: All-hazards disaster behavioral health training.* Tampa, FL: Disaster Life Support Publishing.

Shultz, J. M., Espinel, Z., Galea, S., & Reissman, D. E. (2007). Disaster ecology: Implications for disaster psychiatry. In R. J. Ursano, C. S. Fullerton, L. Weisaeth, & B. Raphael (Eds.), *Textbook of disaster psychiatry* (pp. 69−96). Cambridge, UK: Cambridge University Press.

Shultz, J. M., Espinola, M., Rechkemmer, A., Cohen, M. A., & Espinel, Z. (2016). Prevention of disaster impact and outcome cascades. In M. Israelashvili, & J. L. Romano (Eds.), *Cambridge handbook of international prevention science.* Cambridge, UK: Cambridge University Press.

Shultz, J. M., & Forbes, D. (2013). Psychological first aid: Rapid proliferation and the search for evidence. *Disaster Health, 1*(2), 1−10. Available from http://dx.doi.org/10.4161/dish.26006.

Shultz, J. M., Galea, S., Espinel, Z., & Reissman, D. E. (2016). Disaster ecology. In R. J. Ursano, C. S. Fullerton, L. Weisaeth, & B. Raphael (Eds.), *Textbook of disaster psychiatry* (2nd edition). Cambridge, UK: Cambridge University Press.

Shultz, J. M., Garcia-Vera, M. P., Gesteira Santos, C., Sanz, J., Bibel, G., Schulman, C., ... Rechkemmer, A. (2016). Disaster complexity case study: Disaster complexity and the Santiago de Compostela train derailment. *Disaster Health, 3*(1), 1−21. Available from http://dx.doi.org/10.1080/21665044.2015.1129889.

Shultz, J. M., Marcelin, L. H., Madanes, S., Espinel, Z., & Neria, Y. (2011). The trauma signature: Understanding the psychological consequences of the Haiti 2010 earthquake. *Prehospital and Disaster Medicine, 26*(5), 353−366.

Shultz, J. M., McLean, A., Herberman Mash, H. B., Rosen, A., Kelly, F., Solo-Gabriele, H. M., ... Neria, Y. (2013). Mitigating flood exposure: Reducing disaster risk and trauma signature. *Disaster Health, 1*(1), 30−44.

Shultz, J. M., Muschert, G. W., Dingwall, A., & Cohen, A. M. (2013). The Sandy Hook Elementary School shooting as tipping point: "This time is different." *Disaster Health, 1*(2), 65−73. Available from http://dx.doi.org/10.4161/dish.27113.

World Health Organization. (2015). *Health in 2015: From MDGs, millennium development goals to SDGs, sustainable development goals.* Geneva, Switzerland: World Health Organization. ISBN 978 92-4-156511-0. Available at: http://www.who.int/mediacentre/news/releases/2015/mdg-sdg-report/en/.

Through an Emergency Management Lens

Marianne C. Jackson

AN EMERGENCY MANAGEMENT PERSPECTIVE

Through the years and much trial and error, the EM community, in partnership with colleagues in DBH, has begun to develop experience in building integration between these two essential disciplines. The EM communities in some locales have extensive experience in integrating these two disciplines.

In an ever-changing world, a look at lessons learned from different types of incidents helps illustrate ways to build and sustain EM and DBH integration. This section will present four case studies that demonstrate how EM and DBH did successfully integrate and how that partnership resulted in positive outcomes. Each of the situations described here indicate the diversity of the types of situations that can, and will, arise beyond more typical types of disasters. In addition, they illustrate integration with DBH that reflects need for, and value of, innovation and creativity.

INTEGRATION IN AN URBAN EVACUATION: YONKERS MUDSLIDE

In this author's opinion and experience, on March 11, 2015, the rapid response to the Yonkers, New York mudslide illustrates how local government, in this case the City of Yonkers in Westchester County, and the American Red Cross (ARC) of Greater New York, partnered to help residents of two senior citizen high-rise apartment buildings evacuate when a mudslide threatened the structures. The site had been a source of concern for local officials because of known threats. The geology of the area, surrounding development, recent rain, and melted snow had raised awareness of the potential danger.

Both the city of Yonkers and Westchester County, directly north of New York City, are large entities with strong staffing. The American Red Cross of Greater New York, with a large number of volunteer licensed mental health experts with DBH experience, was available on short-notice. The actions of all responding organizations were managed at an onsite Incident Command Post. This is where the integration between EM and DBH staff initially occurred.

In addition, many of the responders from the various organizations had experience with other area emergencies and had worked together in the past.

They had also participated in disaster exercises together. They were able to build off their existing relationships and integrate their activities. *They were not strangers to each other.*

> The scene of an incident is not the ideal time for responders to meet face-to-face. There is an underlying need, in whatever jurisdiction, for essential staff to know each other prior to an event. These relationships are an essential part of integrating EM with DBH professionals.

The 109 seniors, some of whom had mobility issues but were largely independent, were evacuated in the afternoon, leaving personal property behind, including medications and clothes. With Yonkers EM and Westchester County officials, a wheelchair accessible reception center was set-up by the ARC while hotel rooms were located. The ARC team of licensed mental health professionals, led by a seasoned ARC mental health expert, provided a compassionate presence to the displaced residents. The Westchester County Commissioner of Health was onsite, writing prescriptions for residents, which were delivered by hand messenger to a nearby pharmacy, filled, and hand-carried back to the reception center.

Within 12 hours, beginning with the mandatory evacuation order, all residents were safely lodged, either in hotels or with family and friends. Throughout the event, the residents were protected from the media. This was accomplished because no responding agency released any names or relocation information. This ensured the survivors' privacy and eliminated the need for anyone to relive or retell their stories of going through something they could neither control nor respond to themselves. Because temporary slope stabilization was not feasible, the residents were eventually permanently relocated to other senior living facilities.

INTEGRATION IN A MULTICULTURAL MASS VIOLENCE EVENT: BINGHAMTON NEW YORK SHOOTINGS

The Binghamton New York shooting took place in the afternoon of April 3, 2009. A young man killed 13 people, took hostages, and wounded others before committing suicide at the American Civic Association (ACA) Immigrant Center (McFadden, 2009). Binghamton, a city with a population of 50,000, is in Broome County, which is about 180 miles northwest of New York City. It is home to Binghamton University. The shooter was a naturalized US citizen from Vietnam and had taken English language classes at the center. The police responded within minutes of receiving a 911 call from a wounded victim hiding under a desk.

This incident required coordination with many entities from the city, Broome County, and New York State—especially from the law enforcement community. Language and cultural differences were a significant factor in both immediate and ongoing communications. Immediate access to language translation services, especially for the coroner's office, was essential. The on-scene commander immediately called in a Professor from a local college who was fluent in Vietnamese to help communicate with the shooter in the event of contact. Following the incident itself, most information was initially provided in English only. The fatalities, the wounded, and the survivors were from Pakistan, Brazil, Haiti, China, and Iraq. Binghamton University was able to provide translations services to help deal with the language challenges.

Crime victim assistance agencies at the state and federal levels, along with the Broome County Mental Health Department, mobilized to assist those impacted by the horrific event. Broome County opened a Victim Assistance Center which offered assistance, such as counseling, legal advice, and translator services. After the center closed, a Victim Assistance Hotline was established that was staffed by local and state mental health specialists and social workers.

This incident was primarily a law enforcement response and required the mental health counselors to coordinate with the officers as survivor interviews were conducted. This coordination required the quick development of a new relationship between law enforcement and DBH groups. While the response to this incident was appropriately led by law enforcement, not EM, the Broome County Office of Emergency Management (OEM) initiated an after action review process, which included all parties involved in the incident. The After Action Report (AAR) contained several key recommendations for all of the responding agencies. They make the case and point the way for EM/DBH integration.

Summary of After Action Recommendations

- Increase capability to communicate with those for whom English is not their first language
- Host and promote training in the Incident Command System (ICS) to ensure complementary response and recovery actions
- Predesignate inter-agency points of contact, and develop and share a list that includes a DBH component as well as sources of translation services
- Participate in discussions and exercises with law enforcement and EM that involve all of the agencies who may need to provide resources

---

All of these recommendations present opportunities for the EM and DBH professions to reach out, collaborate, develop joint resource lists, and ensure everyone knows whom to call, as well as when to make the call.

INTEGRATION IN A HUMAN EXPLOITATION CASE: FORCED LABOR OF DEAF MEXICANS

The prior case examples were both no-notice events. They were visible, dramatic, and the types of events that, when preparedness occurs, are not unusual. They are examples of natural disasters and human-generated events. This case example is another less typical no-notice event requiring law enforcement involvement and support from DBH and other social service professionals. It involves the 1997 escape and subsequent rescue of enslaved Mexicans, all of whom were deaf and forced to sell trinkets in the subways of New York City.

Editors' Note: Deaf Mexicans is a social identity label (Davis, 2016; Goldstein, 2006; Ojito, 1998).

The event began when a couple of men entered the 114th New York Police Department (NYPD) precinct in Queens, appearing to want help, but not able to communicate with police officers. The Mayor's OEM brought in staff from the Mayor's Office for People with Disabilities (MOPD) to help evaluate the situation. By then, the group had grown to 64 people, all deaf, ranging in age from babies to adults. The liaison from MOPD, fluent in American Sign Language (ASL), reported that the Mexicans did not know ASL, and only a few even used Mexican Sign Language. Most used forms of "home sign" systems (National Institutes of Health, 2015) used by deaf children who live in isolation. The victims needed interpreters, were in distress, and indicated that some of the slave ringleaders were in the group pretending to be victims.

With information gathered from the group, police soon discovered two crammed houses where the victims were held. Law enforcement personnel from city, state, and federal agencies quickly gathered, with officers keeping the news media two blocks away. More victims were found inside the houses. Rather than deport the victims back to Mexico, the Mayor declared the victims to be "material witnesses" who were detained in protective custody. A team of interpreters was coordinated by the MOPD and assembled by the Lexington School for the Deaf, including those with knowledge of Mexican Sign Language. To protect the victims from further stress, interviews were conducted by a small team which included interpreters and women police officers for the female victims.

The victims were witnesses in the subsequent trials of the criminal leaders. Eventually, some chose to return to Mexico, while those remaining were given green cards. Special visas were distributed for people who made significant contributions to criminal cases. Social workers and mental health specialists continued to support them during the many years they were in protective custody. The human traffickers were eventually convicted in 2006.

As told by Elizabeth Davis, MOPD liaison and Incident Commander for this incident (Davis, 2016).

INTEGRATION IN A TERRORIST EVENT: THE BOSTON MARATHON

The Boston Marathon Bombing took the lives of three people, injured hundreds, and shook the country. Unlike the earlier case example of the Binghamton shooting where the perpetrator committed suicide on-scene, the capture of the bombers was complicated by the multijurisdictional law enforcement response, resulting in the death of a police officer, the death of one of the two bombers, and the capture of the other.

The marathon was a planned event, and EM had developed an incident response plan and structure in preparation for the Marathon with a broad range of agencies and volunteer organizations standing by. After the bombing, significant additional resources were rapidly mobilized, including the Massachusetts National Guard (MNG) and US Department of Health and Human Services.

Massachusetts Office of Emergency Management coordinated the thorough after action report, addressing delivery of DBH in a number of sections. While there were many substantive recommendations, two were of significant interest to both EM and DBH.

- Develop a disaster mental health coordination plan[1]
- Develop a centralized source that preidentifies disaster mental health specialists

1. Editors' note: As discussed in the Introduction, the editors have chosen to use the term *behavioral health* (BH) rather than *mental health* (MH) throughout the book. Since the report referenced uses the term *mental health*, that term is used here.

Other observations found in the report include:

- Mental health counseling was quickly provided at the Unification Center, primarily used by international runners trying to retrieve personal property and return home
- Law enforcement personnel did not receive adequate mental health support
- Mental health care needs of nonpublic safety personnel and human services providers were not adequately addressed
- Lack of a Disaster Mental Health Coordination Plan resulted in confusion and duplication of efforts amongst the many responding professionals
- Some responding disaster mental health organizations came in from out of town without logistics support to obtain lodging and communications, stressing the overall operation

The entire report can be found at http://www.mass.gov/eopss/docs/mema/after-action-report-for-the-response-to-the-2013-boston-marathon-bombings.pdf.

We can all learn from the recommendations and observations. EM has to ensure that disaster mental health is part of the overall planning and preparedness process, as well as for individual, high-profile events. EM must also have a plan to deal with spontaneous volunteers and uninvited or unprepared groups from all professions, including DBH specialists.

UNDERSTANDING VARIATIONS IN LOCAL, STATE, AND FEDERAL DISASTER AUTHORITIES

Disaster preparedness and response is not based on unconstrained flexibility to meet all needs. There are limits, permissions, and definitions that apply at all levels of government and to nearly all elements of preparedness, response, and recovery. In order to assure an integrated response and a reduction in misunderstandings, all parties benefit when the various authorities are understood and tested in exercises.

Local, state, and federal chief executives have various disaster authorities for imminent or actual disaster events. The disaster authorities of Mayors, county executives, Governors, and the President carry the greatest power. Other government agencies may have more limited disaster authority. For example, at the federal level, the Administrator of the Small Business Administration (SBA) can approve a request from a Governor for low-interest disaster loans for businesses, homeowners, and renters in impacted counties. Similarly, the US Department of Agriculture and US Department of Transportation have disaster programs which can be triggered by agencies heads.

At the local level in New York City, the Mayor can declare a local state of emergency, empowering him or her to issue curfews and open emergency shelters for the duration of the state of emergency, or for such periods of time as is necessary to respond to existing conditions. The Mayor can request state assistance to supplement local efforts to save lives and to protect property, public health, and safety, or to avert or lessen the threat of a disaster.

At the New York state level, the Governor can declare a state of emergency, like all state Governors, in response to a request from a chief executive or on his own initiative. The Governor may direct any and all state agencies (including the State Mental Health Authority and/or Health Authority) to provide assistance and temporarily suspend specific provisions of statutes, rules, and regulations. Additionally, the Governor can activate the National Guard and reach out to other states for help through the Emergency Management Assistance Compact (EMAC). More information on EMAC can be found at www.emacweb.org.

Whenever the Governor finds that a disaster is of such severity and magnitude that an effective response is beyond the capabilities of the state and affected jurisdictions, he/she may request federal assistance from the President through the Federal Emergency Management Agency (FEMA) under the Robert T. Stafford Disaster Relief and Emergency Assistance Act. The President of the United States has the authority, under the Stafford Act, to approve an emergency declaration and/or a major disaster declaration, usually at the request of a Governor.

Local, state, and federal staffs conduct Preliminary Damage Assessments (PDAs) to determine the extent of the disaster and its impact on individuals and public facilities. However, when an obviously severe or catastrophic event occurs, such as the 9/11 attacks, the Governor's request may be substituted for the PDA.

There are two types of Presidential disaster declarations authorized in the Stafford Act—an Emergency declaration and a Major Disaster declaration. Both declaration types authorize the President to provide supplemental federal disaster assistance. An Emergency declaration provides federal assistance to lessen or avert the threat of a catastrophe. The President can declare a Major Disaster for any natural event or, regardless of cause, fire, flood, or explosion. Not all programs are triggered for every disaster; determination is based on needs. The two major programs are Public Assistance, which aids government and nonprofit entities, and Individual Assistance, which is aid given to individuals and businesses. Federal regulations require consideration of certain factors concerning aid to individuals. These factors include considerations of the effects to special populations, such as low-income, the elderly, and the unemployed (Code of Federal Regulations, 44 CFR Part 206.48b).

A program often triggered by a Major Disaster Declaration for Individual Assistance is the Crisis Counseling Program administered by the state in partnership with US Department of Health and Human Services. In addition to government-based local, state, and federal DBH resources, the broad resources from Voluntary Agencies Active in Disasters (VOAD) provide critical support to disaster-impacted communities. Voluntary organizations at all levels can be an integral part of the preparedness, response, and recovery framework, with DBH professionals working closely with social services providers.

REFERENCES

Code of Federal Regulations. 44 CFR Part 206.
Davis, E. (2016, January 22). Personal communication via email.
Goldstein, J. (2006, September 28). *Deaf Mexicans recount enslavement in the city*. Retrieved from http://www.nysun.com/new-york/deaf-mexicans-recount-enslavement-in-the-city/40556/.
McFadden, R. (2009, April 3). *13 shot dead during a class on citizenship*. Retrieved from http://www.nytimes.com/2009/04/04/nyregion/04hostage.html?_r=0.
National Institutes of Health. (2015, June 24). *American sign language*. Retrieved from https://www.nidcd.nih.gov/health/american-sign-language.
Ojito, M. (1998, July 18). *For deaf Mexicans, freedom after slavery and detention*. Retrieved from http://www.nytimes.com/1998/07/18/nyregion/for-deaf-mexicans-freedom-after-slavery-and-detention.html.

Making Integration Work

Brian W. Flynn and Ronald Sherman

The title of this chapter, "Integration Across Levels and Locations of Response: From Local to International," required a great deal from both contributors. Their experiences and chapter contributions range from the hyper-local response perspective to the challenges our international counterparts encounter.

Each of them has provided a unique perspective. Their combined experiences highlight the many differences that disaster incidents possess and how those differences can drive response and recovery activities, especially from the EM and DBH communities.

Emergency managers love checklists. They are a tool they routinely use to make sure they have not forgotten to alert a partner agency or request a resource—whether the resource is a truckload of generators or a team of DBH professionals. There are concrete ways both professions can make sure they are making strides toward integration and collaboration. Based on the authors' contributions in both parts of this chapter, following is a checklist both EM and DBH can use or adapt to begin to frame and guide that process (Table 5.2).

TABLE 5.2 Checklist for Integration of Emergency Management and Disaster Behavioral Health

Planning Action Items	Lead	Due Date
Identify lead agencies/organizations		
Include law enforcement		
Build relationships via meetings/calls		
Identify capabilities at all levels		
Ensure understanding of ICS		
Identify vulnerable populations		
Identify/designate human services lead		
Identify/designate DBH lead		
Create DBH Coordination plan and add to Emergency Operations Plan		

(Intended as a guide to be modified, as needed by users. Partners decide who has the Lead)

Dr. Shultz's section, earlier in this chapter, provides two perspectives that can provide overarching guidance for integration steps:

- Internationally, DBH in emergency settings has developed from a different process than in the United States, and resulted in different approaches. International advances in both EM and DBH can provide fresh and helpful perspectives for those EM and DBH professionals willing to explore these approaches, practices, and lessons learned
- The hierarchical incident management organizational structures are often not familiar to, or ideally suited to, DBH approaches. Integration efforts should focus on identifying where structures can be flexible to accommodate the needs and priorities of emergency managers, while also recognizing the longer-term, nonlinear aspects of DBH

Within DBH, researchers need to think like responders and response personnel need to take on the "mind" of a researcher. Practical and pragmatic researchers who are field-experienced are needed.

Chapter 6

Not All Disasters Are the Same: Understanding Similarities and Differences

Daniel W. McGowan[1] and James Siemianowski[2]
[1]*McGowan Enterprises, Inc., Helena, MT, United States,* [2]*CT Dept Mental Health and Addiction Svcs., Hartford, CT, United States*

Through an Emergency Management Lens

Daniel W. McGowan

The definition of a disaster is relative to those experiencing the event. For the purpose of this discussion, the term disaster will include incidents, emergencies, or disasters. An incident is usually an occurrence that is managed by the local jurisdiction. The local jurisdiction does not necessarily equate to a governmental entity; it could be a school district. In these cases, usually a single first response agency responds to the incident and handles the situation. Limited resources and assistance are needed to manage the effects of the incident. An emergency is more complex requiring more than one response entity, an elevated level of resources, has a greater impact on the community, and possibly requires assistance from outside the jurisdiction, which could include state involvement. A disaster, on the other hand, is much more complex than an incident or an emergency. A disaster involves multiple agencies, requires resources from outside the jurisdiction, and typically involves state and/or federal assistance. Regardless of the type of the disaster, workers and victims can be affected from a disaster behavioral health (DBH) perspective because they are involved in abnormal situations and potentially need help normalizing the effects of their involvement on their lives. By the very nature of these definitions alone, it is easy to see the complexities involved with trying to integrate two or more disciplines and occupational cultures especially in light of the variances among events.

At the onset of any disaster, there are a multitude of issues and circumstances bombarding the initial response and recovery efforts. Some of these elements are connected and others have no real relationship to each other. As the event continues, cascading events often occur and elevate the complexity of dealing with the situation. For emergency managers, there can be a sense of disconnectedness between the original mission and the demands of newly emerging challenges.

The challenge for emergency management (EM) is to provide leadership and a sense of organization resulting in an environment where circumstances can be managed.

At the start of a disaster response, the emergency manager can be likened to an artist. When the artist begins, he or she splatters the canvas with all different colors. As they begin connecting the colors together, a recognizable picture begins to form. The outcome is a picture that is a cohesive unit and the artist has managed to represent a thought, idea, occurrence, thing, or place.

The EM's output is a well-constructed action plan for the response.

The integration of a DBH initiative becomes even more complicated when you consider that no two disasters are the same. It is true that floods, tornadoes, hurricanes, and wildfires are a common occurrence in various parts of the country and have taken place multiple times. Each event, however, differs in size, geographical coverage, magnitude, severity, and complexity.

There are major differences among even frequently occurring event types when they strike different areas at different times. To demonstrate the point, the following is a synopsis of several hurricane disasters (National Oceanic and Atmospheric Administration, 2015):
- The Galveston Hurricane of 1900 was a Category 4 hurricane that buffeted Galveston Island with 8–15 foot waves. The storm was noted as the deadliest hurricane in US history. The death toll reached 8000 and the estimated damages were ~30 million dollars
- The New England Hurricane of 1938, known as the Long Island Express, started as a Category 5 storm and diminished to a Category 3 by the time it struck the North Carolina coast. The storm brought sustained winds of 121 mph with gusts up to 183. The storm surge brought 10–12 foot waves and inundated the coast from Long Island and Connecticut to Southeastern Massachusetts. The storm caused 600 deaths and 308 million dollars in damage
- In 1999, Hurricane Floyd was a Category 2 storm that struck the coast of North Carolina and continued up the coast to New England. The storm was noted for

(*Continued*)

(Continued)

its widespread rainfall of 10–19 inches. Flooding caused the majority of the 3–6 billion dollars in damage and was responsible for 50 of the 56 deaths
- In 2005 Hurricane Katrina struck Miami/Dade area of Florida and moved southeast while gaining intensity to a Category 5 hurricane with 175 mph sustained winds. It turned in the northerly direction and brought destruction to Louisiana, Mississippi, and Alabama. The storm caused an estimated 75 billion dollars in damage in the New Orleans area alone. Storm surge flooding capped 25–28 feet above normal tide and brought anywhere from 8 to 14 inches of rain. Katrina was responsible for 1000 deaths

The text box above represents a subset of 36 different hurricanes. The parameters and effects of each are entirely different with regard to geographical area, severity, magnitude, and complexity. This same principle applies to all disasters regardless of the cause—flood, fire, tornado, winter storm, etc. The only like qualities between and among each incident, emergency, or disaster are that each of them occur in a different manner, cover an entirely different geographical area, and present their own unique set of challenges. The real challenge will be developing a DBH initiative and strategy that encompasses the scope and breadth of almost infinite disaster possibilities.

NATURE OF PREPAREDNESS

Another element to disasters that reminds us of the variety of event similarities and differences is the level of preparedness among and between jurisdictions—local, state, tribal, and federal. Not unlike the variations in disasters, the level of jurisdictional readiness varies across the country in direct correlation to the robustness of the EM program. The EM discipline is one of the most, if not the most complex and diverse areas of professional concentration. Being prepared requires major planning efforts in collaboration, communication, coordination, cooperation, and network development. It is difficult to fully understand the full scope of EM and the degree of complexity required to ensure that a jurisdiction is prepared.

Not only must preparedness, response, and recovery take place, its elements must be able to be communicated in many ways to many audiences. It is especially hard to describe the scope of EM in a thirty-second elevator speech. Instead, I defer to the pictorial in Fig. 6.1. During my tenure as a State Emergency Management Director in Montana, I felt compelled to develop a pictorial of the EM discipline to help explain the dynamics and scope of the complex relationships involved in an integrated EM system. I call the pictorial *The Department of Emergency Services (DES) Spiderweb*.

132 SECTION | II Key Areas of Integration

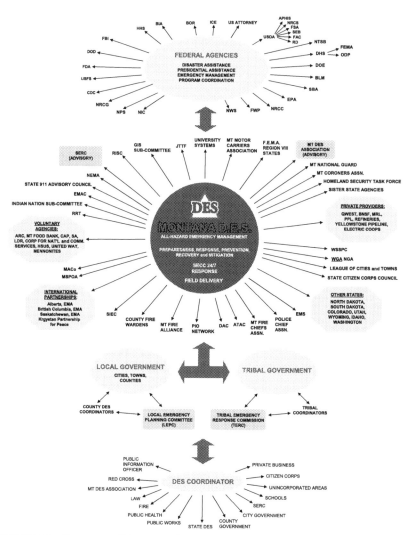

FIGURE 6.1 Department of Emergency Services (DES) Spiderweb.

The diagram depicts the myriad of potential relationships that need to be developed and nurtured. Each contributor at the federal, state, local, tribal, private, and voluntary level has something to bring to the table in mitigating, preparing for, responding to, and recovering from a disaster. These relationships, vertically and horizontally, have some kind of cause and effect relationship affecting the development of the preparedness program or the delivery of response and recovery efforts at the jurisdictional level. The comprehensiveness of the system and the level of preparedness are directly correlated to the level of stakeholder involvement and integration of the initiatives between and among jurisdictions. Further complicating the issue is the fact that the state realizes an expanded responsibility to foster network development. The state is the pivotal partner. In order to implement an effective program, state EM must coordinate preparedness efforts with the federal government, develop the state's preparedness, and assist the local and tribal jurisdictions with their efforts. One may think that the state does not have to worry about further developing relationships at the federal level. The state, however, does work with some federal agencies individually with regard to *agency only declarations*. Some federal agencies, like the Farm Service Agency and the Small Business Administration, have the authority to issue agency only declarations for assistance they can provide to a state or local jurisdiction without a presidential declaration.

In the case of tribal government, there is a quite different picture affecting the complexity of preparedness. The network is affected by how services are performed on the reservation. Services are rendered by compact or contract. Under the compact scenario, the tribe relies on the federal government to provide services on the reservation. The federal government providing law enforcement on a reservation for the tribe is an example of such a compact. On the other hand, some tribes contract with the federal government for so much in funding and take on the responsibility to provide the service themselves. In this case, the tribe would have its own law enforcement structure. Taking the complexity one-step further, not all tribes receive services by one method; it is usually a mix of compact and contract services. The difference in authorities makes the coordination and collaboration at the tribal level much more multifaceted. The implementation of programs will vary based on the level of involvement of the federal stakeholders and the coordination networks that must be developed vertically and horizontally. The variance in jurisdictional authority on a reservation has a direct effect on achieving the desirable outcome of assembling a comprehensive EM program. The tribal picture is further complicated by the fact that there are two avenues to seek assistance. After the Hurricane Sandy Recovery Improvement Act of 2013 was passed, tribes were granted the authority to deal directly with the President to request a declaration. The tribes, however, still have the option to partner with the state and submit a request through the Governor. The avenue of choice depends on what the tribe perceives as the most effective outcome for them (i.e., which approach will yield the most disaster aid in the shortest time).

Keep in mind, the network depicted in the DES Spiderweb is not entirely the same in every state and will vary according to the relationships built among companion stakeholders as well as their degree of involvement toward the preparedness program efficiency and delivery of assistance. The variances in relationships and the degree of organizational development present many challenges to forming a strong foundation for response and recovery. Not unlike disasters, no two response or recovery efforts are exactly the same. Most will have some core elements in common, like the Incident Command System (ICS), Communications, Damage Assessment, and Public Information. Beyond those, the *requirements of the disaster* drive the activation of other response or recovery elements. The robustness of the preparedness element is directly related to the depth of relationship development and the available funding that fuel the effort. Funding for programs across the country varies by jurisdiction. Funding is a major contributor to the robustness of the program. It is important to realize that not all jurisdictions place the same priority on the EM program development. Consequently, there are varying levels of program development across the country. As an example, some local jurisdictions do not have full-time EM Coordinators. Some jurisdictions are lucky to fund quarter-time coordinators. Based on the complexity of the discipline, a quarter-time EM coordinator barely has the time to open the mail. Underfunding the EM Coordinator function directly impacts the level of program effectiveness, degree of relationship development, and the thoroughness of initiative integration.

Variations in the nature of employment for EM management personnel create another level of complexity across the country. On the state level alone, some EM Directors are appointed by and serve at the pleasure of the Governor, others work under the direction of The Adjutant General, and some work under the direction of the Public Safety Commissioner. This scenario creates a real dilemma with respect to program consistency across the country. For example, when the EM function is part of the Governor's cabinet, the respective program direction can be altered each time there is a change in a state's Governor. Likewise, not all program direction will be coordinated between Governor's, Adjutant Generals, or Public Safety Commissioners responsible for the EM development across the country. The key is to bring all of the stakeholders together to develop a collaborative national approach. The outcome can be likened to the development of a major league baseball effort. In order for the league to be effective, the league must collectively work with the teams, the teams must coordinate with the league and among themselves, the finances must be available to develop the program and fund equipment needs, and all the teams have a different subset of players. The common focus is winning. The lack of any foundation elements will weaken the league and create a multitude of variances across the country.

NATURE OF FOUNDATION ELEMENTS

The foundation elements of any program must be aligned in order to develop an effective program and accomplish full integration of initiatives. Homeland Security Presidential Directive 5 (HSPD 5) mandated the implementation of the National Incident Management System (NIMS). Federal grant funding for program development required local, state, and tribal jurisdictions to be National Incident Management System compliant. One of the components of the system is the use of the ICS. The ICS is prescribed as the common management system for any event in whatever jurisdiction is experiencing the disaster. The jurisdictions across the country have integrated ICS into their preparedness efforts. To this point, ICS is institutionalized as the core management architecture for response and recovery efforts at the local, tribal, state, and federal levels of engagement.

> The key to an effective integration of DBH into EM is reaching the point where the initiative is institutionalized as part of disaster response and recovery. In order for the full integration to occur, the foundation elements must be present. If the program elements or policy guidance is not well developed, the integration of a DBH initiative will be ineffective.

At this point in history, all the foundation elements are not aligned, fully developed, or articulated for integrating a DBH component like HSPD 5 and ICS. Remember, these foundation elements are the guidance documents for all local, state, and tribal EM preparedness and initiative development. The critical component of the foundation is the National Preparedness Goal, which is comprised of 32 core capabilities. These core capabilities are distributed over five mission areas: Prevention, Protection, Mitigation, Response, and Recovery. The capabilities are the backbone elements for preparing our nation to deal with disasters. Many of the capabilities are integral parts of various frameworks like response and recovery. The Public Health core capability is structured to "provide lifesaving medical treatment via emergency medical services and related operations and avoid additional disease and injury by providing targeted public health and medical support and products to all people in need within the affected area" (U.S. Department of Homeland Security, 2011, p.15). The DBH component is not even mentioned in any of the core capabilities.

Presidential Policy Directive 8 is the federal document that establishes the foundation for preparedness and mandates the framework for how the nation responds: National Response Framework (NRF) (The National Response Framework, 2013). The NRF includes Emergency Support Functions (ESFs) that outline the responsible entities for various elements of assistance delivery. The NRF outlines the purpose, scope, policies, concept of operations, and actions necessary for the identified initiative. The document also establishes the ESF Coordinating Agency, Primary Agency, and lists the support agencies.

The NRF references the term mental health in ESF #8—Public Health and Medical Services Annex. It is interesting to note that the construct of DBH for this publication mirrors the tenants of ESF #8. According to the

Federal Emergency Management Agency (Federal Emergency Management Agency, 2008), "Public Health and Medical Services include responding to medical needs associated with mental health, behavioral health, and substance abuse considerations of incident victims and response workers" (p. 1). The scope of ESF #8 provides for the assessment of public health and medical needs as well as behavioral health care. The concept of operations includes the responsibility for the assessment of public health needs to determine the appropriate response capabilities. The primary agency, US Department of Health and Human Services (HHS), may task other supporting agencies and ESFs to assess mental health and substance abuse needs (Federal Emergency Management Agency, 2008, pp. 2—6). The very word "may" provides that implementing a DBH component is optional. Interestingly enough, the American Red Cross (ARC) is a supporting agency to ESF #8. The American Red Cross "provides supportive counseling for family members of the dead, for the injured, and for others affected by the incident" (Federal Emergency Management Agency, 2008, p. 15). The only way that the American Red Cross can provide these services is to be aware of the need. Current procedures do not provide for a thorough needs assessment. Without a structured comprehensive needs assessment and follow-up program, the needs will only surface if the individuals are referred to the American Red Cross or they are identified in a Disaster Recovery Center. The real concern with ESF #8 is that the system is set up to help the victims and does not address the behavioral health component for workers as referenced above in the Federal Emergency Management Agency ESF #8 purpose statement. It is true that the Worker Safety and Health support annex coordinated by the Department of Labor is referenced in ESF #8, but it is geared toward occupational safety issues and does not address the behavioral or mental health needs of workers. *The salient point is there are several incongruities within ESF #8 that prevent it from fully establishing the policies and guidelines for an integrated DBH component.*

The National Disaster Recovery Framework (NDRF) "focuses on how best to restore, redevelop and revitalize the health, social, economic, natural and environmental fabric of the community and build a more resilient Nation" (Federal Emergency Management Agency, 2015). The National Disaster Recovery Framework is comprised of six Recovery Support Functions (RSFs). The RSF for Health and Social Services, led by HHS, is responsible for "the ability to restore and improve health and social services networks to promote the resilience, health, independence and well-being of the whole community" (Federal Emergency Management Agency, 2011, p. 2). This foundational recovery element makes no mention of any DBH elements and focuses on the social service network restoration.

The Target Capabilities List (TCL) is another federal effort comprised of 37 core capabilities. The document is used to assess the level of our nation's preparedness. The TCL, at least, acknowledges the need for mental health services for victims and workers under the planning assumptions section. The TCL elements, however, fail the litmus test for a comprehensive DBH initiative. From the worker perspective, the Responder Safety and Health Capability acknowledges, "providing behavioral health and substance abuse services as a critical task under demobilization of responders" (U.S. Department of Homeland Security, 2007, p. 255). Demobilization is after the fact for circumstances that could have drastically impacted active efforts during response and recovery. Community members are addressed under the response Medical Surge Capability as providing "short-term mental health and substance abuse behavioral health services to the community" (U.S. Department of Homeland Security, 2007, p. 455). In this case, the provision only addresses the short-term needs. Neither of these approaches is comprehensive. The worker element only addresses the mental health needs at demobilization and does not address the response or recovery phases or related issues that surface long after the disaster. The community element is only short-term and affected victims may surface months after the disaster.

The ESF #6—Mass Care, Emergency Assistance, Housing, and Human Services Annex is the closest to providing for a mental health component with some sense of action. ESF #6 is designed to provide crisis counseling services to victims and training for EM professionals. The main emphasis is placed on children and those who require assistance with daily living. This ESF relies primarily on the National Voluntary Organizations Active in Disaster to provide services. Once again, there is not a comprehensive approach to DBH, given the fact that services are rendered on a short-term basis. There is no provision for a thorough needs assessment, follow-up process, or consideration for issues that develop long after the disaster.

Federal Emergency Management Agency does have a preparedness guide. According to Federal Emergency Management Agency (2010),

> *"Comprehensive Preparedness Guide (CPG) 101 provides guidance for developing emergency operations plans. It promotes a common understanding of the fundamentals of risk-informed planning and decision making to help planners examine a hazard or threat and produce integrated, coordinated, and synchronized plans"* (p. Intro-1).

This guiding document describes guidance for ESF #8 with regard to mental health as "identify and describe the actions that will be taken to assess and provide mental health services for the general public (including individuals with disabilities and others with access and functional needs) impacted by the disaster" (Federal Emergency Management Agency, 2010, p. 20). The guidance here is incomplete, provides a general scope of the initiative, offers no real details toward developing a comprehensive initiative, and only pertains to the general public.

The foundation elements formed by the major federal guidance documents are filled with inconsistencies and have no coordinated in-depth connectivity. The focal points are not comprehensive in nature. There is no consistency in providing services for the workers and victims in any kind of structured format. Each element hints at only part of the necessary functions toward developing an integrated DBH initiative.

NATURE OF CURRENT SERVICES

The governmentally employed and private professionals are the most common providers for mental health services. In the case of tribal government, Indian Health Services provides for mental health initiatives. Most state systems provide services through an institution or contract with regional mental health centers or private providers. The institutional setting has mental health professionals, but state statutes require them to provide service to the resident clients. The caseloads in most institutional facilities are at or beyond capacity, and therefore are unable to reassign professionals for other purposes, such as DBH disaster initiatives. On the other hand, state public health agencies contract with local mental health centers and private practitioners to provide services for eligible clients. The scope of eligible clients is often limited to those who qualify for government sponsored health programs, excluding the majority of workers and victims affected by a disaster. Like their institutional counterparts, these provider organizations are often operating at capacity and have waiting lists for service.

States and local governmental jurisdictions do have the option to refer affected workers to their Employee Assistance Program (EAP). The administration, however, can only make the referral and cannot make any further inquiries as to progress, unless offered by the employee. They do have the option to design Human Resource policies to enable them to require employees to seek assistance. Even in these cases, information sharing is limited. The limitations of EAPs often make their use as a resource challenging. However, constructive personnel actions aimed toward a positive outcome for an employee can occur if appropriate policies exist.

Critical Incident Debriefing Team development has been around for some time. In 1994, I coordinated an initiative to establish a statewide Critical Incident Stress Management Network in Montana. The network is still active as of 2016 and consists of 17 teams. The network is comprised of trained private practitioners and peers that volunteer their involvement on a team to help first responders deal with abnormal situations through a Critical Incident Debriefing. The sheriff of a particular county understood the value of such debriefings. By policy, the sheriff required all employees engaged in a critical incident to take part in a debriefing.

> Author's Note: Not all response agencies have applied the same philosophy. The use of these single session debriefing strategies varies depending on the approach, history, and resources of the department and their network of relationships. It should be noted the use of Critical Incident Stress Debriefing (CISD) has become an increasingly controversial and a less frequently used intervention tool. This has resulted from research emerging over the years that questions its efficacy and its potential to cause harm. This topic is addressed in other chapters in this book including Dr. Shultz' section of Chapter 5, Through a Disaster Behavioral Health Lens. This topic highlights the importance for integrating EM and DBH to assure that the important psychosocial needs of workers are addressed with evidence-based interventions.

It is not uncommon for private organizations and practitioners to have developed services to help victims and workers of localized events, such as plane crashes or train derailments. These services are normally on a contract basis with the transportation company experiencing the serious incident. In these circumstances, there is a defined universe of primary service recipients. The manifest identifies the victims. The workers are on location. Family outreach can be coordinated. Area security is often manageable with minimum interruptions. Large-scale disaster settings, on the other hand, present an entirely different dynamic. The audience is not a contained group due to the geographical expanse covered by the disaster and the workforce varies exponentially as a function of the magnitude and type of the event.

The services of private behavioral health practitioners are available to any worker or victim through direct payment or health insurance coverages, assuming that the recipient has these resources and these services exist in the area where the recipient works or lives. The key is the individual's acknowledgment that they need help. In some cases, a referral can be the catalyst for seeking help, but that means some type of assessment occurred to identify the need. In the EM world, DBH assessments are not common practice, so identification of those in need is solely based on the robustness of the local, state, or tribal programs or policies.

Voluntary agencies also have mechanisms to provide limited DBH services. Typically, the primary reason for voluntary agency involvement in a disaster is to provide shelter, housing, feeding, and temporary housing repairs. DBH services are not always provided by or through voluntary agencies. Often, if they exist, they are usually only employed if a victim is identified through a recovery center or through an assistance caseworker.

Nationally, the Department of HSS does have a mechanism in place for providing information, referral, and mental health and substance abuse services related to disasters. The website of the Substance Abuse and Mental Health Services Administration (SAMHSA) outlines their services. Their web page provides information for a national suicide hotline and access to a crisis center, a disaster distress hotline that directs viewers to information and contact information, a national helpline for treatment referral of mental health and substance abuse issues, and a behavioral health treatment locator. SAMHSA also provides a grant, under a presidential declaration, for outreach and educational services. The agency can provide training on the behavioral health effects of disaster to jurisdictions (U.S. Department of Health and Human Services, 2015). These resources provide much valuable information to support and facilitate establishing DBH integration initiatives.

The DBH resources that are available during disasters have great value, but the mechanism for providing services is not fully developed or integrated into EM. Adequacy is also questionable as not all provisions address the needs of both victims and workers. There is no mechanism for conducting a comprehensive needs assessment or implementing a follow-up plan for those affected. There are no comprehensive mechanisms providing assistance beyond the debriefing possibilities for first responders. The guidance and policy documents are not aligned to proved for a coordinated and seemless integration into EM and the relationship network necessary to drive integration is far from developed.

REFERENCES

Federal Emergency Management Agency. (2008, January). *Emergency support function #8 — Public health and medical services annex*. Retrieved from http://www.fema.gov/pdf/emergency/nrf/nrf-esf-08.pdf.

Federal Emergency Management Agency. (2010, November). *Comprehensive preparedness guide (CPG) 101 — Developing and maintaining emergency operations plans*. Retrieved from http://www.fema.gov/media-library-data/20130726-1828-25045-0014/cpg_101_comprehensive_preparedness_guide_developing_and_maintaining_emergency_operations_plans_2010.pdf.

Federal Emergency Management Agency. (2011, September). *National disaster recovery framework*. Retrieved from http://www.fema.gov/pdf/recoveryframework/health_social_services_rsf.pdf.

Federal Emergency Management Agency. (2015, July 16). *National disaster recovery framework*. Retrieved from http://www.fema.gov/national-disaster-recovery-framework.

National Oceanic and Atmospheric Administration (2015, December 15). *Hurricanes in history*. Retrieved from http://www.nhc.noaa.gov/outreach/history/#katrina

The National Response Framework. (2013). *National response framework*. Washington, DC: U.S. Department of Homeland Security.

U.S. Department of Homeland Security. (2007, September). *Target capabilities list*. Retrieved from http://www.fema.gov/pdf/government/training/tcl.pdf.

U.S. Department of Homeland Security. (2011, September). *National preparedness goal*. Retrieved from http://www.fema.gov/pdf/prepared/npg.pdf.

U.S. Department of Health and Human Services. (2015, December 28). *Substance abuse and mental health services administration*. Retrieved from http://www.disasterassistance.gov/disaster-assistance/forms-of-assistance/4506/1/6.

Through a Disaster Behavioral Health Lens

James Siemianowski

Disasters have wide-ranging psychological effects that significantly impact individuals and communities, highlighting the importance of behavioral health to response and recovery. Disasters like the Boston Marathon bombing, Sandy Hook School shooting, and Hurricane Sandy are examples of disasters that had pervasive and lasting psychological effects. "Unfortunately governments have seldom emphasized the psychological consequences of disaster as a critical part of disaster preparedness even when plans to deal with other issues are in place" (Gerrity & Flynn, 1997, p. 105). This statement, which was made almost 20 years ago, remains true today and serves as a reminder that DBH is still not well integrated into disaster management. While some progress has been made, the role of DBH is not understood well within the larger behavioral health community or by emergency managers.

> The quote below related specifically to integration following Hurricane Katrina. However, it is applicable to the current relationship between DBH and EM.
>
> "Whenever we think of disaster response and responding to protect the public's health it will require the integration of our public health system, medical care system, and emergency response system. And as we all know, those do not fit together well. At best, they are a patchwork quilt when they are working well" (Ursano, 2006, p.23).

The National Response Plan (NRP) places behavioral health under the health support function, which may obscure the unique contributions the discipline may make in a disaster (U.S. Department of Homeland Security, 2013). Emergency managers are often unfamiliar with the essential role behavioral health plays in response and recovery. At the same time, behavioral health assets may not be clearly defined or well organized. On the local level, disaster plans often ignore the role of behavioral health or bypass mental health resources that may exist within the community. Instead, they may look to organizations like American Red Cross to meet this need without developing their own behavioral health resources. Regardless of the explanation, greater emphasis must be placed on more fully integrating DBH into EM.

For behavioral health, these challenges to integration increase or decrease based on the type and scope of a disaster. Each disaster is unique, affected by the degree of warning, which community or communities are affected, traumatic effects, the extent of the physical damage, specialized needs of victims and survivors, the geographical size of the impact zone, and the scope of the needed response. The kind of disaster, where it occurs, and its magnitude can dictate who manages the incident. It can also affect what DBH assets exist in the community, remaining infrastructure, interventions used, and the funding available to support response and recovery efforts. The nature of the disasters impacts the degree to which the event response expands, adding complexity as the response becomes multijurisdictional and multiorganizational. Table 6.1 below illustrates how some recent disasters differ in key areas.

TABLE 6.1 Unique Aspects of Recent Disasters

Event Characteristics	Hurricane Sandy	Sandy Hook School Shooting	Boston Marathon Bombing
Presidential declaration	Yes, across multiple states	No	Yes
Warning/preparedness	Days in advance	None	None
Incident command	Emergency Management Authority (EMAs) and Federal Emergency Management Agency	Law enforcement	Law enforcement
Impact zone	Multiple countries, states, regions, locales	Confined largely to one town	Boston, suburbs, national and international
Victims	Spanned the life cycle	Mostly young children	Local, national, multinational
Fatalities/injured	147 direct deaths (Blake, Kimberlain, Berg, Cangialosi, & Beven, 2013)	26 dead, 2 injured	3 dead, 286 injured
Psychological effects	Moderate with long-term effects of relocation, rebuilding	High traumatic exposure, small community with high degree of social relatedness	High traumatic exposure, prolonged due to search for bombers

(Continued)

TABLE 6.1 (Continued)

Event Characteristics	Hurricane Sandy	Sandy Hook School Shooting	Boston Marathon Bombing
Damage to infrastructure	Considerable damage to infrastructure (homes, roads, and power)	Minimal damage to infrastructure (school)	Businesses in bombing area
Available behavioral health resources	Multiorganizational, multijurisdictional, across many states	Limited public behavioral health resources in town, Scope led to increased need for DBH assets	Substantial behavioral health resources in Boston
Funders of response	Federal Emergency Management Agency, Crisis Counseling Program (CMHS)	Department of Justice (DOJ) Victims of Crime, Federal Dept. of Education School Emergency Response to Violence (SERV)	Federal Emergency Management Agency, DOJ, and Massachusetts (MA) Victims of Crime

Mass shootings like the Sandy Hook shooting in Newtown present unique challenges for behavioral health responders because of the sudden nature, the traumatic effects, and the size of the response. Compared with the response to natural disasters, command and control are more fragmented and resources are less integrated for mass shootings. This complicates the implementation of behavioral health interventions (Shultz et al., 2014). Tragedies like this include sanctioned responders representing local, state, and national interests and many uninvited mental health personnel who self-deploy.

> Events like the Newtown shooting present a problem of *multiples*: multiple responders, representing multiple jurisdictions and organizations, providing services at multiple community sites, and using multiple models of intervention.

Unlike natural disasters, the nature of these types of events place law enforcement in the lead role for the incident. In Newtown, there was no Presidential Disaster Declaration so mental health response and recovery efforts were largely supported by the Federal Department of Education and the Federal Department of Justice. These funding sources were unfamiliar to the state's mental health authority. The lack of familiarity creates delays in securing funding as states "learn" new funding requirements on the fly.

Hurricane Sandy challenged behavioral health responders in different ways. The hurricane affected the Caribbean, the continental United States, and Canada over a 1-week period, with clean-up and recovery efforts spanning years. While communities had significant warning, the hurricane devastated local infrastructure over multiple states on the eastern seaboard, killing at least 147 people (Blake et al., 2013). The hurricane led to massive relocation of affected residents. Presidential declarations were approved in a number of states, triggering provisions of the Stafford Act (Robert T, Stafford Disaster Relief and Emergency Assistance Act) that placed Federal Emergency Management Agency in a lead role for response and recovery efforts.

The Boston Marathon bombing created other challenges for behavioral health responders because of the terrorist nature of the act and subsequent manhunt over the following days. There were three deaths, but local hospitals treated over 260 injured requiring trauma level care (Massachusetts Emergency Management Agency, 2014). Emergency Medical Services (EMS) played a key role due to the high numbers of wounded. The postincident behavioral health response was made even more challenging because victims and witnesses were not only going to be returning to their Boston area homes, but also returning to other states or countries. Victims from other states or countries did not receive the same level of community support that locals did and creative solutions needed to be implemented to build a recovery network for these individuals in their distant communities. One example included reaching back to local running clubs to enlist supports. Within Massachusetts, a Presidential Disaster Declaration was issued with funding of long-term behavioral health efforts supported by Federal Emergency Management Agency, the Federal Department of Justice, and State Office of Victim Assistance.

The unique aspects of these disasters highlight how DBH and emergency managers may be challenged. Challenges may include any combination of the following:

- Providing psychological support based on the magnitude of the traumatic effects
- Responding across huge geographic areas
- Addressing specialized needs of victims
- Having limited DBH resources immediately available
- Managing the influx of invited and uninvited responders from multiple jurisdictions
- Establishing clear lines of command, control, and ongoing responsibility
- Coordinating the funding of response and recovery activities with funders with distinct eligibility requirements and program restrictions

These challenges impact EM and DBH leadership alike. Basic questions need to be answered by emergency managers and DBH leadership. Ideally, they are considering the same set of questions, but this is often not the case. In order to integrate behavioral health into the broader response, the following questions must be considered:

1. Has the disaster had a significant psychological effect on the impacted community?
2. How can behavioral health support response and recovery?
3. What will the behavioral health mission be?
4. Are there clearly identified behavioral health resources in the community that can assist?
5. Are they sufficient or will other resources need to be identified?
6. If so, how are those assets accessed?
7. Does the town, state, etc. have a DBH plan?
8. Does the plan specify the lead organization for the provision of disaster mental health support?
9. Does the plan specify how to manage an influx of uninvited responders?
10. Does it specify who is responsible for applying for and administering DBH funding for victims and the community at large?

The EM function may be less complicated if a DBH plan exists and identifies a lead mental health agency. Addressing the other questions listed above would be part of the lead agencies' roles and responsibilities following the disaster. Too often, this is sorted out postdisaster. The After Action Report from the Boston Marathon specifically identified the lack of a disaster mental health coordination plan and the negative effects this had on the immediate response (Massachusetts Emergency Management Agency, 2014). The report went on to specify components that should be included in the plan to address some of the questions listed above. The Final Report of the Sandy Hook Advisory Committee contained similar recommendations related to the need to better integrate behavioral health into the Unified Command Structure (Sandy Hook Advisory Commission, 2015).

Common barriers that emergency managers and DBH face as they attempt to integrate a support function like behavioral health into the overall response are found in Table 6.2. The table identifies strategies that can be used by DBH leadership and emergency managers to more effectively blend DBH into broader response and recovery activities. Coordination must be built on structures that formalize the authority and roles of DBH responders and leadership.

Table 6.2 demonstrates that the integration of DBH and EM can be strengthened in a number of ways. Some of the strategies fall under the purview of DBH leadership while other strategies require actions by emergency managers. Integration will not magically occur following a disaster

TABLE 6.2 Integration Barriers and Strategies to Enhance Integration

Barriers	Impact on Integration	Strategy
No DBH plan	Roles and responsibilities are negotiated during or postdisaster creating a delayed/fragmented response	Ensure local and state All Hazards Plans include DBH plan
No preidentified behavioral health assets	EM has no clear options for BH response	Create local and state behavioral health strike/response teams
No established working relationships between EM and DBH	EM has no clear idea about the support role DBH can play following spectrum of disasters	Integrate DBH into local and state disaster planning efforts
DBH not included in disaster drills	No opportunities to drill or practice collaboration between EM and DBH	Include DBH in meaningful role in drills and exercises
No "authority" for DBH response	No point of accountability for oversight of the DBH response	Create/modify legislation to formalize role of DBH in civil preparedness and response
No designated DBH lead agency	Confusion and delayed response as leadership is negotiated during/postdisaster leading to a fragmented response	Designate and formalize in state and local plans the DBH organizational lead and responsibilities pre-post disaster
No DBH representative in incident command post	Decisions with major ramifications for DBH are made without DBH input	Require DBH representation at incident command post
Multiorganizational and multijurisdictional DBH response	"Mission confusion," duplication of efforts, disagreements over response strategy, adverse political consequences	Use DBH lead to form DBH Coordinating Committee, identify needs and gaps, assign roles, obtain additional DBH assets
Limited DBH capacity to sustain response efforts for more than 2–4 weeks	Gaps in DBH service delivery	Immediately begin planning for long-term recovery
DBH Strike Teams do not understand incident command	DBH response is not effectively integrated into broader response	Incident command training for DBH personnel
Unfamiliar "funders" of DBH recovery efforts	May delay long-term recovery and restrict available programs and eligibility	Immediately involve DBH planners in the development of recovery model/program based on unique criteria of "funders"

and must evolve during planning and preparedness phases. These preevent actions provide the foundation for formalizing the DBH structure. They are designed to provide behavioral health with a seat at the EM table, offering behavioral health leadership with an access point to EM. Other strategies relate to actions that can be taken following the disaster. These actions can strengthen integration in the mental health support function and collaboration with EM.

PREDISASTER STRATEGIES

Develop Local and Statewide Disaster Behavioral Health Plans

The Boston Bombing After Action Report (Massachusetts Emergency Management Agency, 2014) recommended the development of a disaster mental health coordination plan. Emergency planners on the local and state level must incorporate DBH into their planning efforts. Local emergency personnel should forge relationships with existing mental health resources, requesting that they participate in local disaster planning efforts. State and local plans should address leadership of the DBH response, roles and responsibilities, credentialing of responders, and development of the incident action plan. By doing this before the disaster, roles and responsibilities are clear and do not have to be renegotiated during a disaster.

Develop DBH Annexes in State All Hazard Plans

At times, the DBH plan may be "lost" under the health support function. Some states have created separate annexes for mental health within the All Hazards Plan, elevating it to a distinct function. This may be important because on many governmental levels, health and mental health are organized in separate departments and do not have significant overlap. While some states or localities may have health superagencies with closer collaboration between these two disciplines, this is not always the reality. A separate annex may place greater emphasis on the role of DBH.

Develop Local and Statewide Disaster Behavioral Health Assets

DBH assets are not always apparent to emergency managers or may be poorly organized. Local and state planners must develop DBH capacity. This is often ignored or difficult to create after a disaster has occurred. State, county, or regional mental health agencies can clarify the assets that can be tapped by locals and they can be instrumental in providing basic DBH training. Training should minimally include Psychological First Aid along with Incident Command training and its importance for disaster management. DBH capacities can also be extended by directing DBH training at funded mental health crisis programs that may be called upon postdisaster.

Provide Basic Psychological Training to Emergency Personnel

Emergency personnel should have a basic understanding of Psychological First Aid in order to better understand basic principle that can applied regardless of roles. The training can provide skills while serving to educate emergency responders and managers about the important role DBH can play following a disaster.

Include Behavioral Health Representatives in Emergency Planning Efforts

Behavioral health is not always included in local and state planning efforts, meaning important relationships do not exist predisaster. Leadership within behavioral health organizations or state agencies must commit staff to participate in planning efforts. Active participation by behavioral health in these planning efforts can educate local and regional planning entities on how they can bring local resources into their response. DBH representation serves to familiarize emergency planners with how DBH can support response and recovery. Successful EM is driven by collaborative relationships and these relationships are developed, nurtured, and formalized during the planning process.

Incorporate Disaster Behavioral Health into Exercises and Drills

Too frequently, mental health is minimally involved in disaster drills, if at all. Even when DBH is involved in drills, the scope of their "play" is not consistent with their actual response roles. Greater attention should be focused on identifying missions or roles for mental health in exercises. This strengthens relationships and increases awareness about the ways mental health can contribute to the response. This is another tool to foster integration prior to the disaster. The manner in which familiarity enhances integration was evident in Connecticut shortly after the Sandy Hook shooting. The local health director and the state's regional emergency administrator immediately contacted one of the leaders of the state's disaster mental health network to request assistance (author's direct experience). The rapid activation and deployment of mental health assets was the result of relationships that had been nurtured over a period of years in regional planning activities and exercises.

Formalize the DBH Role Through Legislation

Integration may suffer because DBH teams do not have a formal state sanction. Legislation can be used to sanction the DBH support function or to formalize the role of DBH Strike Teams. Connecticut codified their state's Disaster Behavioral Health Response Network (DBHRN) in legislation as civil preparedness forces (Civil Preparedness and Emergency Services). This legislative action has increased the integration of mental health into overall EM because the request for DBH assets and authorization to deploy is a function of the state's EM agency. The legislation formalized not only the role of DBH but also the relationship between the state's EMA and DBH.

POSTDISASTER STRATEGIES

Integrate Senior Mental Health Officials into the Command Structure

The Sandy Hook Advisory Committee Final Report recommended that mental health representatives must be included in the Unified Command Structure (Sandy Hook Advisory Commission, 2015). A *senior* mental health official from the lead agency must be integrated into the incident command center. Since state governmental structures vary considerably, this individual in many cases may be part of an agency that is not organized under Public Health. *Senior* is underscored because this person must have decision-making authority and must also be at an organizational level that allows him or her to freely assert their positions about the mental health disaster response. This individual becomes the bridge between EM and the DBH response.

Use Mental Health Experts in a Consultative Role

Placement of mental health personnel in the command center is important for another reason. It creates opportunities for emergency managers to use mental health experts in a consulting role—a role that is often underutilized. As subject matter experts, mental health personnel can provide valuable information regarding issues that impact the mental health of individuals, communities, and responders. Integration into the onsite command center provides mental health responders with opportunities to impact on those decisions that are certain to have psychological implications. This consultative role that mental health can play is often overlooked and its absence can lead to decisions that have harmful, unintended effects.

Link Disaster Behavioral Health Leadership and Experts to Other Decision-Makers

While integration at the command center is important, critical decisions are being made outside of the command center. As a response expands, decisions may be made at multiple sites. For example, in Newtown, town and school officials were making decisions about how to reach out to families, where to move the school, when to begin school, how to support students and staff who would be out for a prolonged period of time, and where memorials should be located. These decisions have huge implications for the mental health response and it is essential that senior mental health leadership is integrated into other settings beyond the incident command post or emergency operations center. If mental health leadership cannot be incorporated into all of these sites, it is important to establish regular feedback loops so key decisions are filtered back to the incident command process and structure.

Require the Lead Mental Health Agency to Develop a DBH Incident Action Plan

The designated DBH lead must be responsible for developing a DBH Incident Action Plan that addresses immediate and longer-term needs. This often overlooked when disasters are rapidly unfolding. The DBH plan provides a structure for improving integration within the DBH discipline. A primary function of the mental health lead is to clarify the extent of the community's needs and to identify the behavioral health resources available to meet these needs. This begins by identifying all of the groups that have been affected by a disaster, finding who is tasked with responding to them, and further clarifying potential roles for the mental health responders.

Develop a Disaster Behavioral Health Coordinating Committee

Integration problems are magnified when the response quickly expands to include mental health responders from multiple organizations and jurisdictions and are responding at multiple community sites. A DBH Coordinating Committee, chaired by the designated mental health lead, is the vehicle for managing the DBH Incident Action Plan and the rapid expansion that often occurs in major disasters. The composition of such a committee becomes a critical initial task and should involve organizations that have responsibility for any victims or responder groups. This might include school personnel, local, state, and national mental health resources, health departments, the American Red Cross, and other Voluntary Organizations Active in Disasters (VOADs). This could also include EAPs serving affected communities or responder organizations. There are sometimes common problems that emerge related to the DBH role, missions, and interventions used. Behavioral health needs may shift or response capacities of organizations may diminish over time. A DBH Coordinating Committee becomes the mechanism for identifying or anticipating problems or emerging needs.

THE FUTURE OF INTEGRATION

All of the strategies described in this section can be used to enhance integration between emergency and DBH. Integration is fostered through relationships and structures that help to formalize and document those relationships. Disaster response is often seriously compromised because leadership and roles are unclear. These problems are exacerbated in certain disasters because of the traumatic impacts or the catastrophic proportions of the event. These strategies identified here can be used to better organize the DBH response and align DBH with the broader response.

While disasters are unique and may present distinct challenges to EM and DBH leadership, much can be done to improve the effectiveness of the response. Fostering improved integration must be cultivated over time and should be a shared responsibility of leadership within each discipline. Integration does not magically occur after a disaster. There must be a foundation for integration that begins with planning and preparedness activities and extends into incident command following a disaster. Many states and communities have taken steps to effectively integrate mental health into the broader EM community, but more work remains.

REFERENCES

Blake, E.S., Kimberlain, T.B., Berg, R.J., Cangialosi, J.P., & Beven, J.L. (2013). *Tropical cyclone report Hurricane Sandy*. Retrieved from http://www.nhc.noaa.gov/data/tcr/AL182012_Sandy.pdf.

Civil Preparedness and Emergency Services, Connecticut General Statutes Title 28, Chapter 517, Sec. 28−1 (5).

Gerrity, E. T., & Flynn, B. W. (1997). Mental health consequences of disasters. In E. K. Noji (Ed.), *The public health consequences of disasters* (pp. 101−121). New York, NY: Oxford University Press.

Massachusetts Emergency Management Agency. (2014). *After action report for the response to the 2013 Boston marathon bombings*. Retrieved from http://www.mass.gov/eopss/docs/mema/after-action-report-for-the-response-to-the-2013-boston-marathon-bombings.pdf.

Robert T. Stafford Disaster Relief and Emergency Assistance Act, Pub. L. 93-288, codified as amended at 42 U.S.C. §§ 5121−5206.

Sandy Hook Advisory Commission. (2015). *Final report of the Sandy Hook Advisory Commission*. Retrieved from http://www.shac.ct.gov/SHAC_Final_Report_3-6-2015.pdf.

Shultz, J. M., Thoresen, S., Flynn, B. W., Muschert, G. W., Shaw, J. A., Espinel, Z., ... Cohen, A. M. (2014). Multiple vantage points on the mental health effects of mass shootings. *Current Psychiatry Reports*, *16*(9), 1−17. Available from http://dx.doi.org/10.1007/s11920-014-0469-5.

U.S. Department of Homeland Security. *The National Response Framework*. (2013). Washington, DC: Homeland Security. http://dx.doi.org/10.1002/9780470925805.ch21.

Ursano, R. (2006). *Mental health in the wake of Hurricane Katrina. Mental health in the wake of Hurricane Katrina: The twenty-second annual Rosalynn Carter Symposium on mental health policy* (pp. 21−25). Atlanta: The Carter Center. Retrieved from https://www.cartercenter.org/documents/rc_mhsymp06_katrina.pdf.

Making Integration Work

Daniel W. McGowan

There are multiple challenges on the road toward developing an integrated DBH component with EM. The critical challenge is aligning the foundation elements to provide comprehensive guidance for developing such a system. The current foundation has barely scratched the surface toward completing such an initiative. There is no evidence of horizontal or vertical connectivity. The nature of disasters provides a challenge that requires any system to be dynamic, as no two disasters will present the same circumstances. DBH preparedness plans and their effectiveness will be challenged by the breadth and scope of the EM relationship network. Another challenge is coordinating and developing the robustness of current services to provide a comprehensive mechanism for effective service delivery to affected victims and workers. The ultimate challenge will be involving the private behavioral health sector as a partner to the initiative on either a voluntary basis or some type of fee for service.

While the challenges are many, there are many benefits, to developing an integrated DBH initiative with EM:

- The first is a consistent manner with which to address victim and worker DBH issues regardless of the disaster, as no structure exists now.
- Second, an integrated and consistent approach will leave less room for victims and workers to slip through the cracks leaving critical needs unaddressed.
- Third, the safety of the workforce will be enhanced and stress-related workforce problems can be reduced.
- Fourth, a more coordinated effort toward helping victims and workers will relieve some future pressures on the already challenged mental health system.
- Fifth, the development of a horizontally and vertically integrated network will provide for a more comprehensive DBH assistance delivery mechanism.
- Finally, the health of the whole community will improve enabling those affected to heal more quickly and move on with their lives.

The public health, behavioral health, and EM communities increasingly see the need for an integrated DBH element to disaster response and recovery. The beginning of an integrated effort is in its infancy. As noted at the beginning of this chapter, the ultimate challenge is checking all the boxes. All potential elements with a stake in developing a DBH initiative need to be identified. The landscape of elements then needs to be massaged into an understandable and manageable delivery mechanism. True integration development must include all stakeholders, including private practitioners, with something to offer toward such an initiative. The development of effective horizontal and vertical network relationships between and among the federal, state, tribal, local, voluntary, and private providers will be critical to a successful initiative. Doctrine, policy, and guidance must be aligned and support DBH integration into EM.

> Following are some simple and concrete steps to help emergency managers get started on integration.
> 1. Create a list of potential stakeholders, including private behavioral health practitioners, voluntary organizations, and faith-based groups
> 2. Convene an introductory/brainstorming meeting; capture the behavioral mental health resources each stakeholder can provide and how
> 3. Identify capabilities and gaps
> 4. Establish communications' links and notification protocols
> 5. Establish a subcommittee of the stakeholders present to design a table top exercise that tests what you have assembled
> 6. Review the results of the exercise with the stakeholders and draft DBH protocols for inclusion in the emergency operations plan
>
> DBH practitioners could follow the same six steps and initiate the outreach process.

The program development will need to provide for a thorough assessment process, matching needs to available resources, follow-up case work, and a mechanism to assist those victims or workers who are detected in need long after the disaster. The collaborative development of planning and preparedness efforts is paramount to a successful effort. DBH integration cannot be an exclusively consultative approach where one entity develops the guidance and consults with the stakeholders to finalize the outcomes. The real test of success will be whether the DBH initiative becomes an institutionalized part of disaster response and recovery regardless of the disaster size or the jurisdiction experiencing the event.

Chapter 7

What Can DBH Actually *Do* To Make Emergency Managers Jobs Easier?

April J. Naturale[1] and Lesli A. Rucker[2]
[1]ICF International, Fairfax, VA, United States, [2]Cenibark International, Inc., Richland, WA, United States

Through a Disaster Behavioral Health Lens

April J. Naturale

In the early years after 9/11, many government agencies stressed the need for disaster behavioral health (DBH) interventions to be integrated with emergency management (EM) planning and response activities. There was a lot of discussion about this concept and some funding mechanisms even added a brief reference about requiring integration. However, they did not describe how it was expected to be done, or by whom, thus responsibility remained largely unassigned. Many state disaster EM and behavioral health response coordinators have worked toward this goal, as have regional and local governments. But it remains a challenge. Like much of the disaster preparedness and response activities, local communities are working to forge partnerships with those aimed toward the same goal of supporting those affected by natural disasters, human caused accidents, incidents of mass violence and terrorism. This chapter attempts to address the various aspects needed for successful integration of DBH into EM.

CONSULTATION TO LEADERSHIP

As the psychological effects of disasters are second only to injury and death (Center for Mental Health Services, CMHS, 2001), DBH subject matter experts can provide invaluable supports to EM staff starting at the highest level of leadership in government and including the ground level Incident Command team members. DBH consultation to leadership can have several foci including:

1. informing leadership of the types of responses to expect in grieving victim's family members, those injured, and other directly affected community members based on the size, type and scope of the event;
2. providing basic crisis intervention and crisis counseling techniques allowing leadership to focus on rescue, safety and other urgent public health issues;
3. helping leadership to form the most effective risk communications and public messaging that may assist in mitigating panic, lack of compliance with safety instructions, and the development of serious mental health disorders;
4. conducting needs assessments to inform leadership of the most highly impacted populations requiring priority attention and supports;
5. providing monitoring and surveillance assessing the psychological effects of the events on their staffs determining who may be experiencing an inability to function safely and effectively in their assigned roles.

Concerns for the mental health status of victim's families and survivors are at the core of the field of DBH. DBH specialists are well versed in assessing and addressing victim's families and other loved ones needs by:

1. helping organize safe spaces such as family assistance centers where families can obtain situational updates including information about the status of their missing or injured loved ones, access resources to help with disaster related needs (e.g., shelter, food, clothing, funds, etc.), and receive DBH interventions such as psychological first aid;
2. collecting vital pre- and post-mortem data sensitively;
3. providing support for family members who are accessing services;
4. bringing homogenous groups (those survivors who have experienced similar losses) together to support each other.

In every disaster, risk communications and public messaging are essential tools for the response. This will be discussed in detail later in this section.

A disaster needs assessment is another tool that DBH staff can provide to support the efforts of leadership to design and implement the most effective disaster response program (United States Code 5121, 2013). Key partnerships can include public health and mental health service providers, universities, local and state governments, and the Federal Emergency Management Agency (FEMA) (in a presidentially declared event). Note that the community may already have a disaster response plan in place, and if so, the goal of the needs assessment is to help hone in on the details as they relate to the specifics of the particular disaster. The needs assessment looks at:

1. the size of the event (large/small geographic areas affected)
2. the scope (large/small numbers of people killed, injured, number whose homes were destroyed or partially destroyed; business and infrastructure

destroyed and those otherwise displaced from their homes or communities)
3. the type of event (natural disaster that was forewarned giving people time to evacuate versus a human caused explosion or terrorist attack that occurred without warning to unsuspecting victims) and other high-risk populations in the affected areas.

The needs assessment can help inform each aspect of the incident command system-planning, operations, logistics, and finance/administration providing estimates for creating a response that is appropriate to the effects of the disaster.

The needs assessment will be key in helping emergency managers determine if the required response can be managed by the community or if they need to request help from neighboring communities, the state, or even the federal government. DBH services are mission-driven to address the unique needs of those in the affected areas. Thus, staff are well versed in conducting needs assessments via focus groups, existing public document review, door to door canvassing, and telephone surveys. They can also participate in analyzing how the results will inform the overall response plan. Working collaboratively, EM and DBH can plan where the most intensive staffing may be required, which high-risk groups exist in which geographic areas and what are the most culturally appropriate approaches for each. Collaborative efforts can help to increase community and individual agency buy-in, which, in turn, can ensure that survivor's needs are identified and are more effectively addressed.

SURVIVOR STRESS

The reach of the psychological effects of many types of traumatic events is much broader than solely victim's family members, those injured, and those whose home were damaged or destroyed (Briere & Elliot, 2000). The negative mental health effects can extend to friends, neighbors, and the community at large especially in those incidents that are human-caused either by accident or with malicious intent (Galea et al., 2005; National Institute of Mental Health, 2002; Norris, Byrne, & Diaz, 2001).

Thus, family members of victims, those injured, and responders are populations considered at highest risk for negative mental health outcomes, partially due to their usually high level of exposure to the incident, death, injuries, and grief, as well as their capacity for empathy (Norris, Friedman, & Watson, 2002a; Norris, Friedman, & Watson, 2002b). For some people (particularly responders), their lack of recognition for the need or acceptance of help can cause their distress to worsen over time and risk becoming chronic (Fullerton, Ursano, & Wang, 2004). The research also indicates that over time (e.g., as the timeframe moves toward and then past the first anniversary of the event), a percentage of victims directly exposed to the event (30−40%), rescue workers (10−20%), and the general population (5−10%) are at higher risk of developing a diagnosable mental illness (e.g., posttraumatic stress disorder (PTSD)) or other diagnosable mental disorders such as depression (Galea, Boscarino, Resnick, & Vlahov, 2002; Kessler, Sonnega, Bromet, Hughes, & Nelson, 1996).

These mental health impacts are often psychologically denied in the general population of those who have never seen themselves in need of mental health or substance misuse services. Denying the need for behavioral health services has historically been seen in the emergency response staff (Naturale & Pulido, 2012). They may be at risk of losing their jobs if they access mental health services and substance use and misuse has traditionally been an accepted part of the culture (Fullerton et al., 2004). This denial can serve both as a strength that allows them to continue working at their height of performance when needed in the acute and immediate response. Often, as the response activities slow, the denial can be problematic as it may prevent people from acknowledging and attending to the negative physical and psychological consequences of chronic stress.

RESPONDER STRESS

Emergency managers in charge of the response and the field staff themselves can experience a compounded negative impact when experiencing high levels of pressure related to leadership responsibilities, exposure to life threatening situations, seeing comrades injured or killed, seeing and hearing gruesome events such as dismemberment, people screaming in pain and fear, handling body parts, and recovering bodies in water (Leffler & Dembert, 1998; Marmar et al., 1999). These experiences can worsen when staff members are additionally stressed by the concern for their own family members, or when their own homes and communities are damaged or destroyed as a result of the disaster.

EM, staff, and responders are typically part of a larger collective that create tight, close knits teams as they work together in the recovery process and when responding to events that directly affect them. They are often required to make rapid adaptations to manage the changes in their perceptions and in their sense of meaning and safety in the world around them. These adaptations may appear similar to those in survivors. However, these workers are different because of the necessity that they continue to do their jobs effectively, often without having the time to give conscious thought to themselves or their personal concerns (Stellman et al., 2008; Wee & Meyers, 2002).

The health and productive functioning of leadership at all levels of government, EM, fire and rescue, law enforcement, EMTs, public health/medical staff, crisis counselors, mental health staff, and other first responders is a priority concern. Historically it has been presumed that disaster response leaders and field staff have learned to be immune or at least inoculated against the negative mental health effects of disasters, but more recent research indicate that they too may experience direct or indirect exposure putting them at high risk of experiencing traumatic stress (Fullerton et al., 2004), shared trauma (if they are experiencing the same trauma as victim's families and survivors) (Tosone et al., 2003), and community trauma particularly in large-scale events.

Friends and loved ones of the injured and other survivors, behavioral health response staff, and responders may experience compassion fatigue (CF) or secondary traumatic stress (STS). This is a phenomenon that can occur when one is exposed to another's trauma and fear, usually someone they care about. In the case of responders, it could occur as a result of witnessing the trauma or fear of someone whom they have helped. CF and STS usually involves experiencing the same symptoms as the survivor themselves, and may be accompanied by extremes of fatigue, sadness and for responders, and a loss of meaning in their work (Figley, 1995; Figley, 1999).

Experienced DBH specialists deployed alongside EM leadership, responders, and into the affected community can work collaboratively with EM to provide the needed assistance to family members, the injured, responders in the field, and leadership themselves. This relationship is likely most effective when they already have a working relationship with EM leadership prior to the event. Pre-existing relationships and an understanding of what each has to offer can serve to provide the best and most effective disaster response program that each person has to offer individually and as a team.

UNDERSTANDING THE BODY AND BRAIN RESPONSES TO EXTREME SITUATIONS: WHAT IT MEANS FOR INTEGRATION

The psychological impacts of critical incidents, such as disasters, which are sudden, unpredictable events that have a catastrophic or life-threatening impact, create responses in the brain that significantly impact adaptation processes. These include both conscious and unconscious thinking. It can extend to automatic impulse, judgment, impulsivity, even in the most seasoned EM staff. Instinct may be affected by a previous experience with a traumatic event and the imprint that these events have left on the brain.

Everyone has traumatic imprints in their memory and often these imprints or memories come flooding back when there are reminders or feelings similar to earlier trauma memory. The most recent research tells us that most of the population has experienced at least one traumatic event by 18 years of age (Centers for Disease Control and Prevention (CDC), 2014).

EM staff know the feeling of getting "pumped up" with the recognition or even perception of danger and responding to the threat of injury or death. This is often the result of long hours, days, and months in a highly dangerous environment that keeps the person in a constantly "hyper-alert" state. Over the long term, when a person who has experienced a stress response does not engage in stress reducing or stress releasing activities, they can become ill as a result, with cardiac problems, depression, anxiety, substance misuse, and other health and mental health concerns (Stellman et al., 2008).

Our reactions can result in poor or dangerous behaviors in response to a crisis. For example, a crisis or trauma survivor will often say, "I don't know what happened. I don't even really remember. I just reacted." Thus, while these automatic reactions can serve to protect us against certain threats, many situations require a rapid, but more thoughtful evaluation. Anyone exposed to a dangerous situation, including EM staff, may experience an instinct to run, freeze (find they are unable to move), fight or even act dangerously, if faced with imminent death or other threats.

Here is where DBH can be helpful. DBH specialists can work with EM throughout all phases of a disaster to help with preparing for events and responding with immediate on-site assistance. This can include guiding EM staff in the development of cognitive tools to help them function at peak performance, mitigate the build-up of stress, and offer additional crisis intervention techniques to help EM staff to cope and control their reactions.

> A brief example of the use of cognitive shifting in the field:
> A behavioral health staff member came upon the scene of a disaster survivor who was highly agitated and was focusing his anger on a nearby responder by getting "in his face" and challenging the responder to hit him- and to fight him. The disaster environment was particularly dire with many responders working for days with very limited food and water and no rest due to a lack of sufficient staffing. Survivors were criticizing responders and taking their anger out on them despite the help offered by these hard working staff. Everyone was frustrated. The responder stopped what he was doing and stared angrily at the survivor. The behavioral health staff, staying a close but safe distance away, speaking directly behind the responder, stated quietly enough for only the responder to hear, "Tell yourself that you will stay focused on your job. This angry survivor's challenge is not worth losing your dignity or your job. You are here to help. Tell yourself to walk away and get back to work." While a supervisor or co-worker may have been able to provide support to the responder in such a circumstance, none were available to help. Behavioral health staff were on the scene and able to support both the responder and then address the survivor directly as well.

Everyone who experiences a disaster will be impacted in some way (Substance Abuse and Mental Health Services Administration (SAMHSA), 2015a; Substance Abuse and Mental Health Services Administration (SAMHSA), 2015b). For some, responses can have profound and lasting effects interfering with such things as their:

- sense of safety
- ability to control their emotions
- sense of who they are and their beliefs
- self-efficacy (their sense of their ability to take care of themselves)
- intrusive and/or disturbing memories
- relationships (the way they respond to people in their lives).

Many people become fearful that their responses mean they are becoming mentally ill. One of the priorities for DBH responders is to provide psychoeducation to provide reassurance that many of these responses can seem like mental illness, but they are not the same and that they are expected to resolve in a reasonable amount of time, especially with good social supports and coping skills (Kaniasty & Norris, 1995; Norris & Kaniasty, 1996; Hamblen et al., 2009).

DBH staff who work behind the scenes, and sometimes side by side, with the EM staff can monitor workers to determine if they are functioning well cognitively and if they are experiencing strong reactions that could potentially create unsafe or even dangerous situations for other responders and survivors. Additionally, DBH staff can train EM Incident Command liaisons, team leaders, and supervisors in using techniques to rapidly identify team members who may be experiencing a serious stress or traumatic stress response in several domains (such as physical, cognitive, emotional, behavioral and spiritual). The following text box highlights examples of responses that can interrupt or compromise the functioning of EM staff.

Negative Cognitive Reactions Can Create a Dangerous Situation for EM Staff and Survivors.

Confusion and disorientation are particularly problematic as they can lead to *impaired thinking, judgment, and decision-making. Poor concentration, poor attention span, and impaired memory* problems can inhibit EM staff from effectively doing their jobs. These stress responses are sometimes difficult to self-identify.

An emergency manager acting as a commander or a team lead needs to be able to recognize these problems in themselves immediately. Ideally, they should collaborate with a DBH person they know and trust. This allows the emergency manager to check-in when such problems arise and obtain help in identifying the nature and severity of the difficulty. In most cases, workers are just overwhelmed, exhausted, and in need of food, drink, sleep or just some down time to repair their cognition. Identifying a trusted DBH professional prior to going into a crisis situation can also serve to relieve an impaired responder with minimal attention or stigma.

A good rule of thumb for all response personnel, including EM and DBH staff, is to establish a "buddy system." For each pair, if your "buddy" says you need to take a break, you do not argue. You remove yourself, and then work out what needs to be done in a confidential space. Typically, only when necessary, the appropriate team lead or commander who has a "need to know" is apprised.

(Continued)

(Continued)

The following tables are examples of what these responses can look like in common terms.

Physical Domain Reactions
- Faintness, dizziness
- Hot or cold sensations in the body
- Tightness in throat, stomach, or chest
- Sudden sweating
- Heart palpitations (fluttering)
- Agitation
- Nervousness
- Hyper-arousal (easily startled by noises or unexpected touch)
- Fatigue and exhaustion
- Aches and pains
- Gastrointestinal distress, nausea, constipation or diarrhea
- Appetite decrease or increase
- Changes in sleep patterns
- Loss of interest in sex
- Headaches
- Lower immune function (including susceptibility to colds and illnesses)
- Exacerbation of pre-existing health conditions

Cognitive Domain Reactions
- Confusion and disorientation
- Poor concentration/poor attention span
- Impaired thinking, judgment, and decision-making
- Impaired memory
- Disorganization
- Sense of powerlessness
- Complete or partial amnesia
- Repeated flashbacks, intrusive thoughts and images
- Obsessive self-criticism and self-doubt
- Preoccupation with protecting loved ones
- Questioning of spiritual or religious beliefs

Emotional Domain Reactions
- Shock, disbelief, minimizing the experience
- Anxiety, fear, worry about safety
- Intrusive thoughts of the trauma (nightmares, flashbacks—feeling like the trauma is happening now)

(Continued)

(Continued)

Attempts to avoid anything associated with the event
Tendency to isolate oneself
Feelings of helplessness
Panic feeling out of control/need to control everyday experiences
Sadness
Grief longing and pining for the deceased
Feeling numb or detached
Disorientation
Denial
Feeling helpless, powerless, and vulnerable
Disassociation—feeling in a dream-like state
Outbursts of anger, rage, and desire for revenge
Irritable, short-tempered, and restlessness
Feeling hopelessness and despair
Blaming self and/or others
Guilt/shame
Unpredictable mood swings (such as crying then laughing)
Re-experiencing pain associated with previous trauma

Behavioral Domain Reactions

Sleep disturbances and nightmares
Jumpiness, easily startled
Hyper-vigilance/scanning for danger
Crying and tearfulness for no apparent reason
Conflicts with family and co-workers
Irritability
Angry outbursts
Fatigue
Unresponsiveness
Avoidance of reminders of trauma
Inability to express feelings
Hysteria or inability to control emotions
Diminished desire for activity
Using excessive sick leave
Isolation or withdrawal from others
Increased use of alcohol or drugs

Spiritual Domain Reactions

Questioning faith, a higher power, and beliefs
Challenging or anger at a higher power
Promising
Bargaining with a higher power

(Continued)

(Continued)
Loss of faith/spiritual crisis
Sense of betrayal and unfairness
World feels unsafe
Separation from the rest of humanity
Loss of meaning
Realization of mortality
Withdrawal from faith and religion
Questions about good and evil
Questions about forgiveness
Redefining moral values and intangible priorities
Concerns about vengeance

In addition, for many disaster survivors, circumstances such as legal and economic issues (e.g., ongoing interactions with insurance or legal entities related to their involvement with the event) can add to stress and distress as well as drain time and emotional reserves, resulting in an exacerbation of their stress responses. EM staff too, can have the layered concerns of caring for their own family and community members when their local community has been affected by a disaster. Their work often mandates them to be in the field and not attending to the needs of their family members as they would like. This is an added stress that can be more difficult than the EM work itself.

Some survivors and responders have stress indicators that cross all of the domains while others' responses will only show up in one or two areas. Many signs are tied to what is culturally acceptable. For example, many men are taught very early in life that "men don't cry." Thus, despite great sadness or depression, they might resist crying at all costs. This sadness will sometimes be concealed through anger, acting out, substance misuse, or silence—all of which are more socially acceptable for men in most Western societies and is historically, part of the responder culture. In some cultures, expressiveness is expected, or else one may be seen as inhumane-cold and uncaring. People in expressive cultures may cry together, with some members of the community being charged with wailing to ensure there is sufficient honoring of a crisis, loss, or death. Still, across the globe, the human responses to crisis are the same: these domains are seen in every culture but they may just be played out differently by different genders, disciplines, communities, or societies (Naturale, 2006; Stamm & Friedman, 2000).

The experienced and skilled DBH responders can act to provide the following information to those affected to help them understand:

- These responses are quite human and expected because these are reactions rather than signs of illnesses.

- These responses will likely decrease with time and that most people will return to their prior level of functioning or experience sufficient adaptation to their disaster affected environment.
- Information about reaching out for assistance if distress remains and is significantly bothersome and/or interrupts functioning.
- The reality that they may need more than crisis intervention services to assist in the mitigation of the development of a diagnosable mental health problem (in the cases of those at high risk who have been assessed for indications).

The provision of this type of information, while vital in helping support and control extreme reactions in those exposed, is secondary to the issues of immediate safety that the EM staff have to address. As psychosocial factors are a priority for the DBH staff, they can help influence the emotional environment and promote calm, provide reassurance, and help create a supportive environment. This will relieve emergency managers from this task, enabling them to focus on priorities of safety and security.

Another important aspect of how human react in the face of danger that is important to note involves the part of the brain that processes and stores memories. During high stress situations, that area of the brain is essentially turned off. As a result, memories of the traumatic event are retained in a very fragmented way, resulting in a lack of logical sequence or details. What tends to be most easily recalled are traumatic sights, sounds, and smells that elicit very strong emotion. Because of this, trauma survivors often have a hard time retelling their experiences. This is very important in emergency situations that are considered crimes and larger offenses, such as terrorism that involves victim interaction with the criminal justice system. Affected survivors may not be able to explain or describe the event or their experiences in a logical way. The importance and credibility of their telling of events should not be disregarded because of their sporadic or initial lack of clear memory.

DBH specialists can work with law enforcement to calm survivors, decrease their stress responses, and help them to access whatever information may be available to them of ultimate value to any criminal investigation. These same problems around impaired thinking, decision-making, and memory can affect EM staff in high stress situations. DBH professionals can help emergency managers by assisting them in mitigating, decreasing, and addressing stress reactions. DBH techniques can help EM staff to focus and regulate their responses by making the cognitive shifts necessary to access the information they need to do their job, react safely, and be able to report events accurately.

RISK COMMUNICATION/MEDIA INFORMATION

Risk and crisis communication is discussed in detail in Chapter 11, Risk and Crisis Communications, yet it is important to highlight in this chapter as well. In some disaster environments, public communications may be the only intervention that is provided to the community. There are two types of public communications that are as important as any other aspect of DBH services. The first of these is public risk communications, which refers to the provision of information that tells the disaster affected community what the potential hazards or risks are that remain as a result of the disaster. This may involve the lack of water, the loss of electricity in certain areas, or the risk of exposure to viruses, disease, hazardous materials, or biochemicals (Substance Abuse and Mental Health Services Administration (SAMHSA), 2002).

In such situations, survivors may create significant challenges for emergency managers. These challenges can include demanding information, creating and spreading rumors, overwhelming police stations or emergency rooms, and possibly delaying care for those with the most acute needs. Ill-informed and highly agitated survivors can also cause disruption in the supportive environment of the family assistance centers where family members have gathered for a peaceful, private place to meet together and grieve.

Alternately, fear can cause survivors to avoid expressing their distress and concerns, leaving them isolated (which can make their distress worse), and inhibiting or delaying them from seeking behavioral health services should their symptoms continue.

Planned, scheduled, and carefully designed public risk communications delivered by leaders who have established legitimacy, especially emergency managers, with confidence and caring, can reduce problematic responses to difficult and highly emotional situations (such as reports of death, injury and coping with the unknown). EM leadership through designated liaison officers (from communications or operations command functions) can benefit significantly from working closely with DBH specialists in this realm. Together, they can carefully craft and properly deliver public risk communications materials to identified populations, such as victim's families, families of missing persons, direct survivors, and the affected community at large. Emergency management services (EMS) and DBH can also work with the media to provide information and guidance, helping ensure that accurate information is made available to specific and general target audiences in a timely manner.

There is a high cost to be paid when accurate communications do not take place. An example of flawed public communications (resulting in increased fear and anxiety rather than decreasing it) was evident in the Anthrax event of 2001, when misinformation was provided to the media and postal workers who were given different medical treatment than what Senate members received. The result was unrest, widespread anger, and perceptions of racial discrimination around the inequality of treatment provided (Quinn, Thomas, & Kumar, 2008).

The second type of public communications involves the provision of psychoeducational information that helps survivors of disasters, responders, and the general public to understand the nature and meaning of what they may be experiencing. This includes factors such as the nature of the stress response, the likelihood that these will subside over time without the development of a mental illness, or receiving formal mental health treatment (SAMHSA, 2015a; SAMHSA, 2015b). Psychoeducational information typically includes identification of what to expect in terms of basic responses and simply-stated, practical coping tips to various high risk populations (e.g., children and their care-takers, frail elderly, or responders). This would help them manage distress and provide a sense of self-efficacy that may facilitate recovery (National Child Traumatic Stress Network (NCTSN) & National Center for Post Traumatic Stress Disorder (NCPTSD), 2005). Targeted and normalizing messaging can reduce stigma associated with behavioral health problems. An additional suggestion includes the use of a disaster or crisis hotline or other resource that leads to more information. Including this information provides an opportunity for the public to ask questions or find out how to access a referral. These services can strengthen psychoeducational messaging and facilitate treatment when indicated.

Comprehensive media plans can be a highly effective component of a comprehensive behavioral health response plan serving to provide the messages of support, hope, recovery, and resiliency, as well as information about accessing available hotlines and referral resources (Norris et al., 2006). There are several good examples of sound media strategies, such as during the aftermath of 9/11 in New York City where various commercials for the DBH response program tailored to specific high-risk populations (e.g., responders, children and parents, and elderly survivors) increased the calls to the crisis counseling program's hotline, as suggested in the messaging (Draper, McCleery, & Schadle, 2006).

New York 9/11/01 media campaign Ad aimed at seniors.

New York 9/11/01 media campaign Ad aimed at caregivers of children.

Callers expressed that the messaging in the commercials was helpful and made them feel like they were not "crazy" and that they were not alone. Fortunately, it is increasingly common in large disasters to find DBH specialists integrating with EM leadership to develop comprehensive media outreach messaging to help provide support and link survivors with appropriate resources.

DBH involvement in the development of media messaging can ensure that messaging is appropriately targeted, addresses the needs of the affected community, and "does no harm." For example, some media messaging can include information, images, or sounds that can "trigger" survivors and their family members. "Triggering" means that the person who experienced the original trauma can be brought right back to the same negative feelings they had at the time of the incident by seeing, hearing, smelling, or somehow being reminded of the trauma. Military members and their families too commonly report this phenomenon in the context of PTSD: "I felt like I was right back at the front, seeing my buddy get blown apart, hearing the screams, and smelling the smoke." Sometimes, media triggers can be so strong that some people can suffer the intense trauma responses similar to being physically present when seeing an incident on television (Dugall, Berezkin, & John, 2002).

The 9/11 media plan specifically included EM responders as one of the targeted groups for the messaging in order to acknowledge their concerns as well as offer support (Fig. 4). As a result, many responders and their family members contacted the disaster hotline to access resources for themselves and their children.

What Can DBH Actually *Do* To Make EMs Jobs Easier? Chapter | 7 **169**

New York 9/11/01 media campaign Ad aimed at responders.

The psychoeducational messaging developed by the Massachusetts Office for Victim Assistance (MOVA) in response to the Marathon bombing terrorist attack was aimed at the general population. This included people from all over the state and the country, since many of whom participated in the Boston Marathon were from a large number of varied locations. Many others witnessed the bombing live on television. The "Boston Strong" tag line and "strong" theme was seen as a positive public message in many communities along with variations (e.g., "Boston's Getting Stronger," "Boston's Still Strong," and "Boston's Strong When We're Together"). After conducting a needs assessment, MOVA created additional messages to address the concerns from survivors who reported that they did not feel strong and did not want to feel like there was something wrong with them because they were still suffering. Many EM staff also reported that they felt like the media messages applied to them and that it was helpful too for their family members to hear about what the experience was like for their loved one. A media campaign by MOVA, called "AskMOVA," stated that "reactions to violence are normal" and encouraged the public to "know the signs of trauma" as a means to normalize the distress many were still feeling, along with providing a website and hotline number to call for more information (L. Lowney, personal communication, July 12, 2016).

Massachusetts office for victim assistance media campaign.

Spontaneous survivor response message.

Providing psychoeducational information that reassures people in the affected areas that having distressing responses to a disaster are quite common and can serve to decrease concerns that there is something wrong with them or that they are alone in their feelings. Effective messaging can also reassure people that their distress is likely to be short-lived and that the symptoms will subside over time, especially by using simple coping skills. These skills might include elements such as reaching out to social supports, talking to people who are willing to listen, understanding and accepting how the survivor feels, and practicing gentle exercise and breathing techniques. DBH professionals can provide assistance to EM staff to tailor messaging that can speak directly to the EM staff as well as survivors and the general public.

Various modalities can be used to distribute DBH information, including:

- television commercials which have the capacity to reach the largest number of people in the general public;
- radio advertisements which are accessed by varied cultures speaking languages other than English or aimed at a specific homogenous group;
- print advertisements which serve to provide resource information;
- pamphlets and brochures which are useful for situations where it is necessary to leave information for reference at a later time or during community meetings and gatherings that do not allow a sufficient amount of time for personal exchange.

Social media has become a necessary mode of communicating with youth as well as with adults across the age and gender spectrum. As social media use has increased in the past decade, the field of EM has embraced its popularity using texting, blogs, websites, and smartphone apps to provide preparedness information, emergency alerts, and instructions as to how to be safe and stay in touch with loved ones in an emergency. Substance Abuse and Mental Health Services Administration's (SAMHSA) Behavioral Health

Disaster Response App has been developed to provide quality support to survivors by both survivors and responders. Users can navigate pre-deployment preparation, on-the-ground assistance, post-deployment resources, and can also can share resources, like tips for helping survivors cope, finding local behavioral health services, and self-care support for responders (Substance Abuse and Mental Health Services Administration (SAMHSA), 2014).

EM staff can work closely with DBH professionals to create and deliver media messaging as a means of expanding their efforts to provide information and support to survivors. This can be accomplished by referring to existing public messages, distributing informational brochures, providing information about accessing the social media sites and apps that are engaging survivors in dialog around preparedness, awareness, and communications during times of disaster.

TRAINING AND EDUCATION

DBH is a unique sub-discipline of mental health with a strong focus on normalizing and not pathologizing the typical stress reactions of survivors. This is quite different than what traditional mental health and substance abuse service administrators require of skilled providers. Typically, mainstream work in behavioral health requires rapid diagnosis of an identifiable disorder (often for reimbursement purposes as well as treatment planning) and implementation of a treatment plan as soon as possible. With the recent initiatives in the mental health field to create trauma-informed providers, many mental health professionals mistakenly believe that because they have learned about implementation of trauma-informed care, they can apply psychodynamically based principles to disaster response work. For this reason, DBH service delivery plans should include training requirements for providers to assure that appropriate crisis response counseling and disaster-specific interventions are being used at various phases of the response and recovery and that interventions are applied in a consistent manner (Young, Ruzek, Wong, Salzer, & Naturale, 2006).

DBH should also be a segment of EM staff training to assure that they are well informed of the DBH specialists' role and know the variety of roles DBH staff can play. Examples include:

- Addressing various levels of distress in the affected population.
- Monitoring and surveying the affected areas while being ready to intervene with survivors whose distress levels are severe or interruptive of safe and effective rescue and recovery operations.
- Supporting emergency managers as they interact with victims and survivors and being ready to help with necessary interventions and resources.

Conversely, as discussed in detail in other chapters, DBH staff needs training and education in the roles, structures, terminology, and culture of EM. Clearly, training and education is a two-way street.

Providing what is referred to in the field as psychological first aid (PFA) is an increasingly popular approach for many who have experienced highly stressful or traumatic events and may be helpful training for emergency managers. Basic PFA training can be easily and efficiently provided to emergency managers at no cost via web-based training. This training will also provide EM staff with an enhanced understanding of the role of DBH professionals, such as assisting rescue and recovery efforts by addressing the distress responses of the survivors and their families. The training will also provide EM staff the skills to address survivors and family members when necessary, in a manner that is neither a formal clinical approach nor does harm (Young et al., 2006).

Many DBH staff receive additional training in a number of interventions that can assist in performing their functions in areas such as:

- Monitoring and surveying disaster survivors, responders, and their loved ones
- Helping support those who are grieving
- Creating outreach strategies
- Making referrals for services and/or treatment
- Providing more intensive interventions as appropriate.

Specifically, this type of trainings may include but is not limited to:

- Assessment
- Suicide prevention
- Stress management and self-care
- Talking to children about traumatic events
- Grief and bereavement
- Survivor guilt in a post-disaster environment
- Substance misuse and abuse
- Working with high risk populations
- Building resiliency
- Cognitive behavioral therapy (CBT) for post-disaster distress
- Trauma-focused cognitive behavioral therapy (TF-CBT) for children and parents
- Prolonged Exposure/stress inoculation
- Eye movement desensitization and reprocessing (EMDR).

Several of the intensive treatments noted above, specifically CBT, TF-CBT for children and parents, prolonged exposure and EMDR, are specialized, short-term treatment modules delivered in an average of 8–10 sessions. They have been used with EM staff to assist with continuing distress symptoms related to working in or witnessing a traumatic event (M. Freire, personal communication, June 14, 2016). These interventions have been well received by emergency managers due to their focus on the current symptoms rather than past familial or relational issues, their short duration, and their level of effectiveness in helping staff get back to their jobs.

PROGRAM EVALUATION, MEASUREMENT, AND MONITORING

Documentation of interventions with survivors by the DBH service providers needs to be collected in a manner that maintains confidentiality and accurately records (coded) identification and demographics of the recipient. This generally includes items such as:

- Demographic information
- Screening scores
- Referral details
- Determination of acceptance by the recipient
- Record of all service provided with the dates and results
- Any follow-up interactions conducted.

In Presidentially-declared disasters, this information is used by FEMA and the Substance Abuse and Mental Health Services Administration (SAMHSA) to improve the design and implementation of their DBH supported activities. It also informs both funders and program leaders so they can determine where to focus service delivery for which population at each phase of the disaster response.

Confidentiality of information is an important consideration to both emergency mangers and DBH service providers. Both parties have a legal and ethical need to appropriately protect certain types of information. At the same time, effective integration of effort requires regular and ongoing information sharing. This has been an area where EM and DBH have had differing views and sorting out those differences has sometimes resulted in valuable time lost and distraction from other priorities. Both EM and DBH will benefit if, *before an event*, they agree on the specifics of what information can/cannot and should/should not be shared, by and with whom, and for what purposes.

Sound documentation of the interventions provided to EM staff, while respecting confidentiality can be a significant help in program debriefing and review activities, developing lessons learned reports and improving future responses. While respecting appropriate confidentiality, sharing aggregate DBH documentation can help identify trends in complex EM issues specific to a particular event types (e.g., flooding vs tornadoes), staffing concerns, roles and responsibilities, time management and staffing levels.

TRACKING, DOCUMENTING, AND PROJECTING BEHAVIORAL HEALTH CONSEQUENCES

The EM incident action plans (IAP) can help inform the DBH staff of the numbers of survivors and family members that EM staff have engaged with and who may need immediate or priority attention. The IAP may also be used to help DBH staff plan for monitoring of EM staff and their responses to a particular incident.

DBH goals, completed activities, planned activities, and the resources brought to the response may assist emergency managers in providing a more comprehensive reporting on the IAP. Some examples of the types of DBH activities that can be integrated into EM plans to assist EM staff include:

- identifying the DBH liaison for EM staff to refer survivors
- family members or responders in need
- providing the number of DBH staff assigned to the incident
- identifying the number/percentage of survivors who have been paired with DBH staff and those who are still outstanding/waiting for services
- describing what resources and referrals are available so that EM staff can appropriately refer if necessary.

REFERENCES

Briere, J., & Elliot, D. (2000). Prevalence, characteristics, and long-term sequelae of natural disaster exposure in the general population. *Journal of Traumatic Stress*, *13*(4), 661–679.

Centers for Disease Control and Prevention/National Center for Injury Protection and Control/Division of Violence Prevention (last updated 05/13/2014). *The Adverse Childhood Experiences Study*. Retrieved from <http://www.acestudy.org/home>.

Center for Mental Health Services (2001). *Disaster Preparedness, Response, and Recovery*. Rockville, MD: Center for Mental Health Services. Retrieved from http://www.samhsa.gov/disaster-preparedness.

Draper, J., McCleery, G., & Schaedle, R. (2006). Mental health services support in response to September 11th: The central role of the Mental Health Association of New York City. In Y. Neria, R. Gross, & R. Marshall (Eds.), *9/11: Mental Health in the Wake of Terrorist Attacks*. New York, NY: Cambridge University Press.

Dugall, H. S., Berezkin, G., & John, V. (2002). PTSD and TV viewing of World Trade Center. *Journal of the American Academy of Child & Adolescent Psychiatry*, *41*(5), 494–495.

Figley, C. R. (1995). *Compassion fatigue: Coping with secondary traumatic stress disorder in those who treat the traumatized*. New York, NY: Brunner/Mazel.

Figley, C. R. (1999). Compassion fatigue: Toward a new understanding of the costs of caring. In B. H. Stamm (Ed.), *Secondary traumatic stress. Self care issues for clinicians, researchers & educators* (2nd ed., pp. 3–29). Baltimore, MD: Sidran Press.

Freire, M. Personal communication, June 14, 2016.

Fullerton, C. A., Ursano, R. J., & Wang, L. (2004). Acute stress disorder, posttraumatic stress disorder and depression in disaster or rescue workers. *American Journal of Psychiatry*, *161*, 1370–1376.

Galea, S., Boscarino, J., Resnick, H., & Vlahov, D. (2002). Mental health in New York City after the September 11 terrorist attacks: Results from two population surveys. In R. W. Manderscheid, & M. J. Henderson (Eds.), *Mental health, United States 2001*. Washington, DC: US Government Printing Office.

Galea, S., Nandi, A., Stuber, J., Gold, J., Acierno, R., Best, C. L., ... Resnick, H. (2005). Participant reactions to survey research in the general population after terrorist attacks. *Journal of Traumatic Stress*, *18*(5), 461–465.

Hamblen, J. L., Norris, F. H., Pietruszkiewicz, S., Gibson, L. E., Naturale, A., & Louis, C. (2009). Cognitive behavioral therapy for postdisaster distress: A community based treatment program for survivors of Hurricane Katrina. *Administration and Policy in Mental Health and Mental Health Services Research, 36*(3), 206–214.

Kaniasty, K., & Norris, F. (1995). In search of altruistic community: Patterns of social support following hurricane Hugo. *American Journal of Community Psychology, 23*(4), 447.

Kessler, R. C., Sonnega, A., Bromet, E., Hughes, M., & Nelson, C. B. (1996). Posttraumatic stress disorder in the national comorbidity survey. *Archives of General Psychiatry, 52*, 1048–1060.

Leffler, C. T., & Dembert, M. L. (1998). Posttraumatic stress symptoms among U.S. Navy divers recovering TWA flight 800. *The Journal of Nervous and Mental Disease, 186*(9), 574–577.

Marmar, C. R., Weiss, D. S., Metzler, T. J., Delucchi, K. L., Best, S. R., & Wentworth, K. A. (1999). Longitudinal course and predictors of continuing distress following critical incident exposure in emergency services personnel. *Journal of Nervous and Mental Disease, 187*(1), 15–22.

National Child Traumatic Stress Network and National Center for PTSD. (2005). *Psychological First Aid: Field Operations Guide*, September, 2005.

National Institute of Mental Health (2002). *Mental health and mass violence. Evidence-based early psychological intervention for victims/survivors of mass violence. A workshop to reach consensus on best practices.* Washington, DC: U.S. Government Printing Office, NIH Publication No. 02- 5138.

Naturale, A. (2006). Outreach strategies: An experiential description of the outreach methodologies used in the September 11, 2001 disaster response in New York. In E. C. Ritchie, P. J. Watson, & M. J. Friedman (Eds.), *Interventions Following Mass Violence and Disaster* (pp. 365–383). New York: The Guildford Press.

Naturale, A., & Pulido, M. L. (2012). Helping the helpers: ameliorating secondary traumatic stress in disaster workers. In J. L. Framingham, & M. L. Teasley (Eds.), *Behavioral Health Response to Disasters*. FL: CRC Press.

Norris, F.H., Byrne, C., & Diaz, E. (2001). *Risk factors for adverse outcomes in natural and human-caused disasters: A review of the empirical literature*. Retrieved from <http://www.ncptsd.org>. Accessed on October 27, 2007.

Norris, F. H., Friedman, M. J., & Watson, P. J. (2002a). 60,000 disaster victims speak: Part I. An empirical review of the empirical literature, 1981–2001. *Psychiatry, 65*(3), 207–239.

Norris, F. H., Friedman, M. J., & Watson, P. J. (2002b). 60,000 disaster victims speak: Part II. Summary and implications of the disaster mental health research. *Psychiatry, 65*(3), 240–260.

Norris, F. H., Hamblen, J. L., Watson, P. J., Ruzek, J. I., Gibson, L. E., Pfefferbaum, B. J., ... Friedman, M. J. (2006). Toward understanding and creating systems of postdisaster care: A case study of New York's response to the World Trade Center disaster. In E. C. Ritchie, P. J. Watson, & M. J. Friedman (Eds.), *Interventions Following Mass Violence and Disaster* (pp. 343–365). New York: The Guildford Press.

Norris, F. H., & Kaniasty, K. (1996). Received and perceived social support in times of stress: A test of the social support deteriorations deterrence model. *Journal of Personality and Social Psychology, 71*(3), 498–511.

Quinn, S. C., Thomas, T., & Kumar, S. (2008). The anthrax vaccine and research: Reactions from postal workers and public health professionals. *Biosecurity and Bioterrorsm: biodefense strategy, practice, and science, 6*(4), 321–333. Available from http://dx.doi.org/10.1089/bsp.2007.0064.

Stamm, B. H., & Friedman, M. J. (2000). Cultural diversity in the appraisal and expression of trauma. In A. Shalev, R. Yehuda, & A. Mcfarlane (Eds.), *International handbook of human response to trauma* (pp. 69–85). New York: Plenum.

Stellman, J., Smith, R., Katz, C., Sharma, V., Charney, D., Herbert, R., ... Southwick, S. (2008). Mental health morbidity and social function impairment in rescue, recovery and cleanup workers at the World Trade Center disaster. *Environmental Health Perspectives*, *116*(9), 1248–1253.

Substance Abuse and Mental Health Services Administration (2002). *Risk Communications Guides for Public Officials: Communicating in a Crisis*. Rockville, MD: DHHS Pub. No: SMA02-3641.

Substance Abuse and Mental Health Services Administration (2014). *SAMHSA's Behavioral Health Disaster Response Mobile App*. Retrieved from http://store.samhsa.gov/apps/disaster/.

Substance Abuse and Mental Health Services Administration (2015a). *Disaster Behavioral Health Interventions Inventory*. Disaster Technical Assistance Center Supplemental Research Bulletin, May 2015.

Substance Abuse and Mental Health Services Administration (last updated 6/04/2015b). *Crisis Counseling Assistance and Training Program*. Retrieved from http://www.samhsa.gov/dtac/ccp.

Tosone, C., Bialkin, L., Campell, M., Charters, M., Gieri, K., Gross, S., & Stefan, A. (Eds.), (2003). Shared trauma: Group reflections on the September 11 disaster. *Psychoanalytic Social Work*, *10*(1), 57–75.

United States Code 5121 (2013). Congressional findings and declarations {Sec. 101}. *Title 42. The Public Health and Welfare. Chapter 68. Disaster Relief.* Robert T. Stafford Disaster Relief and Emergency Assistance Act (Public Law 93-288) as amended. Retrieved from <http://www.fema.gov/media-library-data/1383153669955-21f970b19e8eaa67087b7da9-f4af706e/stafford_act_booklet_042213_508e.pdf>.

Wee, D. F., & Meyers, D. (2002). Stress response of mental health workers following disaster: The Oklahoma City Bombing. In C. R. Figley (Ed.), *Treating Compassion Fatigue* (pp. 57–81). New York: Brunner-Routledge.

Young, B. H., Ruzek, J. I., Wong, M., Salzer, M. S., & Naturale, A. J. (2006). Disaster mental health training: Guidelines, considerations and recommendations. In E. C. Ritchie, P. J. Watson, & M. J. Friedman (Eds.), *Interventions Following Mass Violence and Disaster* (pp. 55–79). New York: The Guildford Press.

Through an Emergency Management Lens

Lesli A. Rucker

INTRODUCTION

What are the benefits of incorporating disaster behavior health professionals (DBHPs) in response and recovery operations and how does doing so make an emergency managers job easier from an EM perspective?

Before answering this question, there needs to be a distinction in the use of the title emergency manager and how it is applied. Typically, when you hear someone is an emergency manager, you think of a person, elected or appointed, at the local city/town, township/county, tribal, state, or federal level. Day-to-day responsibilities of the emergency manager involve coordinating and overseeing EM planning and exercising, implementing programs and activities, and leading response and recovery efforts. For the purpose of this discussion, the distinction of emergency manager is broadened to include anyone that has a role in the coordination, planning, exercising, educating, and implementing EM plans, programs, and activities at any level before, during, or after a disaster event. This includes individuals whose emergency manager role is as an additional duty, reassignment, or specifically hired in a temporary, part-time, or full-time position. Responsibilities may be broad based or singularly focused. Therefore, anyone with EM responsibilities that either directly or indirectly impact individuals and families have something to gain by incorporating DBH considerations into the decision-making process.

Just like people, neighborhoods and communities have their own unique characteristics and personalities. How adverse events affect individuals and families living in those neighborhoods and communities will differ. Although the basic cause and effect may be the same (e.g., flood waters impacting homes and businesses), how communities plan for, respond, and recover from these events is unique to the individual community and those individuals and families in the community. Behavioral health professionals can help emergency managers understand these unique characteristics and personalities. Emergency managers will benefit from this understanding as they consider and tailor specific response and recovery actions and activities.

Timely, effective, and efficient response and recovery are operational goals for any emergency manager following an event requiring response and recovery. On any given day, emergency managers are faced with events and successfully achieve these operational goals. What happens, then, when an event exceeds the responsible EM capacities? Depending on the size and scale of the event, EM support may be required beyond the affected jurisdiction, relying on outside resources from neighboring jurisdictions, state and federal agencies, relief organizations, and the private sector. With this outside support comes the increased need for an understanding of the affected entities' character, culture, and anticipated behaviors to aid in making decisions that are timely, effective, and efficient in a compressed timeframe.

TOPIC AREAS

The following six topic areas illustrate the ways emergency managers benefit by incorporating DBH into response and recovery operations. The topic areas include:

- Awareness and Understanding
- Messaging
- Operations
- Sheltering
- Strategic Locations
- Program Implementation.

Awareness and Understanding

To facilitate the delivery of assistance it is important to gain awareness and understanding of a community's complexities. Regardless of location or size, reactions and actions to an event will vary. Working with DBHPs to increase emergency managers' awareness and understanding of the intricacies of the affected communities and neighborhoods will support emergency managers as they move through the various response and recovery operational phases. Following are some examples where having better awareness and understanding can assist the emergency manager.

No Trust in Government

In the emotionally charged aftermath of a disaster, there can be a lack of trust in the government. There may be a perception that information gathered may be utilized for other purposes than relief efforts. If perceptions such as this exist, individuals in need of assistance may not come forth and avail themselves to what is being offered. The perception may be perpetuated and, as such, an attitude of mistrust may prevail in the community. Working early on with DBHPs to understand these existing or developing perceptions, emergency managers can develop and implement strategies for addressing these misperceptions. Energy and focus can then be put toward the ongoing response and recovery efforts.

Fear of the Government

There is any number of reasons individuals and families may have a fear of the government. Understanding those fears will assist emergency managers in determining how to best communicate and provide assistance. The earlier in an operation the understanding that fear exists, to include the basis of the fear, the easier it is to determine how to communicate and provide assistance. As an example, individuals legally present, such as agricultural workers, may fear their credentials will be revoked and they can no longer remain in the United States. There may be a need to work with an entity sponsoring workers to determine if assistance is required and identify ways to provide assistance to the sponsor or directly to the workers. Individuals may remain in unsafe and/or unhealthy situations due to a real or perceived fear. For example, an individual or family staying in a damaged property on a leased site may not vacate the site for fear their lease will not be honored if they leave the property. In this instance, losing their lease means nowhere to live in a limited, affordable housing area. Working with DBHPs to identify existing and developing fears allows emergency managers to develop and implement strategies to address and reduce the fears. Addressing the fear early will keep the fear from spreading, save time, and allow emergency managers to focus on the delivery of assistance.

Community Leadership

DBHPs can also provide an awareness and understanding of respected community leadership, not only of a political nature but also community and social leadership. Understanding the leadership dynamics within a community provides the opportunity to move the community forward in the response and recovery efforts avoiding loss of time, ineffective efforts, and delays in the delivery of assistance, as well as the opportunity to incorporate community leaders as stakeholders.

Community Diversity

Communities throughout the country are made up of people from different cultures and backgrounds. Working with DBHPs to understand and become aware of a community's diversity following an event aids in effectively communicating and implementing the provision of assistance. For example, a community with a neighborhood made up of individuals and families from a country where there is a significant housing shortage. They are legally working in the United States and living in housing units that may be one-, two-, or three-bedroom unit with 10 or more adults sharing the one unit. Although the living situation is perhaps not typical in the United States, understanding this situation early into the operation allows emergency managers the opportunity to evaluate resource needs and communication methods that can reach all affected by the event. In this example, in assessing required resources, it is important to be aware and understand these dynamics as the number of damaged structures may not equate to the number of individuals and families requiring assistance. This is particularly important when temporary housing resources are limited and emergency managers are determining ways to create temporary housing.

An additional example would be a multi-generational area or neighborhood where families have lived for years and real property has been passed down generation to generation. Proving who actually owns and maintains the property may become an issue. Not addressing an issue such as this early on will result in contention, confusion, and frustration as assistance programs with proof of ownership requirements are implemented. In this instance, being aware and understanding the situation early can allow program managers to put in place specific processes and procedures to address legal ownership documentation.

Questions Versus Statements Posed as Questions

Throughout the response and recovery process meetings involving the public may be held to discuss any variety of topics: operational progress of an event, specifics of any given program, explanation of decisions. Community meetings for any number of reasons may be emotional, energy charged and confrontational. The reason behind these emotions may have nothing to do with the topic being presented. However, the forum provides an opportunity to release frustration and anger and be heard. Individuals may be asking questions, but in reality they are making a statement posed as a question. In that instance, while no specific response is necessarily required, a simple acknowledgment may suffice. DBHPs awareness and understanding of the current state of the community can help with the preparation for these meetings, suggestions for conducting the meetings, informal interaction with attendees at the meetings, and help to support and provide feedback on community reaction. Working together, the DBHPs can help those conducting and presenting at meetings understand the root of emotions that may emerge, how best to address statements posed as questions, suggestions for conducting productive meetings and strategies for addressing and achieving positive meeting outcomes. The behavioral health staff can also interact directly, but informally, with attendees. Chapter 11, Risk and Crisis Communications, discusses further how DBHPs can assist emergency managers with preparing for and understanding the psychosocial dynamics of meetings that have potential for being emotionally charged.

Perceptions

There is any number of possible perceptions that develop following an event. Individuals may be too proud to avail themselves of the assistance being offered. Individuals from a specific area believe that assistance is not being provided to their area and that another area is receiving preferential priority. There have been instances where elderly recipients have not wanted to apply for aid as they are worried their children or others need the funds more than they do, or they are worried the funds are going to run out. Similarly, recipients receiving assistance, however, do not cash their assistance check for the same reason. In addition, emergency managers who are responding to and assisting with recovery from an event, at all levels, may bring their own perceptions of what needs to be done, how it needs to be done, who needs to be involved, and when something needs to be done. Working together, emergency managers and DBHPs can identify, understand, and determine the best communication strategies and methods for addressing individuals, organizations and agencies misperceptions. By dispelling inaccurate perceptions, both internally and externally, emergency managers are able to address other emerging issues and focus on the delivery of assistance.

Messaging

As response and recovery operations move forward, relief activities and programs are initiated and a wide array of assistance is made available and provided. If an event is federally declared, federally-funded recovery programs may be authorized and implemented. These programs fall under federal law and regulation[1] and have very specific criteria for recipient eligibility. States and tribes may also administer or have their own disaster aid programs with specific criteria for who may apply for and receive the assistance. Additionally, not-for-profit organizations, private sector assistance, and insurance proceeds may also be available with specific recipient requirements for eligibility to receive funds and for making claims. With the myriad of available assistance, what assistance is available and who qualifies for the assistance can become confusing. Misunderstandings and misinformation develop and emerge. When today's reality of instantaneous communication is added to the mix, inaccurate and incorrect information can spread very quickly. In the absence of information, people will fill in the gap whether or not the information is true or accurate. As misinformation and negative perception develop and grow, the more time emergency managers spend addressing the emerging issue, taking personnel and time away from program delivery. DBHPs working with program managers can help to identify misinformation and negative perceptions developing within the community. Working together, the DBHPs and emergency managers can determine the best methods for clarification, correction, and dissemination of information. This includes not only the method, but also the message itself.

---

1. Code of Federal Regulations, Title 44, Emergency Management and Assistance, Chapter 1, Federal Emergency Management Agency, Subchapter D, Disaster Assistance, October 1, 2002.

Operations

As events unfold and grow, so will the need to expand emergency operations in response to the scale of the event. With this expansion comes the requirement to staff a multitude of positions and, in the case of large-scale events, expand to a 24 hour operation. The staff for these positions comes from a variety of sources: internally from responsible agencies or organizations, neighboring jurisdictions, mutual aid agreements, EM assistance compacts, federal agencies, and those hired locally. Individuals may be brought in from locations all over the country. As the operation continues to grow and the event moves through simultaneous response and recovery phases, decision-making is made in a condensed time frame. The number of emergency managers who have specific experience and knowledge of the affected areas characteristics, culture, and politics becomes diluted, thus increasing the need to include local DBHPs into the operation. It is important to coordinate and integrate DBHPs into the decision-making processes as soon as possible as operational objectives, strategies, tactics, and plans are being determined and implemented. Operational benefits for emergency managers include less duplication of effort, informed decision-making and maximization of limited resources.

Sheltering

As response and recovery operations go forward, the congregate sheltering requirement shifts from an immediate need to provide a safe secure location for individuals and families evacuated and/or dislocated from their homes to a need for individuals and families to transition out of the congregate shelter. Shelter occupants may go back to their homes, apply for and access disaster related housing programs, utilize personal insurance funds, or make alternative arrangements with friends, families, and acquaintances. Some of the circumstances that may be encountered during the transition to closure and how DBHPs can help emergency managers follow:

Internal Shelter Dynamics[2]

Those experienced with shelter operations understand that a variety of dynamics may develop within a shelter. Shelter occupants may not want to vacate the shelter, as they believe they will soon be returning to their home. They may also want to stay, as they have come together as a community within the shelter and they do not want to leave this newly formed community. DBHPs can help with understanding these dynamics, such as what it means to the shelter operations and how to best address these dynamics to

---

2. For additional information regarding shelters contact the American Red Cross. http://www.redcross.org.

help facilitate successful shelter transitions. Emergency managers can then focus their attention on other operational priorities.

Prioritization

Local, tribal, and state officials, including the Governor's or Tribal Leader's office, may be asked to establish priorities for who will receive a specific type of assistance. For example, it may be required to prioritize which shelter occupants will be the first to be offered limited transitional or short-term lodging resources. The DBHPs can assist in the development of a well-thought out strategy for prioritization that will help facilitate a smooth, acceptable transition. Integrating DBHPs as soon as possible in a disaster operation means familiarity with the overall operation, not only singular issues and priorities. By being included in briefings, meetings, and discussions throughout the operation, the DBHPs will have a better understanding of the very dynamic and rapidly changing operational goals, objectives, and priorities.

Transitional and Short-Term Housing Assistance

In large disaster operations, the demand for transitional and short-term housing assistance may be immense. The shortage of short-term housing solutions may require emergency managers to come up with creative alternatives. Although on the surface these alternatives may seem reasonable to the emergency managers, the shelter occupants may not be so accepting of the resource. An example of a solution that on the surface seemed straight forward, required a significant amount of time, communication, funds, and energy to implement involved a cruise ship.

A cruise ship was brought in to house individuals and families living in a congregate shelter. A priority was established for the ship to first house older adults and families with young children. The nearest accessible port was in a neighboring state. The ship was brought in and the priorities were announced. The reaction from the shelter occupants was unexpected to those who were not experienced with shelter operations. Hardly any of the occupants wanted to leave the shelter. In the days that ensued to sort out concerns, it was discovered people did not want to be located several hours from their community, and others thought the ship was going to go set out to sea once they were on board. Those without transportation were worried about how they were going to get around. Some parents were worried about schooling for their children and wanted to remain in their community, albeit significantly impacted, for when school reopened. The sheltering plan that was developed and put into place was very comprehensive and included a variety of counseling support to include disaster mental health counseling. Having DBHPs integrated throughout a shelter operation will help EM at all levels to understand how transitional and short-term housing alternatives may be received. DBHPs can also aid in determining strategies for presenting and implementing this type of nonconventional housing resource from concept to closure. For emergency managers, an effective, efficient, strategy will help alleviate fears, maximize resources, and move individuals and families into an improved situation.

Closing a Shelter

There comes a time when it is time to close shelter operations. Shelter occupants may return home, move in with family and friends, receive additional living expense funds from insurance, be provided with a form of temporary housing assistance, or locate a permanent housing option. Moving the remaining occupants out of the shelter(s) may be a challenge. Those remaining may have been homeless prior to the storm, are afraid to apply for assistance, may not qualify for housing assistance, and as such present a challenge for shelter closure. Working with shelter managers, local DBHPs familiar with the area, and emergency managers a plan can be developed for how to transition remaining individuals out of the shelter. A well-planned transition will not only be of benefit to emergency managers, focusing on emerging issues, and continuing operational priorities, but it will also benefit shelter managers and shelter occupants with a smooth transition to closure.

Strategic Locations

To provide a variety of services, such as information dissemination, commodities distribution or provisions of assistance, publicly accessible mobile or fixed sites and centers will be identified and established. If there is not an awareness of the dynamics of a community, sites and centers may be established in less than optimal locations. Individuals may not go to the location due to social, cultural, or historic divides that are largely invisible to outsiders, but are very real to the individuals within the community. Individuals from one area will not conduct business or go to a neighboring area, even if it is just across the street. The location is out of the way or unknown, so individuals may have a difficult time finding the location. The site or center location is important to reach the total population, those affected may have specific transportation, communication, and/or access needs. For emergency managers, establishing sites and centers that will not be accessed or utilized to their maximum may require the site or center to be relocated or closed, and as such result in a loss of valuable time, energy, and expending of resources—including people, materials, and dollars.

In addition to the intended purpose, these locations may serve unintended purposes, such as central gathering points for those affected or become central meeting sites to gather information and answer questions. DBHPs can help to identify the significance of these unintended purposes and work with emergency managers as they determine a site or center's purpose, importance and timing for closure.

Program Implementation

Following an event, and in particular an event that has significant impacts such as loss of life and high degree of destruction, numerous programs will be authorized and implemented from the federal, state, tribal, and local governments. As assistance programs are authorized and implemented, program managers will encounter a variety of challenges working with individuals and families as they go forward with program implementation, delivery, and closure. Working together, DBHPs and emergency managers can develop and implement strategies that will not only benefit emergency managers but also those impacted and receiving assistance. How working together can be a benefit is illustrated below:

Interim Housing Solutions

In large events, in addition to financial assistance to rent alternative housing, there may be a requirement to create temporary housing solutions. One solution to a lack of post-disaster housing is to bring in manufactured housing units. Use of the temporary housing units may be provided for up to 18 month or longer.[3] Along with providing a place for individuals and families whose homes have been destroyed a temporary place to live, often perceived or real concerns from past events emerge. If these concerns and fears are not adequately addressed, it may delay the timely provision of the temporary housing assistance. Partnering with DBHPs emergency managers can anticipate, eliminate, or lessen the stress, and concerns prior to them becoming larger and impacting the delivery of assistance.

Property Acquisition and Buy Outs

Following an event, funds may become available for property acquisition or buyouts. Local officials make the difficult decision to apply for and implement a program to remove properties from hazard situations. These decisions may be met with skepticism, unease, and criticism. They may reflect long-standing concerns regarding social justice and, as mentioned earlier, distrust of government. With limited funding, one family's home is bought out; however, the property nearby is not part of the buyout. This may create a resentment as the neighbor(s) might ask, "why did they get it and I didn't?" Homes and land holdings may have strong symbolic values, reflecting family

---

3. Code of Federal Regulations, Title 44, Emergency Management and Assistance, Chapter 1, Federal Emergency Management Agency, Subchapter D, Disaster Assistance, Part 206-110, October 1, 2002.

and community history. This may make emotional responses seem out of proportion to the decisions/actions at hand. Working with DBHPs will not make these decisions easier; however, they may be able to help emergency managers appreciate the psychosocial dynamics involved, develop strategies regarding how to best present the program, and provide perspective and assistance throughout the numerous steps of the process.

Rebuilding

As rebuilding begins following an event, building permits are going to be required prior to beginning construction/rebuilding projects. In instances where significantly damaged properties are located in a special flood hazard area, as part of the permitting process, there may be a requirement to elevate, or otherwise flood proof, the home. As a condition of funding, there may also be specific requirements, such as to purchase and maintain flood insurance. Even in non-flood events, there may be specific requirements due to a properties location in a flood hazard area. As the rebuilding process goes forward and these requirements become known, emotions can run high especially if rebuilding in certain areas is significantly restricted. Working together, emergency managers and DBHPs can help local officials, program managers, and the community understand requirements, decisions, and provide support going forward with the rebuilding process.

Redevelopment

Communities will be faced with a number of post-disaster, redevelopment issues, such as land use, location/relocation of public infrastructure and facilities, economic redevelopment, permanent housing solutions, and restoration of health and social services. Developing and implementing post-disaster redevelopment plans is an all-inclusive process, involving not only emergency managers and DBHPs but also local officials, community stakeholders, and the general public. Redevelopment planning is locally driven requiring leadership, participation, and commitment. As with the rebuilding process, emotions can run high, particularly when not everyone in the community may be accepting of the redevelopment plans. DBHPs can assist with the overall planning process, as well as helping in understanding the specific health and social services issues and impacts of the redevelopment plans.

BENEFITS TO EMERGENCY MANAGERS

The benefits of incorporating DBHPs in pre-event planning, exercising, and post-event response and recovery, including long-term recovery, are that the emergency manager's job will be made easier to include time and energy savings and improved efficiencies and effectiveness in the delivery of assistance. Additional benefits include:

- Understanding of how the two disciplines, EM and DBH, fit together and support one another.
- Early issue identification and resolution.
- Communication efficiencies.
- Distribution of quality information.
- Identification, understanding, and resolution of real or perceived assistance barriers.
- Limiting duplication of effort.
- Informed decision-making.
- Maximizing efforts in the successful implementation and delivery of assistance.
- Making the most of limited resources.
- Smoothly transitioning between operational phases.

Disasters are change-makers, and no amount of money or technology will bring individuals, families, neighborhoods, and communities back to exactly the way they were pre-event. Working together BDH professionals can not only make the emergency manager's jobs easier, but also make it an easier road to travel for those impacted.

Making Integration Work

April J. Naturale

Many organizations use formal programs to help them assist their employees with individual/family and community events involving concerns such as suicide, domestic violence, substance misuse, sudden death, and disasters. The disaster field itself needs a comprehensive implementation strategy to integrate behavioral health into its overall plan of response. This strategy should become part of the disaster and EM organizational culture that is implemented from the top down with modeling, encouragement, and support for the activities targeted to each population that may need general or specialized attention. These include but are not limited to victim and survivor family members, those injured, witnesses, and members of the general population who are experiencing severe levels of stress, physical exhaustion, and even re-traumatization.

A plan to address these issues should include provision of information, education, assessment, and, where necessary, disaster-specific interventions. Such interventions can mitigate the development of consequences such as (US DOJ/OVC & CMHS, 2004):

- Prolonged mourning
- Physical and mental exhaustion
- Substance misuse
- Aggravation or exacerbation of health problems (which often result in greater health care utilization)
- Poorer quality of life and relationships
- Job and/or school-related problems and on both individual and community levels
- Higher long-term disability costs
- Increased personal and financial burdens

Emergency managers will find that DBH staff are more accepted and welcomed because of the legitimacy that emergency manager leaders bestow upon them and when they are included in all communication as well at the Incident Command Center, Family Assistance and Information Centers, and in the respite sites for responders. If the behavioral health staff can blend more easily with all of the emergency response staff and survivors, they are more likely to be approached as part of a comprehensive and cohesive team. This enhances the potential for DBH professionals to avoid standing out as separate from the primary effort and to reduce the potential that they will be seen only as staff who are seeking to identify "troubled" survivors or responders. The more casual and integrated the initial interactions between behavioral health staff, other responders, and survivors are, the more likely their information about managing traumatic stress reactions and later recommendations for specific interventions will be accepted.

One of the best examples in recent times comes from one of the worst hurricanes in the history of the United States, Hurricane Katrina, followed by two almost equally devastating storms within just a couple of weeks, Hurricanes Rita and Wilma. These storms effected the five Gulf states for several years. An integration strategy was born out of necessity. The Mississippi Emergency Management Agency (MEMA) needed all the trained personnel, it could find from every agency available. DBH staff were assigned to every shift and included in every Incident Command briefing from the start of the response to the end. This inclusiveness provided the DBH staff opportunities to design and integrate their specific activities as part of the response program in an ideal structure, with representatives designated to address survivors, family members, responders, and the general public from the most highly affected coastal areas of the state to the more northern counties with different types of damage. The EM staff had the benefit of input from the DBH staff in planning and implementing a comprehensive response. This plan included assessing needs and strategically detailing staff across the state using information from the MEMA to geo-map the areas with populations at highest risk. EM staff was relieved of the burden of addressing severely distressed survivors because the DBH staff immediately assumed those tasks. They were pleased to find that DBH staff could assist in problem-solving unusual concerns brought to the Federal Coordinating Officer. For example, early in the response, New Orleans EM leaders called to say that 90 nurses from Charity Hospital were just laid off and would disperse, possibly leaving the state to find opportunities for income. The DBH program was able to hire these staff into the crisis counseling positions and build their capacity to provide supports in a way that supplemented their existing knowledge. Not only were these survivors able to remain in their home area reducing their distress, but then were then able to return, later to local nursing positions. Additionally, the disaster intervention techniques they learned as part of the DBH team remained a part of their skill set.

A simple exercise for any state EM agency would be to go through each of their checklists including Incident Command Briefings, Safety Briefings, IAP, Communications Plans, and all others, making sure to add a DBH designee to each of these EM activities. This would allow for representation of DBH throughout the disaster recovery, response, and resiliency-building phases.

REFERENCE

U.S. Dept. of Justice, Office for Victims of Crime and the Center for Mental Health Services, the Substance Abuse and Mental Health Services Administration (2004). *Mental health response to mass violence and terrorism: A training manual.* Rockville, MD: DHHS Pub. No SMA 3959.

Chapter 8

Expanding the Tent: How Training and Education Partnerships with Other Professions Can Enhance Both EM and BH

Laurence W. Zensinger[1], Gerard A. Jacobs[2], Brian W. Flynn[3], and Ronald Sherman[4]

[1]*Director of the Recovery Division for the Federal Emergency Management Agency (Retired), United States,* [2]*University of South Dakota, Vermillion, SD, United States,* [3]*Uniformed Services University of the Health Sciences, Bethesda, MD, United States,* [4]*Independent Consultant, FEMA Federal Coordinating Officer (Retired), United States*

Through and Emergency Management Lens

Laurence W. Zensinger

A focus on training and education for both emergency management (EM) professionals and colleagues in other professions with whom emergency managers work can be an effective means of expanding the understanding and appreciation of the role of behavioral health in the EM process. It is my guess that most practicing emergency managers, not to mention members of other professions that have EM responsibilities in their job description, have had little or no formal training or education that has addressed behavioral health, particularly as that topic intersects with the planning for and management of disasters. Nevertheless, the behavioral health impacts of disasters have been well recognized in the United States since the inception of national level disaster programs. Many emergency managers may be familiar with the Crises Counseling Program (CCP) administered by the Federal

Emergency Management Agency (FEMA), or if such a program has at some point been activated in their state or local jurisdiction in the aftermath of a disaster. Emergency managers are rarely part of the delivery mechanism of this program.[1] Even an awareness of this specific program is not very helpful for emergency managers in understanding the importance of disaster behavioral health (DBH) in the execution of their emergency response rolls. Emergency managers need to know about DBH well beyond the Crisis Counseling Program (CCP).

In this chapter, I hope to accomplish three objectives. First, I will address what I believe to be the critical role emergency managers play in influencing the stress levels disaster survivors experience in the aftermath of a disaster or emergency. There are a number of specific disaster response and recovery functions, which, depending upon how they are executed, can make the difference between exacerbating or alleviating the inherent stress of a disaster on significant portions of the affected population. Second, there are some related professions which, if included in the EM/Behavioral health process, can help to enhance the overall disaster behavioral objectives and thereby "expand the tent." Finally, I will offer some thoughts on the current status of training for emergency managers in topics related to DBH, and suggest potential resources that are or could be available to "expand the tent" of understanding in the emergency community to enhance partnerships between emergency managers and disaster behavior health professionals.

THE COURSE SYLLABUS: DISASTER BEHAVIORAL HEALTH FOR EMERGENCY MANAGERS

Most emergency managers understand that emergencies and disasters can adversely impact behavioral health for those individuals and families who have the misfortune of experiencing direct or indirect effects. These impacts are often obvious and, as discussed in other chapters, well-documented. Damage to public infrastructure can cause the loss of basic community services that everyone takes for granted, such as transportation systems, electric power, water supply, and health and medical facilities. Disasters impact the lifelines upon which all depend, such as food supplies and other basic commodities. Loss of personal property, one's dwelling, income, or social networks can disrupt lives significantly and often for extended periods of time. These effects alone create very high levels of stress, leading to issues impacting behavioral health, in the lives of people who otherwise cope effectively.

---

1. FEMA's Crisis Counseling Program (CCP) underwrites grants, following federally declared disasters, for community based counseling and referral services focusing on the unique effects of the disaster on the general population. The program is administered through the Substance Abuse and Mental Health Services Agency (SAMHSA, 2016), part of the Department of Health and Human Services.

What is less understood or appreciated throughout the extended EM community is the extent to which the actions of response and recovery agencies in the aftermath of a disaster contribute to behavioral health of both individuals and the community as a whole. They do not always contribute to the behavioral health "healing" process but in many instances actually amplify the traumatic effects of the disaster. There will always be a sense of disillusionment among the affected population following a disaster,[2] no matter how well the response and recovery plans were executed and no matter how well available forms of assistance match the needs created by the disaster. That said, it is still always possible to make things worse. In the aftermath of one of the great U.S. disasters of the late 20th century which overwhelmed not just the Federal Emergency Management Agency but federal disaster preparedness capabilities as a whole, one observer, who did not want to be identified, noted "first there was the earthquake, then there was the federal response, then there was the disaster."

How can certain EM functions and activities lead to greater or lesser adverse DBH effects? The answer to this question revolves around how EM officials relate to and communicate with the public in the aftermath of disasters.

> I believe there is no greater role for emergency managers than to understand that the specific actions they take and how they characterize the evolving disaster situation in communication with the public following disasters can significantly influence DBH outcomes.

There is an old military adage that says "no battle plan ever survived contact with the enemy." This similarly can be said for disaster response and recovery plans. Emergency managers invest a lot of their time in developing plans to address the response to the range of potential hazards that have been identified in their communities. Nevertheless, no plan can anticipate every situation that will arise. A certain degree of improvisation will always be necessary, and, even then, things will not always go as smoothly as hoped. The question is not whether things will go wrong, but how we handle unanticipated problems that invariably arise.

It is difficult to say if the response and recovery to a specific major disaster or significant emergency has been "successful" from a DBH perspective. However, a case can be made that the overall response and recovery from some disasters has been less harmful than others in impacting behavioral health impact. Based upon my personal observation of the outcomes of nearly eight hundred federally declared disasters over the course of 25 years, I believe it is possible to generalize a list of common themes or factors

---

2. See "Common Stages of Disaster Recovery", North Carolina Cooperative Extension Service, 1999. The four phases are Heroic, Honeymoon, Disillusionment, and Reconstruction.

which can help to move the disaster response and recovery process in a positive direction along the scale of success in terms of DBH.[3] These common themes are summarized in the following questions:

Who Is in Charge?

Just as the concept of "unified command" is essential to get a range of entities to work cooperatively following disasters, it also essential that someone assume the mantle of "comforter-in-chief." This person gives reassurance to as many people as possible that there is a process in place, in capable hands, aggressively addresses needs, and moves the recovery process forward. Whether it is at the local, regional, state, or federal level, a leader must be a positive presence in the eyes of the public, preferably a chief-elected or appointed official with the most authority and responsibility for the disaster response and recovery.

As the face of the government response structure, the leader needs to demonstrate empathy, offer hope, transmit useful information, and do all of this consistently, as well as at regular intervals for as long as necessary. In the aftermath of the 9/11 attacks on the World Trade Center, for example, New York City Mayor Giuliani filled this role in a manner which many believe to be the classic model. Mayor Giuliani conducted daily press conferences during which he gave updates on key activities, such as the search and rescue program, which became the most anticipated and most trusted source of information available. He was authoritative and accurate in the information he reported. He inspired confidence in the eyes of most New Yorkers that the response and recovery process was in good hands. As a result, even though this event resulted in a great many deaths and destruction of an iconic part of the city if not the nation, there was much less of the stress induced adverse outcomes to this disaster than for many others of less significant impacts.

Do Expectations Reflect Reality?

If not realistic and met to some tangible degree in response to the disaster, the expectations of the public will increase disillusionment and distrust and result in adverse DBH effects. Elected officials and their EM staff, at every level of government, walk a very tight line between projecting competence and setting reasonable expectations.

---

3. While there is no tested and calibrated quantitative means for measuring the DBH outcome of a disaster, it has never been difficult for senior FEMA staff to discern which operations had the most egregious outcomes. Widespread dissatisfaction among the population of the disaster affected area results in negative media coverage. Extensive and justified adverse media reports get the attention of the national political leadership, congressional oversight committees, Inspectors General and the Government Accountability office (GAO). Congressional hearings are held, remedial actions are demanded and occasionally agency heads lose their jobs.

> During my 26-year tenure with Federal Emergency Management Agency, the expectations of the public with respect to Federal Emergency Management Agency's response to disasters grew exponentially. The Federal Emergency Management Agency evolved from an organization that most people had not heard of to the organization that many people believe has the primary responsibility for all disasters throughout the nation. While rising expectations are not bad if they are the motivation for concomitant advances in capabilities and performance, they can be problematic if they reach levels that cannot be practically achieved. Even though the Federal Emergency Management Agency's statutory role has never changed, expectations for what it should and could do have significantly risen in the minds of the public.

This rise in expectations is probably the result of a variety of factors, one of which is the notion that a President and his or her administration are responsible for what occurs on their watch. In actuality, governors of the 50 states retain the primary responsibility for responding to disasters in their states. Yet, an increasing range of disasters and emergencies can be politically damaging if a President fails to show up at the scene.

The presence of high-level public officials at the scene of a disaster can be very reassuring and the "comforter in chief" can have very positive impacts on DBH. Nevertheless, in this process, there is also a tendency with many governmental officials at all levels, often inadvertently, to create unreasonably high expectations. If the chief-elected official promises that something will occur, or some problem will be solved in a certain way or by a certain date, the EM director and staff will do everything within their powers to meet the objective, even if they know it is impossible to do so. Misguided withholding of accurate information that might reflect badly on the recovery effort and those in charge seldom succeeds; the truth usually prevails. When disclosed, intentional misinformation increases disillusionment and erodes trust, thereby increasing stress in disaster survivors. Unfortunately, bad news does not go away if it is ignored or worse yet, if it is disguised as something else.

> The challenge is this: in communicating with the public during disaster operations, how can you strike a balance between honestly discussing the severity of the situation and the enormity of the recovery task at hand while, at the same time, offering hope and optimism for the long-term outcomes? For the emergency manager, their staff, the leaders of all the relevant emergency support functions, private nonprofit agencies, and private commercial enterprises engaged in the recovery process, one of the most important and challenging jobs is to help strike this balance. It is essential to do so for support of positive DBH outcomes.

Disasters and emergencies create innumerable situations in which expectations are set. When will the power be restored? How much of my losses will be covered? Will disaster assistance "make me whole?" When will I be able to leave the shelter and get into temporary housing? When will the roads be open? When will the mobile homes be ready for occupancy as temporary housing? Emergency managers must be expert at not only knowing how these programs work. Accuracy and credibility matter. Emergency managers must also be expert at ensuring that the standard answers to these questions are adjusted to reflect the impact of leadership credibility and guard against the tendency to over-promise when the only feasible outcome will be to under-deliver. Credible leadership that delivers what it promises and promises only what if can deliver has a positive impact on the psychological health of survivors and impacted communities.

What Is Your Approach to the Media?

We are blessed (and some would say cursed) in the United States with the right to freedom of the press. It is especially important for emergency managers to understand two things about the media. First, most reporters and other representatives of the media have very little, if any, understanding of how disaster response and recovery processes work. Many of them are reporting about a disaster for the first, and perhaps last, time in their careers. They may bring a range of preconceptions about disaster processes to the job. If these notions become part of the narrative espoused by reporters, they can have a very negative effect on the psychosocial impact on the general public. It is important that emergency mangers and behavioral health experts have patience and take care to do the education necessary to improve media coverage.

Second, if the managers of the disaster response and recovery process do not "tell their story" as a narrative for the media to embrace, the media will design its own narrative. Nature abhors a vacuum, and this is certainly true when it comes to media reporting. Any adverse perceptions that develop in the disaster-affected population about the quality or timeliness of disaster relief and recovery efforts, unless countered, will be reflected in media coverage. The degree of accuracy of these perceptions does not matter. Negative media reports will exacerbate the stress of the disaster. Most emergency managers do not have direct responsibility for designing media strategies. By the same token, most public affairs officers do not understand the subtleties of disaster relief or recovery programs or strategies. Emergency managers must ensure that public affairs personnel understand that "good press" and DBH impacts are two sides of the same coin. Much more detail regarding how integration of communication can enhance both EM and DBH is contained in Chapter 11, Risk and Crisis Communications.

BRINGING ALLIED PROFESSIONS INTO THE TENT

EM as a profession is a relatively recent phenomenon. As a result, higher education in the field of EM is still emerging. As in other professions, there is considerable variability among the programs in EM offered at many colleges and universities. There is ongoing debate around the question of what constitutes the central discipline of EM. However, one attribute shared by many degree programs in EM is a highly interdisciplinary approach. Higher education programs in EM tend to reflect the reality that emergency managers need to have at least some degree of understanding of a very wide range of core knowledge, but are not expected to be experts in all of them. There are already a great number of professions, of which their subdisciplines contribute to the recommended core knowledge for EM. The list is still growing.

The variety of professions or that contribute to EM constitute a starting point for identifying those allied professions that would benefit from a greater understanding of EM and behavioral health. This takes the concept of interdisciplinary training of emergency managers, stands it on its head, and begs other questions: How can the profession of EM contribute to and enhance these other professions? By doing so, how can it bring a more comprehensive and unified understanding of behavioral health?

A workshop held in October, 2003 brought together 55 leaders from the "hazards" community to discuss higher education opportunities in EM and to help promote the "professionalization" of the profession of EM.[4] One of the products of this workshop was identification of 21 professions contributing to EM and which were considered areas appropriate for interdisciplinary research to advance the field of EM. While this list was useful for developing research priorities, it is not of great value in identifying which professions or occupations that are traditionally allied with or work in partnership with emergency managers. I developed my own list of candidate professions. For each of these professions, I have highlighted either the significance of that profession in impacting DBH, or the potential role that a greater understanding of DBH could have in improving outcomes after disasters. This list should be considered a starting point and can be used as a blueprint in identifying additional relevant professions and disciplines (Table 8.1).

---

4. Designing Educational Opportunities for the Hazard Manager of the 21st century. Workshop Report, October 22−24, 2003, Denver, Colorado, By Deborah Thomas, University of Colorado at Denver And Denis Mileti, Natural Hazards Center, University of Colorado At Boulder, Working Paper No. 109, The Natural Hazards Center, University of Colorado, Boulder, Colorado (Thomas & Mileti, 2003).

TABLE 8.1 Associated Disciplines with Emergency Management Roles

Profession/ Discipline	EM/Behavioral Health Role
Business	Importance of business continuity planning, especially as it relates to continuing income streams for employees: corporate programs for helping employees recover from disasters.
Communications	The communications industry is the means by which information is transmitted. It has a critical role in getting accurate and reliable information to the disaster affected pubic.
Economics	Broadly speaking this could include the insurance and financial services industry, among others. These two professions/functions have a critical role in the recovery process by providing the financial resources. They need to perform their services with an understanding of the potential psychological issues their customers may be facing in the aftermath of disasters.
Education	Awareness for educators of emergency managers roles and processes used and potential student behaviors that may be driven by disaster effects and tools for addressing them.
Engineering	Engineers and engineering organizations perform damage assessments, and in many instances, such as earthquakes, play an important role with helping the public to understand whether structures remain safe for occupant. Reliable and well-crafted statements from the engineering community can help to alleviate concerns and stress among disaster survivors.
Environmental management	Environmental Managers may have responsibility for preparing public statements, in the aftermath of disasters, regarding the safety of water supplies or waste treatment systems, or the potential for the spread of contagious disease. They have key role in both providing necessary safety precautions as well as reassuring the public regarding potential health and safety concerns, all of which affects the DBH following a disaster.
Journalism	Specific training in journalism schools on covering disasters and emergencies which includes reference to DBH and the crucial role of the media.
Law	Disaster Legal services, rule of law and support for law practices impacted by disasters.
Law enforcement/ criminology	Recognition of role of DBH in certain criminal behaviors and the need for appropriate, case-specific interventions.
Political science	Political science programs at the undergraduate and graduate level should touch upon the politics of disasters and should

(Continued)

TABLE 8.1 (Continued)

Profession/ Discipline	EM/Behavioral Health Role
	incorporate an understanding of DBH outcomes to enhance the effectiveness of political leadership.
Psychology	Create the curriculum for DBH training and education for the range of EM partners in concert with emergency managers.
Public administration	Most city managers and many other key government officials come out of a public administration background. Many of these individuals ultimately have key responsibilities in disaster response and recovery operations. It is critical that they understand the EM and DBH nexus.
Public health	An importance with feet in both the EM and mental health camps that can help promote awareness of behavioral health issues and messaging to lessen adverse impacts.
Sociology	Training for those working in social services/social work to understand and work with the EM/DBH nexus.
Urban planning	Dual use of public participation mechanisms to include outreach into disaster affected communities to achieve consensus and support for disaster recovery proposals.

EMERGENCY MANAGEMENT TRAINING AND EDUCATION AND DISASTER BEHAVIORAL HEALTH

As higher education in EM has evolved and expanded over the past 25 years, the number of degree programs available has increased substantially, and the focus and specialization of these programs has broadened. According to FEMA (2016),[5] there are 49 colleges and universities in the United States which offer Bachelor's degrees in EM, with additional programs that offer an EM "concentration" in a different degree program or certificate programs in EM. Federal Emergency Management Agency also identifies 39 master's degree and nine doctorate programs currently offered by U.S. colleges and universities.

An analysis of required courses and electives to those with an EM major reveals the extent to which the subject of DBH may be included in the undergraduate studies for EM degrees.[6] I have attempted to be as inclusive as possible in performing this analysis to include any required or

---

5. FEMA Emergency Management Institute, Higher Education Project, "The College List." Information taken from FEMA website (https://training.fema.gov/hiedu/).
6. This analysis is drawn from the description of required and elective courses published for each degree program on the FEMA website (https://training.fema.gov/hiedu/).

TABLE 8.2 Behavioral Health Related Bachelor Degree Required or Elective Courses in U.S. Universities and Colleges

	Number
Introduction to/principles of psychology	4
Psychology of disasters	5
Social media applications to EM	2
Social dimensions of disaster	6
Public information skills for emergency managers	2
Psycho-social aspects of disasters	4
Community psychology	1
Crises and disaster psychology	1
Socio-behavioral foundations of emergency management	3

elective course that could potentially touch upon the subject of DBH. I have not reviewed specific course content to determine the nature or extent of DBH topics covered in these courses, if any (Table 8.2). Nevertheless, if each of these courses (a total of 28) includes at least some relevant DBH material, it still appears that no more than about half of the B.A. or B.S. EM programs require any preparation which connects EM with issues of psychology or behavioral health.[7]

If the topic of DBH gets very little emphasis in formal academic education, how is it treated in the realm of postgraduate or career oriented training? To answer this question, I reviewed material published by the two organizations that do the most to promote training and skill development of EM: Federal Emergency Management Agency's Emergency Management Institute and the International Association of Emergency Managers' Certified Emergency Manager (CEM) program.

Federal Emergency Management Agency delivers the majority of its training course through its independent studies (IS) series of courses. These courses are available on the internet to virtually anyone that wants to take them, completely free of charge. The courses are self-guided and easy to use and offer continuing education credits that can be used for a variety of purposes (including qualifying for CEM training and education requirements). Federal Emergency Management Agency's website identifies ~200 IS

---

7. For an excellent example of a curriculum for an emergency management post-graduate level course which brings together EM and DBH, see "EMGT 607: Emergency Mental Health and Trauma Syllabus," Millersville University of Pennsylvania, Center for Disaster Research and Education, Master of Science in Emergency Management.

covering the full range of EM topics and programs, with the exception of any courses dealing with behavioral health or mental health issues in disasters. Once again, I have not reviewed in detail the course content of each of these courses. One or more of them may touch upon DBH, but the *absence of a single course that deals with DBH issues exclusively is significant.*

The International Association of Emergency Managers (IAEM) confers recognition for two levels of professional experience in EM: Associate Emergency Manager (AEM) and CEM. The CEM is considered the gold standard recognition for professional emergency managers and requires the demonstration of knowledge, skills, and abilities based upon training and education, and professional experience as well as professional accomplishments. The qualification requirements are fairly rigorous and require extensive documentation, including multiple professional references.

The range of categories of training that can be used to meet this requirement is relatively broad and flexible. Nevertheless, in order to qualify under the "emergency management" category, training in a related discipline (such as behavioral health) must be substantially related to EM. Training uniquely related to behavioral health, however, would qualify under the "general management" training requirement. Since they are free and easily accessible, Federal Emergency Management Agency's IS courses serve as a very useful and convenient source of training credit for members of the EM profession or those striving to receive one of the IAEM certifications. Yet, the absence of a Federal Emergency Management Agency IS course serves as a constraint which limits the promulgation of information about DBH for emergency managers as part of their ongoing professional development.

SUMMARY AND FURTHER THOUGHTS

There is very limited emphasis on DBH in the core or elective requirements of colleges and universities that offer undergraduate degrees in EM. There is virtually no DBH content in the single greatest source of continuing education (and professional certification training) for emergency managers. This poses a major challenge expanding the tent for the field of DBH among emergency managers. One of the primary underlying reasons for the dearth of these types of courses is the highly interdisciplinary nature of the subject. Instructors with extensive training and experience in both DBH and EM are fairly rare. For Federal Emergency Management Agency's resident on-campus courses, the scarcity of this combination is a serious challenge. However, developing an IS course that integrates DBH and EM could be easily achieved. Such a course, freely available from Federal Emergency Management Agency's online IS program, especially if it were adopted by the IAEM as eligible for meeting the CEM training requirement, would go a long way toward "expanding the tent" of the DBH/EM nexus.

REFERENCES

Federal Emergency Management Agency (FEMA). (2016). *Alphabetical Listing of Emergency Management, Homeland Security, and International Disaster Management and Humanitarian Assistance, and Related Higher Education Programs in the US*. Retrieved from https://training.fema.gov/hiedu/collegelist/Alphabetical%20Listing%20of%20EM%20HiEd%20Programs.doc?d = 2015-09-08.

Substance Abuse and Mental Health Administration. (2016). *Crisis Counseling Assistance and Training Program (CCP)*. Retrieved from http://www.samhsa.gov/dtac/ccp. Accessed on July 12, 2016.

Thomas, D., & Mileti, D. (2003). Designing educational opportunities for the hazards manager of the 21st century. In *Workshop Report*, Boulder, CO: Natural Hazards Center.

Through a Disaster Behavioral Health Lens

Gerard A. Jacobs

When disasters occur, some people are directly affected (i.e., those who are usually labeled "victims" or "survivors") and some are indirectly affected (e.g., families of those directly affected, witnesses, law enforcement, fire/rescue personnel, emergency medical care providers, and disaster relief staff from governmental agencies and nongovernmental organizations [NGOs]). Those indirectly affected also include some people who are not often thought about, including civic officials, political figures, and emergency managers as well as their staff. Some of those directly or indirectly affected may find the experience overwhelming. Those people are likely to experience overwhelming (traumatic) stress. Traumatic stress itself is not a disorder. It is not psychopathology. Rather, it is an ordinary response to an extraordinary event in life. Being overwhelmed is not a sign of weakness, merely a sign of being human. No matter how good one's coping skills are, or how "strong," tough, or macho someone is, a specific event can be experienced as overwhelming or traumatic. The large majority of those who experience traumatic stress will work through that reaction in 4–6 weeks. A minority of those who experience a traumatic reaction may develop psychopathology.

Many people, when they see someone who is feeling overwhelmed after experiencing a disaster, empathize with the overwhelmed individual and feel that it would be a good idea to comfort that person with some caring support. Disaster relief directors and emergency managers may also be moved by this humanitarian scenario, yet until the 1990s, deliberate psychological support was not included in disaster relief efforts. Disaster relief managers have traditionally been trained to focus on the physical aspects of disaster relief (e.g., how many pounds of food and gallons of water were delivered, how many people were sheltered, how many people cured of illness, etc.). Their focus has been on the logistical needs following a disaster. It is not that they did not care. They just did not know how to provide that support in the midst of chaos when there were so many physical and logistical needs. In fact, the first grant the Disaster Mental Health Institute (DMHI) ever received was an unsolicited contribution from the South Dakota Association of Emergency Managers, who asked DMHI to just keep doing our work.

One Federal Emergency Management Agency regional director, a former urban firefighter, told an audience that when he was a firefighter responding to a house fire, he always thought about what the family must be going through. He said he would remind himself that the Red Cross would take care of them, to focus on fighting the fire effectively to try to minimize the damage, and to keep his colleagues and himself safe. His job was not to help with the psychological support, but he was glad that there was someone designated to provide that support. (However, it is also important to realize that the DBH professionals are also there to support the firefighters, when necessary.)

A Fatal Fire

In a rural Midwestern community, the local fire/rescue service responded to a fire call at a trailer park. Although their response time was impressive, they arrived to find the trailer already 80% involved. They could hear cries coming from inside the trailer, but no entry seemed possible through doors or windows. While someone brought tools to open the trailer wall, firefighters could hear a young boy screaming for help next to them. The boy died before they could rescue him. It was a tough call—a multiple fatality. However, what stuck with the firefighters was hearing the boy die while they stood by helpless. A few months later, several of the wives of the firefighters went to an experienced DBH professional. They told him that their husbands had been having nightmares and struggling with the memories of the boy's screams as he died. They asked if the DBH professionals would try to help their husbands to work through their experience. He agreed to approach them individually, and with the wives' permission, told each of the men that their wives were concerned about them. This took place at a time when it was unusual for a firefighter to get psychological support, and each of them refused, despite their wives' encouragement. The problem with the stigma of accepting psychological support is that traumatic stress can affect decision-making and problem-solving. Thus, this can potentially endanger or impact not only the firefighter, but their colleagues, those whom they serve, and their families.

In 2004, a group of mental health professionals experienced in humanitarian assistance proposed to the U.S. Agency for International Development (USAID) that it include psychological support as part of its foreign disaster assistance. USAID is the branch of the federal government which manages nonmilitary foreign aid. After a day of presentations and meetings in Washington D.C., one of the USAID staff told me that the idea of adding psychological support to their humanitarian disaster assistance sounded advantageous. However, she said, the idea would never be enacted unless the most experienced field member of U.S. Agency for International Development's Office of Foreign Disaster Assistance (OFDA) endorsed the idea.

(Continued)

(Continued)

I was frequently working overseas at that time in the city where this individual was based, and talked with some of my colleagues there to see what they knew about him. They told me that he was well known to them, tough as they come, and able to handle any horrendous situation in the world. They were certain that he would not be interested in talking to a "shrink." When these colleagues heard that he had given me an hour-long appointment, they were shocked that he had even agreed to see me. They warned me that he must be planning to tear me to shreds, because he usually only gave 15 min long appointments.

I approached his office, admittedly with some trepidation. He answered the door. I was shocked when he literally grabbed me by the front of my shirt, pulled me into his office, and physically sat me down in a chair. He then proceeded to stand in front of me lecturing to me for an hour about the importance of psychological support in humanitarian assistance. He told me that in his decades of experience, in many of the worst disasters around the world, he had noted that survivors of disasters who were experiencing traumatic stress reactions slowed the recovery of the entire community. Furthermore, he reported that the whole relief operation was impeded when members of his own disaster relief staff experienced traumatic stress. Therefore, he said he absolutely endorsed including psychological support in humanitarian assistance—both for U.S. Agency for International Development's Office of Foreign Disaster Assistance staff and for the residents of the area who had experienced the disaster. Only a few months later, U.S. Agency for International Development's response to the December 2004 Indian Ocean earthquake and tsunami was the first Office of Foreign Disaster Assistance response to include psychological support.

The perception that workers experiencing traumatic stress impede the disaster relief operation is accurate. Among the effects that result from traumatic stress reactions is an impairment in cognitive functioning. People who feel overwhelmed by the situations they experience may not think clearly and may make poor decisions. This is true even if that individual is ordinarily the most skilled member of the team. It is important for emergency managers to understand the impact of traumatic stress on their own staff, as well as on those whom their operations serve, and how to minimize the incidence of and the impact of traumatic stress. The explanation of this reality has often been the key for our team in convincing first responders to accept psychological support. First responders who make bad decisions can put themselves, their colleagues, and those whom they serve in danger.

The purpose of this chapter is to discuss education in disaster psychology and EM. It will also touch upon the usefulness for mental health professionals and emergency managers to understand the basics of each other's roles and the contributions they make to the recovery of individuals and communities.

NATIONAL BIODEFENSE SCIENCE BOARD (NBSB)

In 2008, the Secretary of Health and Human Services (HHS) formed a Disaster Mental Health (DMH)[8] Subcommittee for the National Biodefense Science Board (NBSB; now the National Preparedness and Response Science Board or NPRSB) in response to a directive from then President Bush. In keeping with the presidential directive, 12 invited experts were appointed to the subcommittee after a national search, and 14 other subcommittee members were appointed by cabinet members. The president asked that, within six months, the subcommittee provide recommendations on preparing for, responding to, and recovering from the mental health consequences of catastrophic health events. The subcommittee's report was presented to the full NBSB in November 2008. It was approved by the full NBSB board and forwarded to the Secretary of HHS.

The Subcommittee's report (U.S. Department of Health and Human Services, 2008) included recommendations across three areas: intervention, education and training, and communication and messaging. Among the conclusions regarding education and training, the report recommended that all mental health professionals receive some formal training in DMH. The conclusions went further:

> *It is important to extend psychological support training beyond mental health (e.g., psychiatry, psychology, counseling, social work, and marriage and family therapy) and health care professionals (e.g., medicine, pediatrics, nursing, and epidemiology) to include the full range of emergency responders (e.g., law enforcement, fire service, emergency medical responders), coroners and morgue staff, disaster relief personnel (e.g., American Red Cross and National Voluntary Organizations Active in Disaster), faith-based professionals and leaders, disaster response leaders (e.g., incident commanders, emergency managers, and civil service and elected government leaders), and educators.*
>
> (U.S. Department of Health and Human Services, 2008, p. 13)

The subcommittee specifically acknowledged that such training could be provided in DMH (particularly for mental health professionals, but also as an alternative for medical staff, first responders, and civic officials). However, in addition, they acknowledged that the education could be provided as "Psychological First Aid" (PFA) for those who were not mental health professionals.

The term PFA has unfortunately not been used in a consistent way in the literature—ranging from using PFA as just another term for DBH, to models

---

8. Editors' note: As discussed in the Introduction, the editors have chosen to use the term *behavioral health* (BH) rather than *mental health* (MH) throughout the book. Since the name of this subcommittee and its products use the term *mental health*, references to it and its work will use that term.

that train paraprofessional teams to go into communities in the aftermath of disasters, to models that seek to train entire communities in basic psychological support. The DMH subcommittee recommended a specific model of PFA that is often referred to as Community-Based Psychological First Aid (CBPFA):

> "Psychological first aid," as used in this context, refers to psychological support that is both used to improve one's own resilience and is provided by non-mental health professionals to family, friends, neighbors, co-workers, and students. Psychological first aid focuses on education regarding traumatic stress and on active listening. The term also incorporates more sophisticated psychological support given by primary care providers to their patients. Properly executed, psychological first aid is adapted to the needs of each group or community (i.e., group of people with shared interests) implementing it, ensuring that the psychological first aid that is introduced in the community does not conflict with the world view of the group. It also emphasizes the inclusion of effective strategies for psychological support that may be specific to that group. This is done in concert with a representative community committee which helps to ensure responsiveness to the specific community. Psychological first aid includes understanding one's role; the difference between anticipated stress reactions and traumatic stress; how to engage in active listening; when and where to refer individuals for additional assessment and intervention; and the importance of supervision, ethical behavior, and self-care.
>
> (U.S. Department of Health and Human Services, 2008, p. 12)

The subcommittee also acknowledged that there was inadequate research on the CBPFA model as a whole, although many of the components of CBPFA have been thoroughly studied. There is more research today on the implementation of this type of CBPFA model, but continuing attention is needed to the validity of the model as a whole.

EDUCATION/TRAINING STRATEGIES IN DISASTER BEHAVIORAL HEALTH

There is presently no accepted standard for what education is necessary or adequate in DBH or disaster psychology. There are many formats available to learn about psychological aspects of disaster. More universities are providing formal training in DBH, ranging from a single academic course to full graduate specializations. In addition to a Doctoral specialization in Clinical/Disaster Psychology, which began in 1997, the DMHI at the University of South Dakota offers an online graduate certificate in DMH. The graduate certificate requires completion of three core courses (DMH, Crisis Intervention, and Serving the Diverse Community in Disaster) and at least one elective course (either Traumatic Stress or Management in DMH). Based on the NBSB's recommendations that DBH could be profitable for first responders, medical staff, and civic officials, the admissions requirement for the graduate certificate were broadened from focusing on existing and prospective mental health professionals to including most applicants with accredited undergraduate degrees and an acceptable grade point average.

The American Psychological Association, in 1992, began its national disaster response network (DRN), which included a group of psychologist volunteers trained in DBH and prepared to respond. (The DRN will soon be renamed the disaster resource network, reflecting changes in communication and service delivery.) Since that time, many state psychological associations have formed their own DRN committees. Some of these committees have developed their own DBH education materials, varying from printed materials, to workshops, to online courses. Other mental health profession organizations have also developed training programs for their members. Some state and federal agencies have also implemented such training.

The American Red Cross offers a 4 h course called "Fundamentals of Disaster Mental Health." This class is focused on how mental health professionals can use their professional skills within the context of the Red Cross organization. It encourages participants to get additional training elsewhere to gain a better understanding of DBH.

An even less formal approach is self-instruction—employing any of a variety of books covering DBH. Because the science in the field continues to develop, and Red Cross policies and procedures for DBH continue to evolve, choosing more recent publications is probably the best strategy, even though a number of older texts have very strong coverage of fundamental principles in DBH.

Educating the Education Community

In community situations that involve potential danger, schools are often immediately locked down. In some situations, schools have been isolated for a fairly lengthy time before the situation is resolved. Teachers have an important role in the life of a child on an ordinary day, but in the midst of a crisis their ability to support the children is critical. It would be very profitable for teachers or education students to be trained to provide psychological support to their students. In the DMHI's undergraduate program in disaster response one of the most popular courses is Children and Traumatic Stress. Many of the students who take the course are future teachers. While this is not training specifically in DBH, it does help future teachers (and future parents) understand traumatic stress experienced in children, and helps them understand how to provide basic psychological support for their children. Similarly, school counselors and school psychologists (who often complete their training in schools of education) could benefit significantly from being trained in DBH. In emergency response and Disaster Response Operations (DRO), children often experience some of the most stressful times of their lives. Moreover, their usual sources of social support, such as parents and friends, may be less available than usual while the parents deal with the demanding process of helping the family recover and the child's friends may be among those evacuated. There is often a shortage of DBH professionals who are knowledgeable and experienced in working with children.

> **Explosion and Gas Cloud**
>
> A number of years ago there was a major explosion in the early morning hours at a chemical factory on the edge of a Midwestern city. The morning was cool and foggy with very light and variable winds. The explosion resulted in a large poisonous gas cloud. The heavy fog made the cloud invisible, and officials had difficulty determining the location of the cloud as it moved around with the varying breezes. The weather conditions also prevented the cloud from dissipating. The school day began as normal, and much of the public was unaware of the poisonous gases. Officials determined that one of the schools was being enveloped by the gas cloud and ordered that the school be evacuated to a shelter not far away. The children were evacuated through the cloud and urged to hurry to school buses that would take them to the shelter. The children could smell the chemicals and were aware of the fear of the teachers and administrators and the police officers overseeing the evacuation. The light winds caused the cloud to shift and it soon enveloped the shelter to which the children had been moved. This process repeated three more times. Fortunately, DMH providers were with the children from the time the first evacuation was ordered. Regardless, if it were your child in this situation, wouldn't you prefer that your child's teacher were trained in psychological support in crises?

Educating Criminal Justice and Law Enforcement

It is also profitable for criminal justice programs and law enforcement to become familiar not only with CBPFA or DBH, but with crisis intervention as well. Police officers are beginning to appreciate the benefits of crisis intervention skills. For example, situations can frequently be deescalated when police officers are skilled in formal crisis intervention strategies (James & Gilliland, 2013).

Psychological First Aid Training

PFA training is more variable. As mentioned earlier, the term itself is defined in various ways by different organizations. The training that is probably most widely available is that of the National Center for Posttraumatic Stress Disorder and the National Children's Traumatic Stress Network

(NCPTSD/NCTSN) (Brymer et al., 2006). This training seems to be more of a response-team model than a community-based model—the focus is on training paraprofessionals and mental health professionals to respond to affected communities. This is contrasted with the CBPFA model which seeks to train groups of people who live or work together, not depending on outsiders for basic psychological support.

The American Red Cross (2014) has also released a PFA course for the public entitled Coping in Today's World (2014). This 4 h course has not been widely implemented and its usefulness is not yet fully evaluated.

Part of the DMHI's undergraduate minor and psychology specialization in disaster response is a full undergraduate course in CBPFA. Unfortunately, there are few formal resources for community-based PFA. The NBSB's description of CBPFA above is a good general definition. Hobfoll et al. (2007) offered a list of criteria for quality mass casualty DMH interventions. Those criteria have become a widely discussed yardstick for evaluating the development of psychological support programs for disasters and traumatic events generally.

A number of practitioners offer training workshops prepared for specific populations. Our own team (DMHI) has offered CBPFA workshops for health professionals and first responders, workshops for emergency medical technicians and paramedics, and hospital staff throughout the state. Fortunately, one text on CBPFA has recently been published (Jacobs, 2016).

An Introduction to PFA

One Regional Health Education Center sponsors a disaster training day for all the health professions once a year. Students from all the health professions at regional universities from specific years in their training gather for the day to learn a number of skills for disaster response, including point of distribution (POD) procedures. One of the sessions students attend is an introduction to PFA. It is little more than a brief explanation of what they need to learn, but at least students are given an introduction to the importance of psychological support.

Learning about Each Other

In order to optimize the efficacy of emergency managers and DBH professionals, it is important for them to have a better understanding of each other's fields. This may be achieved by taking courses, workshops, and/or trainings in each other's areas. There can also be elements of one field taught in the education programs of the other. Some of the basics of each field that can profitably be learned by professionals in the other professions are noted in Table 8.3.

TABLE 8.3 Elements to Know About Each Others' Fields

Things DBH Professionals Could Profitably Know about EM	Things EM Professionals Could Profitably Know about DBH
Incident command system	Nature and effects of traumatic stress, particularly effects on cognitive functioning
The overall National Response Framework (NRF)	Effects of traumatic stress on community members
The role of DBH within NRF, and how DBH can help maintain or improve the emergency response or relief operation	Effects of traumatic stress on first responders and relief workers
	Effects of traumatic stress on emergency managers
	Role of DBH in maintaining or improving the emergency response or relief operation

Mental Health Professionals

Mental health professionals will be more effective in emergency responses and disaster relief operations if they understand the philosophy and structure within which they need to function. The federal government has established an overall National Response Framework (NRF) that guides the nation's all-hazards response (U.S. Department of Homeland Security, 2013). Most communities use the National Incident Management System (NIMS) (U.S. Department of Homeland Security, 2008), which is related to the NRF and provides more detailed specifics on the operation of the emergency response. NIMS replaced the Incident Command System (ICS) in 2008. The importance of this aspect of EM is indicated in the NIMS document:

> *NIMS provides a consistent nationwide framework and approach to enable government at all levels (Federal, State, tribal, and local), the private sector, and nongovernmental organizations (NGOs) to work together to prepare for, prevent, respond to, recover from, and mitigate the effects of incidents regardless of the incident's cause, size, location, or complexity.*
>
> (U.S. Department of Homeland Security, 2008, p. 1)

The general course introducing NIMS is "IS-100.B: Introduction to Incident Command System, ICS-100" (U.S. Department of Homeland Security, 2011). Additional related training is also available at the NIMS website (https://training.fema.gov/nims/).

Emergency Managers

For emergency managers the most important DBH information to garner is to have a clear understanding of traumatic stress, that it is an ordinary reaction, is not psychopathology, and what the symptoms are for those directly affected by incidents, for responders, and for the managers themselves. This information is often covered in PFA training, and emergency managers would usually receive more detailed information in DBH courses.

It would also be very helpful for emergency managers to understand how DBH professionals can assist both individuals and communities within the emergency response or DRO, as well as to consult with emergency managers on the psychological aspects of the incident management. This information is best acquired through DBH training, as described earlier.

Disaster Drills

Formal education is not always the most practical way for professionals to learn about each other's work. It can also be quite useful for emergency managers to include DBH within emergency response and disaster drills. All of the agencies and professionals in various roles have the opportunity to better understand the necessary interactions involved in an effective incident response. When one sees other professionals performing their role within the drill, it becomes easier to understand how those professionals fit within the structure of the response.

School Disaster Drill

A small city organized resources in the region and held a full-scale, real-time disaster drill. The scenario was an active shooter in the local high school. Several law enforcement entities from the region were involved. The local hospital was participating and ready to drill on receiving mass casualties. Both regional helicopter medical evacuation units were involved. Many ambulance units from the region participated, as well as the local fire/rescue department. Local high school students had volunteered to play the role of wounded and dead victims of the shooter(s). The local EM office for the first time had invited DBH professionals to participate as well.

The DBH staff were pleased to participate, but anticipated a very boring afternoon. Before the drill began the police officers, EMTs, and paramedics were milling about quite casually and looking forward to the exercise. Many of them spoke with DBH staff, a bit uncertain about what DBH's role was in this kind of incident. When the signal was given for the drill to begin, the police units entered the high school, who in the scenario were the first to reach the scene, followed by others, on a timed schedule, estimating response time in a real situation.

(Continued)

(Continued)

What the police officers and emergency medical staff did not know was the reality of the drill had been augmented by a sophisticated audio presentation, with broadcasting of localized sounds throughout the building of students screaming or crying in fear, gunshots, and of the wounded crying out in pain. The "wounded" had been prepared by professional moulage makeup artists and moulage blood pools had also been staged where the wounded were located. Some of the wounded students had dragged themselves on the floor, leaving blood trails in the hallways. In keeping with standard procedure, the emergency medical teams were allowed to enter and recover the wounded when a section of the building had been cleared.

The DBH staff were surprised when some police officers emerged from the building and asked to speak with a member of the DBH team. At first, the DBH staff thought it was a staged part of the drill to make the DBH staff feel welcome, but the police officers quickly assured them that their requests for help were real. Officers reported being overwhelmed by the sights and sounds. Many of them found themselves thinking about the fact that their own children attended that high school, and what such an incident would be like for them. Some of the emergency medical teams also asked for support from DBH staff. They reported being shocked by the sounds, what they saw, and by the horror of realistic wounds on the staged victims of the shooting.

In the after-action review, the emergency manager said that one lesson that was clearly learned was that DBH from that day forward would be a standard part of their response.

Emergency response and disaster relief work drills can be extremely helpful. In the DMHI's undergraduate course on Introduction to Disaster Response, students for many years have been required to participate in a sheltering drill. Each student has their unique role and assignment to accomplish. Situational updates complicate their strategies or are designed to be distractions such as those that often arise in real DROs. Students have consistently reported that they never understood the other roles in the shelter until they needed to work together with others in those roles to solve problems.

Decontamination Unit

At a regional terrorism preparedness conference in a far northern state, a city's hazmat unit proudly presented a full-scale demonstration of their new portable decontamination unit. The unit had plasticized canvas floors, walls, and ceilings. The unit had two lanes, one for men and boys and the other for women and girls. In each lane, those thought to have been exposed would enter

(Continued)

> **(Continued)**
>
> a room and remove all their clothes. They would then enter a group shower area shielded from outside view. Once properly scrubbed they would step out into an open exposed area. When the hazmat staff finished their explanation, a mental health professional in the audience asked if he understood the situation properly, "So those who have been decontaminated step out of the gender-isolated lanes into a common area. Everyone has just been told they may have been exposed to something deadly, and told to strip naked in front of others (all of the same gender), and, when they finish decontamination, both genders emerge naked and wet into the same open area. What if this happened in the winter? Then they would be naked in front of each other, wet, potentially terrified, and in subzero temperatures. Have you made provisions for warm clothing or for mental health professionals to be part of the decontamination team?" The hazmat team looked stunned and said they had not thought about clothing or blankets, or even towels after the showers, or about the need for DBH. It is very important for those involved in disaster preparedness to understand the basic human dignity and the psychological support needs of those being served.

OBSTACLES IN CROSS TRAINING

While it is easy to describe training opportunities that might be beneficial for DBH and/or emergency managers, there are problems.

Stigma About Receiving Psychological Support

Perhaps the most prominent is a perception among many emergency managers, first responders, and many political and civic officials that accepting psychological services may be experienced as an embarrassment or perceived by others, particularly superiors, as a sign of weakness.

The NBSB DMH subcommittee's recommendations clearly called for DMH and/or PFA training for mental health professionals, emergency managers, first responders, civic and political leaders, and medical professionals. Rudolph Giuliani, the mayor of New York City at the time of the terrorist attacks on the World Trade Center in September 2001, addressed mental health policy makers gathered at the Carter Center in Atlanta in 2002. He asserted the importance of good psychological support in a crisis. He had learned, following the crash of Flight 800 in 1996, about the importance of DBH both for his own well-being and to ensure he made good decisions for those affected by that aviation disaster. He said that as soon as he realized the scale of the September 11th event, he loudly called for one of the senior DBH staff in the city. Giuliani told the DBH professional that he wanted that professional to stay directly behind Giuliani's right shoulder throughout the operation—to help ensure that Giuliani was functioning well. The DBH professional was also instructed to advise Giuliani on how his decisions might affect the public. Giuliani said that when he openly demanded that psychological support, his senior response staff looked around and began to ask where *their* mental health professionals were. When the senior decision maker in the room demanded continuing mental health support, it was easier for others to ignore any stigma of asking for psychological support themselves. This same openness will flow down to the front line providers who often look to their leaders for guidance. If the leaders of the various professions endorsed training in DBH and/or PFA, as called for in the NBSB recommendations, this destructive stereotype might begin to wane.

Costs

Formal education in these fields and specialized training such as a custom CBPFA class can be expensive. Fortunately, the federal training both in CBPFA and in EM described above can largely be done online without cost, although more formal academic training in EM can also be expensive.

Time

Given the existing demands of working in both fields, professionals are unlikely to seek cross-training unless they are either personally invested in the training or if there are licensing or certification requirements that mandate DBH, CBPFA, PFA, or EM training. The NBSB recommendations included a call for such requirements.

Aviation Disaster Drill

A large international airport held a full-scale real-time disaster drill. A team of national observers had been chosen to watch the drill and comment in the after-action review. One of the observers was a DBH professional experienced in aviation disasters.

The setting for the exercise was impressive—directly on a large body of water and directly adjacent to an active runway, with wide-bodied jets landing or taking off nearby every few minutes. The script for the drill was that a large passenger jet had crashed in the water alongside the runway and sunk. Divers played the role of passengers in the submerged fuselage. Rescue divers needed to extract them from the wreckage and bring the passenger divers to a barge at the surface. The diver passengers "tagged" high school students waiting on the barge, who took over the role of a designated passenger. The students had received professional moulage makeup representing their physical condition. As the student passengers, whose condition was designated as ambulatory, were brought to shore, they were to be transferred to buses and taken to a special

(Continued)

(Continued)

area in the terminal for appropriate nonemergency medical care and DBH support.

The drill went quickly awry, however, as members of the response team helped the ambulatory off the boat and directed them to go to the evacuation bus—some 150 feet away in the direction of the runway. Some of the students did not understand the directions, and were wandering about confused in the chaos of the scene. They were also unnoticed by the response team. The situation posed a genuine and immediate danger because the nearby runway was still very active. There were also a large number of emergency vehicles moving about. The observer team, prohibited from commenting during the actual drill, nevertheless felt the need to intervene early and point out the problem to prevent some of the students from becoming genuine casualties.

In the after-action review, the DBH observer noted that these students had not even gone through the traumatic experience of an aviation incident, but were still confused by the rescue team's instructions, as well as by the chaotic scene in which they found themselves, and therefore had difficulty following the team's directions. How much more difficult would it be for someone who had just experienced the traumatic experience of a violent aviation disaster? Instructions were reviewed and procedures were added to escort the ambulatory survivors to the designated buses. The disaster response plan was also changed to include DBH professionals to accompany the ambulatory survivors on the buses.

REFERENCES

American Red Cross (2014). *Coping in today's world.* Washington, DC: American Red Cross. Retrieved from https://intranet.redcross.org/content/redcross/categories/our_services/disaster-cycle-services/dcs-capabilities/individual_clientservices/disaster-mental-health-toolkit/coping-in-today-s-world.html. (NOT AVAILABLE TO THE PUBLIC).

Brymer, M., Jacobs, A., Layne, C., Pynoos, R., Ruzek, J., Steinberg, A., ... Watson, P. (2006). *Psychological aid: Field operations guide* (2nd ed, Retrieved from http://www.nctsnet.org/nctsn_assets/pdfs/pfa/2/PsyFirstAid.pdf). Los Angeles: National Child Traumatic Stress Network and National Center for Post-Traumatic Stress Disorder.

Hobfoll, S. E., Watson, P., Bell, C. C., Bryant, R. A., Brymer, M. J., Friedman, M. J., ... Ursano, R. J. (2007). Five essential elements of immediate and mid−term mass trauma intervention: Empirical evidence. *Psychiatry, 70*(4), 283−315.

James, R. K., & Gilliland, B. E. (2013). *Crisis intervention strategies* (7th ed). Belmont CA: Brooks/Cole.

Jacobs, G. A. (2016). *Community-based psychological first aid: A practical guide to helping individuals and communities during difficult times.* Amsterdam: Butterworth-Heineman.

U.S. Department of Health and Human Services (2008). *Disaster Mental Health Recommendations: Report of the Disaster Mental Health Subcommittee of the National Biodefense Science Board.* Retrieved from http://www.phe.gov/Preparedness/legal/boards/nprsb/Documents/nsbs-dmhreport-final.pdf.

U.S. Department of Homeland Security (2008). *National Incident Management System*. Retrieved from https://www.fema.gov/national-incident-management-system.

U.S. Department of Homeland Security (2011). *IS-100.B: Introduction to Incident Command System, ICS-100*. Retrieved from https://training.fema.gov/is/courseoverview.aspx?code = IS-100.b.

U.S. Department of Homeland Security (2013). *National Response Framework* (2nd ed.), Retrieved from http://www.fema.gov/media-library-data/20130726-1914-25045-1246/final_national_response_framework_20130501.pdf.

Making Integration Work

Brian W. Flynn and Ronald Sherman

The previous sections in this chapter have provided detailed and valuable information about how to expand the range of expertise and contribution to not only the individual professions but to their effective and sustained integration. The content of this section draws key content from the information and perspectives in those sections and identifies practical ways the ideas can be operationalized.

"Through a Disaster Behavioral Health Lens" section begins to lead us in into a discussion of key elements of what the two professions can learn from each other. "Through and Emergency Management Lens" section adds important information on the current status of available training in the EM field. The Table 8.4 brings together their ideas of practical ways to realize the goals identified in this chapter.

TABLE 8.4 Suggestions for Making Integration Work

Recommended Priorities	Behavioral Health	Emergency Management
Share the general knowledge base	• Know and effectively communicate the knowledge base (e.g., What the NBSB, IOM, ARC others recommend regarding disasters) • Become part of ongoing EM training, professional meetings • Jointly author publications	• Know and effectively communicate the primary guiding principles, documents, and knowledge base (e.g., NRF, ESF structure, ICS, etc.) • Become part of ongoing BH training, professional meetings • Jointly author publications
Share the intervention knowledge base	• Be up to date on current evidence of various intervention strategies • Share this knowledge with EM in understandable and relevant ways	• Through work with DBH professionals, be aware of the state of the science regarding disaster BH interventions • EM could benefit from orientation to and training in PFA

(Continued)

TABLE 8.4 (Continued)

Recommended Priorities	Behavioral Health	Emergency Management
Explore/expand opportunities for enhancing professional training and education	• Expand disaster related content in BH training and education • Expand opportunities for EM to participate in DBH training	• Expand behavioral health content in EM training and education (Specifics provided in "Through and Emergency Management Lens" section) • Expand opportunities for DBH professionals to participate in EM training
Expand collaboration with other allied professions	• Specific examples are identified in "Through and Emergency Management Lens" section • Use academic connections to facilitate these linkages	• Specific examples are identified in "Through and Emergency Management Lens" section • Use academic connections to facilitate these linkages and increase opportunities
Enhance school preparedness	• DBH professionals working with/in schools should promote emergency and disaster preparedness and response capability	• As EM works with schools in preparedness and response, highlight the importance of integrating DBH elements
Expand the capacity of law enforcement and the criminal justice system	• DBH professionals should continue to provide crisis intervention training • Ensure that this training includes disaster and the importance of these systems' integration with EM	• As EM works with these systems in disaster preparedness, EM professionals should promote the importance of enhancing crisis intervention skills
Integrate drills and exercises	• Become part of the planning process so that when drills and exercises take place, DBH is integrated, and	• Assure that DBH professionals are integrated into the preparedness process

(Continued)

TABLE 8.4 (Continued)

Recommended Priorities	Behavioral Health	Emergency Management
	DBH scenarios are realistic • Actively participate in drills as players, designers, and judges	• Integrate DBH into drills and exercises as players, designers, and judges.
Anticipate and strategize regarding sigma related to BH issues	• Acknowledge the role of stigma and identify ways in which it might compromise both DBH and EM goals	• Acknowledge the role of stigma and identify ways in which it might compromise both DBH and EM goals
Utilize existing online training from the Emergency Management Institute (EMI) and other sources	• Seek out EM training opportunities through contact with EM counterparts	• Promote EMI online courses to DBH partners

Chapter 9

Linking with Private Sector Business and Industry

Diana Nordboe[1] and Susan Flanigan[2]
[1]Emergency Manager, Independent Consultant, Carter Lake, IA, United States
[2]Missouri Department of Mental Health, Jefferson City, MO, United States

The authors believe so strongly in collaboration and professional partnership that this chapter is presented as a joint blend of our philosophy and practice rather than as separate sections. Collaborating locally is the best starting point when integrating emergency management, behavioral health, and the private sector. Disaster response is relationship-driven, even in today's data-driven, social-media focused, "there's-an-app-for-that" society. Human-caused and natural disasters disrupt infrastructure at all levels. Local leaders and decision makers launch initial efforts, control access to the impacted area and request resources from a regional, state, or federal partner when the local needs exceed local resources. Before disaster strikes, get to know the community leaders and decision makers, join disaster response groups sanctioned to launch when disaster strikes, and become educated on incident command and DBH.

Collaboration Basics

Readers are advised to identify and engage with the following key contacts in their community.
1. Emergency manager and staff
2. Lead public behavioral health agency/authority
3. Community business leaders and partnerships
4. Community disaster response collaborative

HISTORY AND OVERVIEW

Since 1974, the Center for Mental Health Services (CMHS) and its predecessor agency, the National Institute of Mental Health (NIMH), have administered the Crisis Counseling Program (CCP) in coordination with the Federal Emergency Management Agency (FEMA). The CCP provides supplemental funding to the State Mental Health Authority (SMHA) for short-term crisis counseling services to victims/survivors of presidentially declared disasters. Legislative authority is given to the President under Section 416 of The Robert T. Stafford Disaster Relief and Emergency Assistance Act of 1988 (Public Law 100−707) to provide training and services to alleviate mental health problems caused or exacerbated by major disasters.

DBH theory and practice evolved with each jurisdiction's response to natural disasters, but sharing this collective knowledge was difficult pre-internet when final reports were submitted in paper to the federal agencies. In 1990, the coordinators of seven county projects addressing the emotional needs following the California Loma Prieta Earthquake CCP realized their collective knowledge would be helpful to other mental health managers responding to community-wide disasters. The Emergency Services and Disaster Relief Branch, then at the National Institute of Mental Health, worked with an informal, California-based group of experienced disaster mental health professionals to publish *Disaster Response and Recovery: A Handbook for Mental Health Professionals* (Myers, 1994). This publication, which became commonly known in the field as the *Purple Book*, became the go-to resource for those interested in the behavioral health factors in disasters. The publication provided guidance on managing an event and outlined 14 key concepts of disaster mental health, which most consider basic foundational knowledge in the field of DBH. They are especially important knowledge elements for those new to the DBH field.

1. No one who sees a disaster is untouched by it.
2. There are two types of disaster trauma—individual and collective.
3. Most people pull together and function during and after a disaster, but their effectiveness is diminished.
4. Disaster stress and grief reactions are normal responses to an abnormal situation.
5. Many emotional reactions of disaster survivors stem from problems of living caused by the disaster.
6. Disaster relief procedures have been called "The Second Disaster".
7. Most people do not see themselves as needing mental health services following disaster and will not seek out such services.
8. Survivors may reject disaster assistance of all types.
9. Disaster mental health assistance is often more "practical" than psychological in nature.
10. Disaster mental health services must be uniquely tailored to the communities they serve.
11. Mental health staff need to set aside traditional methods, avoid the use of mental health labels, and use an active outreach approach to intervene successfully in disaster.

12. Survivors respond to active interest and concern.
13. Interventions must be appropriate to the phase of the disaster.
14. Support systems are crucial to recovery.

A 15th item is suggested and added by the authors of this chapter: *All disasters are political events*. Although never officially listed as a key concept, it is consistently acknowledged as an important consideration in each event.

This chapter contains several case examples of integration among emergency management, behavioral health, and private sector organizations and groups. Readers will note that all of the examples cited above reflect the changing status and needs of victims across the disaster event cycle. Needs change over time and the outcome of the types of collaborations noted need to change also. A graphic illustration of the trajectory of individual, family, and community response is presented in Fig. 9.1.

As an aid to readers in understanding the various important variables and considerations by disaster phase, Table 9.1 is provided. The table defines description of the major characteristic presents at each phase described in Fig. 9.1.

The CCP program changed again in 1995 when The Alfred P. Murrah Federal Building in Oklahoma City was bombed on April 19, killing 168 people, including 19 children from the on-site day care center. For the first time, a CCP program was activated for a human-caused event. Sadly, human-caused events continued to impact the nation, ranging from the Columbine High School shooting in 1999, to the September 11, 2001 terror

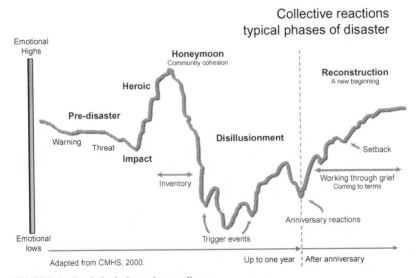

FIGURE 9.1 Psychological reactions to disaster.

TABLE 9.1 Description of Disaster Event Phases and Responses

Phase	Characteristics
Predisaster phase	• Disasters vary in the level of warning communities receive. When there is no warning, survivors may feel more vulnerable, unsafe, and fearful of future unpredicted tragedies. • When people do not heed warnings and suffer losses as a result, they may experience guilt and self-blame.
Impact phase	• The impact phase can vary from the slow, low-threat buildup (floods) to violent, dangerous, and destructive outcomes (F-5 intensity tornado or 9/11). • The greater the scope, community destruction, and personal losses associated with the disaster, the greater the psychosocial effects. People's reactions range from constricted, stunned, shock-like responses to the less common overt expressions of panic or hysteria. • Typically, people respond initially with confusion and disbelief. They tend to focus on the survival and physical well-being of themselves and their loved ones. • When families are separated during the impact (e.g., children at school, adults at work), survivors will experience considerable anxiety until they are reunited.
Heroic phase	• In the immediate aftermath of a disaster event, survival, rescuing others, and promoting safety are priorities. • For some, postimpact disorientation gives way to adrenaline-induced rescue behavior to save lives and protect property. Activity level may be high but actual productivity is often low. • The capacity to assess risk may be impaired and injuries can result. • Altruism is prominent among both survivors and emergency responders.
Honeymoon phase	• During the week to months following a disaster, formal governmental and volunteer assistance may be readily available. • Community bonding occurs as a result of sharing the catastrophic experience and the giving and receiving of community support. • Survivors may experience a short-lived sense of optimism that the help they will receive will make them whole again. • When DBH workers are visible and perceived as helpful during this phase, they are more readily accepted, providing a foundation from which to provide assistance in the difficult phases ahead.

(Continued)

TABLE 9.1 (Continued)

Phase	Characteristics
Disillusionment phase	• Over time, survivors go through an inventory process during which they begin to recognize the limits of available disaster assistance. • They become physically exhausted due to enormous multiple demands, financial pressures, and the stress of relocation or living in a damaged home. The optimism can give way to discouragement and fatigue. • As assistance agencies and groups begin to phase out, survivors may feel abandoned and resentful. Survivors calculate the gap between the assistance they have received and what they will require to rebuild their lives. • There are stressors abound—family discord, financial losses, bureaucratic hassles, time constraints, home reconstruction, relocation, and lack of recreation or leisure time. • Health problems emerge/intensify due to ongoing, unrelenting stress and fatigue. • Ill will and resentment may surface in neighborhoods as survivors receive unequal monetary amounts for what they perceive to be equal or similar damage, undermining community cohesion and support.
Reconstruction phase	• The reconstruction of physical property and recovery of emotional well-being may continue for years following the disaster. Survivors realize that they will need to solve the problems of rebuilding homes, businesses, and lives largely by themselves and gradually assume responsibility for doing so. • With the construction of new residences, buildings, and roads comes another level of recognition of losses. Survivors are faced with the need to readjust to and integrate new surroundings while they continue to grieve losses. • When people come to see meaning, personal growth, and opportunity from their disaster experience, despite their losses and pain, they are well on the road to recovery. While disasters may bring profound life-changing losses, they also bring the opportunity to recognize personal strengths and to reexamine life's priorities. • Individuals and communities progress through these phases at different rates, depending on the type of disaster and the degree and nature of disaster exposure. Progression may not be linear or sequential, as each person and community brings unique elements to the recovery process.

Source: Adapted from SAMHSA (SAMSHA, 2000).

attacks on the Twin Towers, Pentagon, and airline industry, to the 2012 Sandy Hook School shooting. These events became game changers in psychological response because public and private sector work environments such as schools, office buildings, and the transportation industry became intentional targets of destruction. Human-caused events propelled the need for further integration between the public and private sector response to an event. Not only do human-caused events create greater psychological stress, but, historically, private enterprise focused on business continuity as opposed to human continuity. In the authors' personal experience, understanding trauma and the impacts to the most valuable asset—personnel—led to collaborations and better alignment among emergency management, behavioral health, and public and private sector agencies.

The evolution of the 24-hour news cycle coupled with citizen journalists posting, viewing, and commenting on videos as well as photos, and frontline accounts of global traumatic events expanded and complicated the psychological impact of events. At times, fortunately, and simultaneously, behavioral health, emergency management, and the private sector align during crises to share intelligence, maximize resources, and execute strategies to address community gaps. In the authors' experience, collaboration depends upon local leadership, resources, and politics. Successful responders are natural networkers and collaborators who integrate lessons learned from previous disasters into their response plans. Good leaders understand that natural disasters and human-caused events affect not only the communities directly impacted but also the families, friends, colleagues, satellite offices, and global citizens following the story well beyond the impacted area. Positive outcomes are relationship-driven.

ROLE OF PRIVATE SECTOR AND NONGOVERNMENTAL ORGANIZATIONS (NGOs) IN EMERGENCY RESPONSE AND RECOVERY

The private sector and Nongovernmental Organizations (NGOs) have long histories of response and recovery activities following disasters, and their roles have been integrated into the national planning process. The Federal Emergency Management Agency facilitated a cooperative planning process with other federal departments and agencies in 1992 to establish the Federal Response Plan (FRP) that outlined federal roles and responsibilities following large-scale disaster events. Following the terrorist attacks of September 11, 2001, and the creation of the Department of Homeland Security, the FRP was replaced in 2004 with the National Response Plan (NRP), which recognized the importance of incorporating the private sector and NGOs in national planning.

The National Response Framework (NRF) (Federal Emergency Management Agency, 2016a, 2016b, 2016c), introduced in 2008 and revised in 2013, recognized that government resources alone cannot address all of the needs of those impacted by terrorist attacks, natural disasters, and other catastrophic events. It also established guidelines for incorporating the whole community into all phases of a disaster event. The NRF defined the roles of NGOs and the private sector through a focus on the whole community and core capabilities. The National Incident Management System (NIMS) (The Federal Emergency Management Agency, 2016a) provides the nationwide template on how the whole community works together to not only respond to incidents, but also prevent, protect against, mitigate, and recover from the impact.

The private sector is defined by the NRF as nonprofit and for profit organizations that are not part of the government structure. The private sector is comprised of businesses and industries of all sizes, commerce, and private cultural and educational institutions. Emergency Management Agencies at the local, state, and federal levels have established partnerships with private sector to share and coordinate prior to, during, and after incidents. The NRF outlines response roles for organizations within the private sector that impact economic recovery (local, regional, or national), infrastructure (including hospitals and health facilities), hazardous operations, and response resources (donated or compensated).

Private Sector Responsibilities under the National Response Framework

The NRF identifies responsibilities of the private sector to include addressing the needs of employees, infrastructure, and facilities and maintaining continuity of business operations as (The Federal Emergency Management Agency, 2011, pp. 10−11):
- Addressing the response needs of employees, infrastructure, and facilities
- Protecting information and maintaining the continuity of business operations
- Planning for, responding to, and recovering from incidents that impact their own infrastructure and facilities
- Collaborating with emergency management personnel to determine what assistance may be required and how they can provide needed support
- Contributing to communication and information sharing efforts during incidents
- Planning, training, and exercising their response capabilities
- Providing assistance specified under mutual aid and assistance agreements
- Contributing resources, personnel, and expertise; helping to shape objectives; and receiving information about the status of the community

---

NGOs are nonprofit, voluntary groups that provide service or humanitarian functions on a local, state, national, or international level. The American Red Cross and the National Voluntary Organizations Active in Disasters are NGOs designated within the NRF as support elements to national response capabilities (The Federal Emergency Management Agency, 2013a, pp. 8−9).

NGOs' interests and values provide for specialized services for the whole community and populations within the community, such as:

- Children
- Individuals with disabilities, access, or functional needs
- Diverse religious, ethnic, and racial backgrounds
- People with limited English proficiency

The contributions of NGOs include:

- Volunteer resources
- Sheltering
- Emergency commodities and services
- Search and rescue
- Transportation
- Logistics services
- Identifying unmet needs in the community
- Coordinating and providing disability-related assistance and functional needs support
- Providing health, medical, mental health, and behavioral health resources

Behavioral health organizations that are private sector or NGO entities are a crucial component within the whole community for recovering from all types of incidents. The Federal Emergency Management Agency, in coordination with the Substance Abuse and Mental Health Services Administration (SAMHSA) of the Department of Health and Human Services (HHS), awards grants for behavioral health services to state and county departments of behavioral health through the Crisis Counseling Assistance and Training Program of the Stafford Act. The major accomplishment of the CCP over the decades has been to promote resilience within communities and support individuals and families with educational and counseling services following large-scale incidents. NGOs and private sector organizations have played integral roles in the CCP grants. However, not all disasters qualify for such federal funding. Behavioral health services and local emergency management often create memoranda of understanding (MOU) typically in collaboration with the American Red Cross and private sector Employee Assistance Programs (EAPs) to provide scalable psychosocial support and services to local communities and organizations when disaster strikes.

EVOLUTION OF DISASTER COMPLEXITY AND PRIVATE SECTOR COLLABORATION

Collaboration between emergency management, behavioral health, and the private sector is complex. Yet, when operationalized, it can provide rapid, creative programs and services that can be tailored to emerging and changing survivor needs in ways few other entities can. This portion of the chapter will highlight several examples of this type of collaboration. Examples have been selected that highlight not only the power of these collaborations but the wide variety of situations in which they can be enormously helpful.

The Power of Water

Private sector involvement in disaster response continues to evolve in sophisticated ways as the disasters morph from natural disasters into human-caused events. Consider delivery of fresh water to impacted locations.

Case Example: Missouri Floods

The Anheuser-Busch (AB) Brewery headquartered in St. Louis has packaged and delivered more than 72 million cans of drinking water for disaster relief efforts since 1988 (Anheuser Busch, n.d.) (http://anheuser-busch.com/index.php/our-responsibility/community-our-neighborhoods/natural-disaster-relief/). During the Floods of 1993, AB provided beverage cans of water, affectionately known as "Floodweiser," delivered to responders and communities in need. Volunteers, including crisis counselors affiliated with the local community mental health centers, chatted with and handed out water to Missourians sandbagging levees, relocating families out of harm's way, and removing debris from flood-damaged businesses and homes. According to the authors' experiences, early, helpful presence forged strong community relationships and connected impacted citizens to resources and emotional supports well after the floodwaters receded.

Case Example: Lead Contamination Flint, Michigan

Hurricane Katrina lessons learned from the direct impact to and volunteer involvement of their associates resulted in strategic philanthropic policy for Walmart in 2015. Wal-Mart with the Wal-Mart Foundation contributed $25 million to disaster relief worldwide plus grants to US nonprofits to improve disaster collaboration and community resiliency with the best practices and technology. In January 2016, Wal-Mart, Coca-Cola, PepsiCo, and Nestle, through Good360, a not-for-profit based in Alexandria, Virginia, pledged to meet the daily clean water needs of over 10,000 school children affected by the water crisis in Flint, Michigan. Up to 6.5 million bottles of water or 176 truckloads will supply water through the end of the school year (Walmart, 2016).

The disaster recovery arm of Good360 (Good360, n.d.) connects on-the-ground nonprofits with corporate donors to ensure that the right goods get to the right people at the right time during all stages of a disaster (Walmart, 2015). This is exceptional logistic and philanthropic alignment as is the national United Way 211 system (United Way, n.d.). Private, public, and not-for-profit donated goods and services are triaged to citizens and groups requiring disaster resources during an active event and with human services and other social services agencies daily, a relatively simple and effective solution for those in need.

The complications of the Flint water crisis continue to evolve. The event received a presidential emergency declaration on January 16, 2016, with the Federal HSS as lead agency. But, as of this writing, no federal behavioral health program has been enacted as the public debate continues. Experts suggest relocating children and families, politicians debate plans to replace pipes, and the media highlights the ongoing impact of poor government decisions (Redlener, 2016).

Although, at this writing, no formal federally supported behavioral health program is ongoing, Flint does have emotional supports embedded in the community. Good360 connected with the creative and effective programs delivered by The Crim Fitness Foundation, a respected community agency of long standing. "We are part of the DNA of Flint," shared Christina Ferris (Ferris, Personal communication, n.d.), Crim development director. Crim brings the community together by working with schools and community members around gardens, helping children to achieve 60 min of physical activity daily and teaching mindfulness. Crim states, "Mindfulness builds the brain's prefrontal cortex which is the area impacted by lead poisoning. Lead can affect impulse control and lower IQs. Mindfulness can exercise the brain and mitigate the impact." Crim has provided mindfulness training to more than 700 teachers as well as thousands of students, and is continuing to expand these programs. Mindfulness improves the ability to concentrate, reduces impulsivity, fosters better decision-making, and brings a sense of peace and calmness to those who practice. In fact, it has been shown to increase brain size and rebuild pathways in the prefrontal cortex—exactly what the children of Flint affected by lead poisoning need now and for years to come. Nutrition education is another tool for healthier life, such as promoting foods rich in calcium, iron, and vitamin C to prevent absorption of lead. Crim states, "We are so embedded in the community and committed to our message of help, hope and healing."

The Crim staff take care of one another and practice what they preach as they address this public health crisis. "The water situation has thrown every work plan out the window," Ferris explains. "We are committed to one another and try to practice good work/life balance" (Ferris, Personal communication, February 24, 2016).

Case Example: Sandy Hook School Shootings

Private sector goods, services, and logistics skills are typical areas for public and private collaboration. In the aftermath of the Sandy Hook shooting, such a collaboration stands out as an example from which other communities can learn (Rallo, Personal communication, October 30, 2015). Jeff Immelt of

General Electric (GE) received more than 100 ideas from the 150 employees who were residents of Newtown, Connecticut, regarding ways they could assist the community following the December 2012 Sandy Hook Elementary School shooting. Immelt approached First Selectman Pat Llodra to discuss what would be most beneficial.

Llodra told company executives that the demands on the government, schools, and community were so "extraordinary" that skilled management personnel were needed to cope with a broad spectrum of needs. Everything from public relations and fund and event management to long-term recovery was needed. "We thought things would quiet down in January, and we'd start to feel more in control, but that was not happening," Llodra said. "We were at a loss at how to manage it all" (Hutson, 2013).

GE executives created three strategic positions and invited company employees to apply. More than 40 were interviewed (Hutson, 2013).

Elizabeth Rallo brought her skills in corporate finance, quality training, facilitation, and process improvement to Newtown. It was an unscripted assignment and the first thing she learned was is that the rules of engagement and behavior are different. "In a typical job, you generally have goals, deliverables, a plan and a timeline. I personally had to look beyond task oriented and move towards doing what was needed, even if that was just listening. The role was not the standard 'job', it was a servant of the community: one head, hand and heart surrounded by people who each had a story to tell and a part to play in the community rebuilding" (Rallo, Personal communication, October 30, 2015).

The GE team, based in the Municipal Center, reported to the First Selectman and became the go-to facilitators on everything from community resources and services for families to answering media inquiries and requests for donations. They managed correspondence, gifts and scheduling of appearances, and meetings of officials for all things related to the shootings. They coordinated public information leading up to the Sandy Hook Elementary School task force's decision to raze the existing school and build a new facility on the site (Hutson, 2013).

In April 2014, Ms. Rallo returned to her job at GE Capital. One year later GE announced that GE Capital would be sold. She was prepared for the emotional spectrum that ensued. Her experience in Newtown fostered a new personal resilience and self-awareness. She practiced active listening and peer support with her colleagues as they collectively moved through a range of emotions to transition from being a corporate family to the next phase of their professional lives.

"My role in Newtown sparked within me a desire to provide greater benefit for society as whole. GE has given me formidable skills, tremendous relationships but most importantly this unique opportunity to assist this precious community and find my true calling" (Rallo, Personal communication, October 30, 2015).

> **Integration Lessons Learned from Newtown, Connecticut**
>
> When blending public and private leadership, consider Elizabeth Rallo's lessons learned (Rallo, Personal communication, October 30, 2015):
> - **Start by building trust.** Fragile communities are wary of outsiders. Build personal relationships and be reliable in making progress
> - **Honor community decision-making.** Understand the balance between driving toward goals and allowing people to come to a decision or conclusion
> - **Stay neutral.** Community factions arise so remain centered
> - **Communicate.** Practice active listening and communicate in a compassionate way
> - **Get media training.** The media was omnipresent

Case Example: Omaha Metropolitan Medical Response System

The examples above focus on *responses* to particular events. Cases in which entire systems, municipalities, and regions integrate their efforts in *preparation* for extreme events are also instructive. This section focuses on the experience of one metropolitan area (Dutton, Personal communication, October 10, 2015).

Throughout the United States private and public sector partnerships have been formed to support the community response to disasters. Partnerships develop at all levels to maximize resources and deliver coordinated services. The Omaha Metropolitan Medical Response System (OMMRS)/Healthcare Coalition (HCC) (Omaha Metropolitan Medical Response System, Health Care Coalition, 2015) is an illustration of private and public sector leaders forming partnerships to integrate medical response to disasters and to foster preparedness within their own organizations. The representatives within OMMRS include private and public hospitals, behavioral health, public health, fire service, law enforcement, emergency management, major businesses, volunteer agencies active in disasters, and community organizations of the metropolitan area. The OMMRS crosses state lines and includes counties in both Nebraska and Iowa. As an aid to other communities interested in promoting such collaborations, Table 9.2 identifies key elements of system development and potential approaches to system development.

The diversity of the organizations participating in the partnership has provided the local Behavioral Health Authority the opportunity to integrate DBH response with the larger community. The OMMRS/HHS Behavioral Health Committee oversees the planning for the behavioral health response for the medical community in the metropolitan area (Omaha Metropolitan Medical Response System, Health Care Coalition, 2015). Following a disaster, the committee conducts an assessment to determine behavioral health needs of the medical community. The primary

TABLE 9.2 Key Elements in the Development of the Omaha Metropolitan Medical Response System (OMMRS)/Healthcare Coalition (HCC)

Element	History/Approach
Key event	Joint Department of Defense (DoD)/Omaha Fire Department chemical exposure exercise and after action report
Motivation	Identified need for hospital/medical integration
Funding	Federal funding for Metropolitan Medical Response System
Key consequence of funding	Expand multiple disciplinary/jurisdictional involvement Development of plans and exercises
Key stakeholder involvement	Development and sustained committee structure including: Alternative Care Facilities, Badging, Behavioral Health, Communications, Communications Recruitment and Training, Community Plan, Drill/Exercise Design Team, Emergency Coordination Center, Equipment and Training, Law Enforcement, Long Term Care, Mass Care Support, Mass Fatality, Media/Public Relations, Pharmacy, Public Health, Transportation, Volunteer Processing, Steering, and Executive
Key to success	Longevity and active participation through a productive, well-organized structure

agency responsible for implementing the OMMRS/HCC Behavioral Health Plan is Region 6 Behavioral Health Care, the local behavioral health authority appointed by the State of Nebraska to provide services to the five western counties.

The Behavioral Health Committee is comprised of a large number of organizations that work collaboratively to respond to disasters. The scope and comprehensiveness of this coalition is a key factor in its productivity and sustainability. Committee membership includes faith-based behavioral health and family services organizations, EAPs, children's behavioral health services, county-based community mental health centers, local American Red Cross chapters, the Medical Reserve Corps, medical centers and hospitals, Veterans Affairs, county health departments, emergency management, state-wide critical incident management, academic institutions, and the OMMRS/HCC Coordinator.

OMMRS/HCC activates an Emergency Coordination Center (ECC) at the request of the medical community to respond to disasters. Behavioral Health is represented at the ECC and is responsible for activating and coordinating specialties such as clergy, Region 6 Behavioral Health (local mental health authority), school systems, American Red Cross, and Critical Incident Stress Management (CISM). The Medical Behavioral Health Section Leader in the ECC may activate behavioral health resources and Behavioral Health

Committee members. Other roles of the Medical Behavioral Health Section Leader in the ECC are to obtain preidentified data required for the Federal Emergency Management Agency Crisis Counseling Assistance and Training Program, assess the stress level of the OMMRS/HCC representatives in the ECC, work with the Pharmacy Leader regarding psychotropic medication needs, and provide other support as needed (Omaha Metropolitan Medical Response System, Health Care Coalition, 2015).

OMMRS/HCC participates annually in a variety of community exercises ranging from tabletop to full-scale exercises. The exercises provide behavioral health centers, hospitals, medical centers, and other health facilities and organizations the opportunity to test their internal plans during a large-scale event and integrate response capabilities with emergency management as well as the community.

OMMRS/HCC is a partnership that engages the whole medical community, coordinates with a variety of response organizations, and crosses both county and state lines. The planning completed by OMMRS/HCC demonstrates how DBH can be included in the planning, training, and exercising in all hazard preparedness for community incidents.

Missouri's Evolving Practice: From Floods to Ferguson

A challenge for any location responding to an event is coordinating the public and private sector funding, grants, and donations. Following the Great Flood of 1993, at the state level, Missouri's Governor convened the public and private sector responders to ensure nonduplication of efforts and maximum impact of funds received following the flood. The Governor's Flood Recovery Partnership was facilitated by the Director of the Department of Mental Health (DMH). The DMH, through the Federal Emergency Management Agency CCP, worked with communities and responding agencies to facilitate emotional recovery from this statewide disaster. State Emergency Management embraced and collaborated with the behavioral health programs. Not only were resources streamlined, but issues were addressed and policy agreed upon collectively too. Now known as The Partnership (see The Parnership, Missouri State Emergency Management Agency, http://sema.dps.mo.gov/programs/the-partnership.php for more detail regarding structure, membership, and function) this collaborative model of State departments and private organizations, aligned with front-line community responders and businesses, continues to evolve as an active and successful framework in Missouri.

Case Example: Ferguson, Missouri

On August 9, 2014, recent Normandy High School graduate Michael Brown was killed by a police officer outside the Canfield Green Apartments (Bogan & Moskop, 2015). There is general agreement that the civil unrest that erupted in the Ferguson community ignited due to decades of racial inequity, excessive municipal fees, as well as anger with the lack of minority representation among local elected officials. "Black Lives Matter" became a global focus. For weeks, the unrest became a regular event in St. Louis County and was covered 24/7 by the international media. Local law enforcement and the Missouri National Guard were deployed to North St. Louis County. A state of emergency was declared but this event was unlike anything seen before in Missouri. A behavioral health response was activated through local agencies that were already embedded in the community. New partnerships were formed with law enforcement and economic development groups. Existing relationships with behavioral health partners and emergency management leaders were leveraged. The following text box contains details of the collaborations supporting DBH goals. These details are provided to assist readers in examining and exploring relevance to their own locations and preparedness and response strategies.

Behavioral Health Framework Creates the Resilience Coalition

In the immediate aftermath of the unrest, the Missouri Department of Elementary and Secondary Education (DESE) contacted the DMH to develop and deliver training for 200 teachers, school counselors, resource officers, social workers, and administrators on how to assist children and staff dealing with the emotional impact of what became known as *Ferguson*. DMH collaborated with internal experts and academics, and 2 days later delivered training with facilitated breakout sessions hosted at Harris Stowe State University. DESE selected the site and it was an important choice to host this gathering in the City of St. Louis at a Historically Black College and University (HBCU). From this collaboration, the Directors of Missouri DMH, DESE, and Social Services sanctioned the creation of "The Resilience Coalition," comprised of local behavioral health providers, the United Way, and other not-for-profits, law enforcement, social justice organizations, private foundations, the faith community, academic experts, and others. This coalition was to address the complex issues and toxic stress in North St. Louis City and County. DMH provided in-kind leadership, meeting space, and support to foster alignment with the behavioral health supports already embedded in the community. Using a crisis counseling and community outreach model, the Resilience Coalition was formed and member agencies volunteered key staff to address the historic and ongoing stress in the targeted area. Simultaneously, The Ferguson Commission, appointed by Governor Jay Nixon, held public meetings and looked fully at the issues impacting the St. Louis Metropolitan area. Their final report was widely distributed and posted online in December 2015 (The Ferguson Commission, 2015).

(**Source**: Author (Flanigan, 2014) first hand experiences) Content compiled from author participation in ongoing conference calls and Resilience Coalition report to Missouri Mental Health Commission Dec. 11, 2014.

The collaborations developed and described above were creative, cohesive, and targeted. Key goals of the Resilience Coalition are detailed in Table 9.3.

TABLE 9.3 Key Goals, Activities, and Partners of the Resilience Coalition

Goals	Activities and Partners
Coordinate Outreach and Response	
Facilitate discussions on race	Racial Equity Learning Exchange (RELE) provided historical context of race and the connection to trauma and toxic stress. RELE sessions were delivered to leaders of State Agencies, Law Enforcement, Public School Safety Officers and more, including the DMH Mental Health Commission and executive management team.
Foster community engagement	Resilience Coalition cohosted Listening Sessions with faith communities and other groups to focus on trauma, toxic stress, and the resulting impact on individuals, families, and communities.
Offer behavioral health support	Resilience Coalition collaborated with Metropolitan Congregations United to organize sanctuary and safe spaces with mental health professionals for the days leading up to and weeks following the grand jury decision. Resilience Coalition partnered with University of Missouri-St. Louis to provide resilience and coping groups to law enforcement and families directly impacted by events in Ferguson. Resilience Coalition teamed with the Association of Black Psychologists to provide clinical support to Ferguson residents.
New Strategic Partnerships	
Urban League	The Resilience Coalition and DMH Human Resource staff partnered with American Federation of State, County, and Municipal Employees (AFSCME) representatives at an Urban League Job Fair to recruit Ferguson area residents for open positions at DMH St. Louis facilities (received 200 applications).
Economic Development	The Resilience Coalition promoted behavioral health resources at "Shop Ferguson" Listening Forums and addressed trauma through providing requested behavioral health services for affected business owners/employees.
United Way	Resilience Coalition collaborated with the United Way emergency assistance to support their Ferguson efforts of food distribution, transportation, and crisis counseling.
Building Trauma-Free Communities	
Bridges to Care & Recovery	Partnered with Bridges to Care & Recovery featuring nine churches that completed training to become "behavioral health friendly churches."
Alive and Well	Resilience Coalition members joined the Executive Steering Committee of **Alive and Well**, the Regional Health Commission initiative to build a resilient and trauma-informed communities.
Youth Violence Prevention (YVP) Partnership	YVP is a multiagency, multidisciplinary approach with involvement in public schools, juvenile justice, criminal justice, health, mental health, and social services to reduce violence in St. Louis City among youth ages 14–24.

Case Example: Virginia Responds to the 9/11 Attack on the Pentagon

Behavioral health has a long history of collaboration and partnership during disaster response, predating the establishment of the NRF guidelines and the NIMS template for integration of the private sector and NGOs in terms of the local response to a disaster described earlier. The following example illustrates how collaboration helps to promote emotional healing within the workplace, the community, and for individuals.

On September 11, 2001, terrorists perpetrated unprecedented attacks on the World Trade Center and the Pentagon that captivated and terrified the world. The 9/11 terrorist attacks left many people across the United States feeling unsafe and vulnerable.

At the Pentagon in Arlington, Virginia, the attack caused major damage and the deaths of 189 individuals. Sixty-four of the fatalities were onboard the American Airlines plane that hit the building. Following the attack, the Commonwealth of Virginia was awarded the Federal Emergency Management Agency Crisis Counseling Assistance and Training Program funding for the Community Resilience Project (CRP) to provide crisis-counseling services in Northern Virginia through local mental health authorities.

As the Metropolitan Washington D.C. area reacted to ongoing terrorist related events, including the anthrax attack, sniper shootings, and military actions in Afghanistan and Iraq, the project reassessed needs and adapted services to respond to the new events. The CRP successfully implemented a wide range of resilience building initiatives during the two and one half years it was operational. A detailed description of the history, structure, services, challenges, and accomplishments of this Project is available for interested readers (Commonwealth of Virginia, 2004) http://www.council ofcollaboratives.org/files/HelpingtoHealManual.pdf.

Of special note are several challenges faced by the CRP that have not previously been experienced by more traditional DBH programs. Key challenges included:

- The Pentagon is a high-security, military facility with controlled access for any civilian programs. The CRP's innovation, persistence, and coordination with the Department of Defense's own programs opened the door for civilian programs to the military community
- News coverage had focused primarily on the World Trade Center, leaving the victims of the Pentagon feeling isolated and overlooked. A media campaign was initiated to increase awareness of the impact on the Pentagon's personnel
- A large number of people in Northern Virginia are of Middle Eastern descent and have been the subject of hate crimes and bias since

September 11, 2001. National and local events, such as the federal raids on homes, further traumatized this community. Crisis counselors with strong ties to the Middle Eastern population were hired to serve this vulnerable group

The complexity of the impact of the terrorist incident required the CRP to outreach the entire area and rely on community partners to assist with (1) identifying the diverse communities in need of assistance and (2) providing perspective on the impact of the ongoing events. NGOs and the private sector played major roles in reaching out to the community and assisting the CRP to be successful.

Each county developed multicultural teams to reach throughout the diverse Northern Virginia region and to respond to the needs of the individuals and groups. The hiring process for outreach workers and crisis counselors emphasized those already affiliated with NGOs serving the community. These groups included Muslim, Vietnamese, Hispanic, and Korean populations.

All of the county CRPs developed partnerships with public and private businesses, educational institutions, public and private businesses, faith-based organizations, human service, and community-based organizations. The CRP was allowed to participate in already-scheduled events offered by the private sector and NGO partners providing an opportunity to promote services, learn about partners' resources, and develop relationships that lasted throughout the project.

Engaging the faith community was essential. The CRP worked with the interfaith group, Faith Communities in Action, to conduct cross-cultural dialogues on the 9/11 experience. The dialogues were well attended and received by people of different faiths and provided a supportive forum for sharing 9/11 experiences and they continued after the CRP ended. This is just one example of the legacy of the CRP that was sustained through the collaboration with NGOs and the private sector.

Other CRP partnerships included an initiative to colocate multicultural and/or multilingual outreach workers within community host organizations to access low income, older adults, and immigrants within the community. Participating partners included health clinics serving low-income residents, government human services, a residential high-rise with at-risk low-income and immigrant residents, and a senior center.

In discussing important partnerships, one CRP leader stated, "..... For mental health, in terrorist-related types of disasters, the State Disaster Mental Health Coordinator, the local Red Cross Chapter, and the Employee Assistance group within your particular organizations are probably the three preestablished relationships, that are most critical in the provision of mental health services" (Community Resilience Project of Northern Virginia, 2004, p. 13).

The Arlington CRP is also an illustration of a local government entity supporting federal resources with unique services. At the request of the Garrison Commander at Fort Meyer, the CRP provided stress management services to civilian and active duty military personnel—many of whom had been among the first to respond to the attack on the Pentagon. In addition, the CRP facilitated support groups on Fort Myer for spouses and children of soldiers in response to active duty members being deployed to Iraq.

Local private sector businesses were very supportive of the behavioral response. The public information campaign for the CRP received 513 free radio spots and 593 free television ad spots to promote CRP services and promote healing. A pizza franchise distributed over 8000 flyers attached to pizza boxes advertising the services offered by the CRP. The CRP worked with the private sector to conduct stress management classes, booths, or displays at varied facilities and community meetings, including public libraries, resource centers, Girl Scouts, senior centers, recreation centers, churches/mosques, service organizations, assisted living facilities, corporations, fitness centers, and local and chain retailers.

The experiences of the CRP demonstrated repeatedly that responding to the behavioral health needs of disaster survivors requires a *whole community* approach to identify and reach those in need and to provide comprehensive and diverse services.

COMMUNITY SUPPORT

No group can accomplish their goals and missions alone. Responders at all levels in all agencies should be engaging in comprehensive and collaborative alignments. The Federal Emergency Management Agency promotes a whole community approach to emergency management that engages private and nonprofit sectors, as well as the general public, into the disaster response. The whole community needs to be engaged in order to meet the multitude of needs of individuals and the community.

Emergency Management Whole Community Approach

Engaging the whole community and empowering local action will better position stakeholders to plan for and meet the actual needs of a community and strengthen the local capacity to deal with the consequences of all threats and hazards. This requires all members of the community to be part of the emergency management team, which should include diverse community members, social and community service groups and institutions, faith-based and disability groups, academia, professional associations, and the private and nonprofit sectors, while including government agencies who traditionally many not have been involved in the emergency management (The Federal Emergency Management Agency, 2011, pp. 4–5).

INTEGRATING VOLUNTEERS INTO DISASTER RESPONSE

Volunteers are essential to the response to large-scale events, including behavioral health services. Local behavioral health resources may become quickly overwhelmed. Augmenting the response with volunteer responders can provide for more expansive services, including services for diverse groups. Many of the state mental health authorities coordinate volunteer DBH response teams to supplement the local behavioral health response.

Organizations within the private sector have also established volunteer DBH teams. The American Red Cross is one of the most well-known volunteer organizations active in disasters (VOAD). They respond to a wide range of events from assisting individual families to large-scale community disasters. DBH services are a major focus of the American Red Cross in their effort to assist families and communities to recover (American Red Cross, n.d.-a). The American Red Cross deploys a wide variety of licensed behavioral health professionals who have completed specialized training to assist both the victims of disasters and the relief workers to deal with the trauma and stresses related to the disaster event (American Red Cross, n.d.-b).

The National VOAD is a gateway for faith-based and nonprofit organizations interested in becoming integrated into disaster response. In May 2015, the National VOAD ratified points of consensus on disaster emotional care for all phases of a disaster. The National VOAD (Volunteer Organizations Active in Disaster, 2015) describes, "Accepted types of disaster emotional care include, but are not limited to:"

- Preparedness activities
- Assessment and triage activities
- Psychosocial support activities
- Early psychological intervention activities
- Recovery activities

State VOADs are typically represented in the State Emergency Operations Centers, providing the opportunity to formally integrate with key partners, including governmental organizations at all levels during the disaster response and recovery. Federal, state, and local governments' recognition of the importance of volunteers has provided better access and support for organizations and individuals wanting to help build resiliency in their communities.

There are many opportunities to volunteer during disaster response and recovery. Volunteering through a VOAD or State DBH Team provides the behavioral health provider access to training resources and a supportive environment.

STRATEGIC STEPS TOWARD ACCOMPLISHING EMERGENCY MANAGEMENT, DISASTER BEHAVIORAL HEALTH, AND PRIVATE SECTOR INTEGRATION

Step One: Educating Individuals and Organization

Integrating stakeholders and including volunteers from the community have been a focus of emergency management for many years. The first step in becoming part of the effort to build community resiliency and respond to disasters is to become educated on the whole community approach to emergency management and the role of DBH. The Federal Emergency Management Agency and many other organizations provide free information and training that can be used to educate individuals or organizations interested in disaster response and recovery and building resilient communities.

A Wealth of Free Information and Training is Available

1. The Federal Emergency Management Agency's A Whole Community Approach to Emergency Management: Principles, Themes, and Pathways for Action at https://www.fema.gov/media-library/assets/documents/23781 introduces the collaborative model of whole community approach to emergency management and provides a strategic framework for integrating this approach into daily practices. In addition, the Federal Emergency Management Agency's website Whole Community at http://www.fema.gov/whole-community provides an overview of the principles of whole community approach to emergency management
2. National VOAD's website at http://www.nvoad.org/ provides guidance on how to help and how to be prepared and provides links to partner organizations that offer volunteer opportunities
3. Citizen Corp provides an opportunity for individuals to volunteer. Information on Citizen Corp is at http://www.ready.gov/citizen-corps
4. Behavioral Health Providers can contact their State Department of Mental Health and/or State VOAD to identify behavioral health volunteer opportunities
5. The Substance Abuse Mental Health Services Administrations (SAMHSA) Disaster Technical Assistance Center at http://www.samhsa.gov/dtac has free information on DBH
6. The Federal Emergency Management Agency's National Emergency Training Center is the home to the National Fire Academy and the Emergency Management Institute (EMI). Online independent study (IS) courses are available at no cost. A complete list of residential and online course offerings is available at: http://www.training.fema.gov/apply/
    a. A first step for any individual or organization interested in becoming part of emergency response is to assure their own and/or the individuals in their organizations personal readiness. The Federal Emergency Management Agency offers online IS courses through its EMI. The IS-22

    (Continued)

(Continued)

course Are You Ready? An In-depth Guide to Citizens Preparedness is an excellent introduction (https://training.fema.gov/IS/courseOverview.aspx?code=IS-22)

b. IS-394a Protecting Your Home or Small Business from Disaster describes protective measures that can reduce or mitigate the negative consequences of disasters on homes or small businesses (https://training.fema.gov/IS/courseOverview.aspx?code=IS-394.a)

c. IS-660 Introduction to Public–Private Partnerships is an introduction to the public–private partnerships and describes roles and responsibilities. This training will introduce the common language used in emergency management and basic principles (http://www.training.fema.gov/is/courseoverview.aspx?code=IS-660)

d. IS 288a The Role of Voluntary Organizations in Emergency Management describes the history, roles, and services of disaster relief voluntary agencies. The course was developed for both the general public and emergency management (https://training.fema.gov/IS/courseOverview.aspx?code=IS-288.a)

e. There are many other courses that may be beneficial to an individual or organization. It is recommended that the IS list be reviewed

Step Two: Preparing Individuals and Organizations

Preparedness begins at the individual, family, and business/organizational level. Education is a powerful tool for facilitating preparedness. Organizations and businesses can challenge members and employees to be ready for all hazards. Being prepared will empower businesses and organizations to contribute to community resiliency.

Businesses form the backbone of the community, state, and national economy. A commitment to planning today will help support employees, customers, the community, the local economy, and even the country. All businesses should have a business continuity plan to ensure the business continues to operate regardless of the interruption. The Federal Emergency Management Agency offers several planning resources and templates, including a business continuity plan, business impact analysis, computer inventory template, and other relevant tools (The Federal Emergency Management Agency, 2013b).

Step Three: Reach Out to Key Partners

After reading the materials related to personal, family, and organizational preparedness, reach out to the group(s) active in the disasters with similar interests to learning more about their roles, goals, and activities.

Step Four: Live Emergency Management for Life

People who have participated in a disaster response frequently tell similar stories. They often say that this work is a life-changing experience that can provide a better understanding of one's community and the vital role emergency management plays in all lives. This experience can lead to many more opportunities to become part of community resilience building. Remember that disasters occur locally, but can still impact the entire nation.

CONCLUSION

All who have worked in the world of disasters understand that no two experiences are alike, and that some experiences personally and professionally impact us more than others. The authors of this chapter are no exception. What follows are reflections by both authors of two of events that have been most significant to us. It is our hope that these examples provide both personal and professional insight and motivation.

A Nation Responds: Diana Nordboe's Reflection on 9/11

September 11, 2001, may have occurred in New York City, Arlington, Virginia, and in a field in Pennsylvania, but it was a national incident. The care, concern, and help provided to the survivors, families, and responders of 9/11 had a profound impact on the recovery of individuals and assisted to build resilience in the impacted communities. The adverse psychological consequences of the disaster were immediate and widespread and individuals and communities across the nation struggled to redefine their daily lives. The impact of a terrorist event is far reaching and requires a whole community response.

In Virginia, the CRP, discussed earlier, found that, in addition to the families and friends of the casualties and employees at the Pentagon, countless others were greatly impacted and in need of crisis counseling services. They include first responders, commuters who witnessed the incident, school children, airport employees, immigrants that had fled war torn countries and terrorism, and many others. Fear and grief were the most prevalent emotions reported by community members to the crisis counseling staff. In Northern Virginia schools and across the nation, our children watched the horror unfold in unrelenting news accounts of the terrorist attacks.

The economic impact was also considerable. The Pentagon is one of the world's largest office buildings with 23,000 employees who live in or proximate to Northern Virginia. Hundreds of workers at the Arlington County-based Reagan Washington National Airport, which brings an estimated $5.6 billion into the local economy each year, have been impacted by the terrorist events. The temporary closure of the airport and reductions in operations after its reopening adversely affected the economy in the region. While the region's 10,000 airport employees were directly impacted by the disaster, an additional 70,000 jobs in the hotel, restaurant, and travel industries were threatened. The 9/11 attack and events following the disaster contributed to a $2.1 billion budget shortfall in the Commonwealth (Fuller, 2004).

Northern Virginia represents the most densely populated region in the Commonwealth, with a population in excess of 1.8 million. Virginia citizens from all nine jurisdictions comprising the Northern Virginia area were directly or vicariously impacted by the terrorist attack. The Northern Virginia area has an extremely diverse population with over 100 languages spoken in the region (George Mason University Center for Regional Analysis, 2011). There are many Northern Virginians who have come to the United States fleeing war in their countries of origin. They came to this country for a new beginning and to feel safe. The 9/11 attack left them fearing for their safety, feeling vulnerable, and worrying that there would be additional terrorist attacks. Despair over the tragedies, exacerbated by fear and isolation, created an extended need for crisis counseling services. The response is Virginia was successful because it immediately reached out to others within the community for support and help. Faith-based groups, businesses, NGOs, and individuals were available to help as needed and become part of the emergency management and behavioral health response.

The nation has learned many lessons during and since 9/11. One of the most important lessons is that it takes a community to respond to a disaster. Emergency management cannot respond alone. The needs of a community are complex prior to a disaster, and even more difficult to address after a significant incident. Conducting outreach in a community following a disaster is like peeling an onion—the more layers you peel the more you want to cry. The disaster so often pushes people who were marginally getting by to a point where they feel that obstacles are too overwhelming to overcome. The partnership between behavioral health and other organizations providing assistance in disasters helps those impacted to regain control over their lives. As the nation continues to develop the whole community approach, our ability to respond will grow exponentially.

Diana Nordboe's Lessons Learned

1. Know your village. People live, work, shop, and commute through your community daily. Design programs and outreach to address anyone residing in or passing through the area when the disaster occurred
2. Address the economic earthquake. The disaster impacts the economic epicenter of the community. Consider the financial aftershocks. Support

(Continued)

> **(Continued)**
>
> local businesses and services and identify second and tertiary businesses impacted by the disruption to the economic drivers
> 3. Cultural competency is a foundation to any program. Know your population and hire respected members of the local community and agencies to advise your staff and deliver programs
> 4. Adopt the whole community approach

Susan Flanigan's Reflections: Ferguson

I was at the Chase Park Plaza for the August 10, 2014 matinee of "Get on Up," the movie biography of James Brown, and everyone in the theater was talking about it. Michael Brown died the day before in Ferguson, a suburb 10 miles from my St. Louis City neighborhood. We filmgoers, strangers yet neighbors, collectively recognized the complexity of the event. The next morning, I drove 2 hours to my office in Missouri's capital, Jefferson City. I asked my boss, Missouri's Director of Mental Health, "What are we going to do about the unrest in Ferguson?" He answered, "We've not been asked for anything." To which I replied, "It's not going away." One week later, the Missouri Commissioner of Education asked for our help in creating a training for teachers, staff, and administrators returning to the schools that would have children directly impacted by the events in Ferguson.

Crises do not fit into neat categories. An incredibly political landscape existed. Plus, generally speaking, the St. Louis area preferred to use their deep pool of local experts who understood the metropolitan area as compared to public servants temporarily deployed from Central Missouri.

The crisis counseling model, Missouri's collaborative strength, St. Louis' strong behavioral health community, and understanding the process to complement existing resources led to the creation of the Resilience Coalition. However, the event that brought us together was not a natural disaster, but was rooted in a history and context of racial inequity, municipal practices, and law enforcement responses infused with socio-economic disparities. All this came together in the perfect storm of civil unrest.

In greater St. Louis, specifically North County, behavioral health collaborations were already embedded in Ferguson and Florissant, as well as nearby school districts. A back-to-school training delivered in 48 hours revealed the need for something more. A core team came together, volunteering their time to focus on the emotional needs of the children, families, and communities most impacted. Outreach consisted of listening sessions primarily hosted with the faith community to discuss toxic stress. As the formal coalition phases down the behavioral health quiet presence continues through Bridges to Care and Recovery (Behavioral Health Network, n.d.), a faith-based organization that continues the legacy and applies lessons learned. Several members of the Resilience Coalition sit on the steering committee of Alive and Well, which is addressing the impacts of toxic stress in St. Louis by striving to build a trauma informed community (Alive & Well, n.d.).

Susan Flanigan's Lessons Learned

Having worked in public and private organizations at the local, state, and national level directing public and legislative affairs, crisis communication, business continuity, and CCP and DBH programs and services, Susan shares these insights:

1. Know what you do not know. Have a "go-to" list of experts, prescreened, both formal and informal, for consultation. Experts should be based within your state and local area, and as well as out-of-area. Out-of-area experts help with perspective and in-area experts help with access and local credibility
2. Identify key community leaders and resources and align with them early. Be inclusive
3. Understand the internal and external political landscape. All disasters are political events and there are always those trying to enhance their reputation, whether they have the needed skills and resources or not. Find meaningful ways to keep the key decision makers positively engaged and know the chain of command and nuances to prevent blunders which could mar program efforts
4. Communication, communication, communication. What you say and how you say it is important. Understand risk communication. Review your communication for content, clarity, empathy, truth, and reassurance. Offer concrete actions when possible. Do not contribute to the ambient noise
5. Sharing a meal brings people together. Yet, public entities often cannot pay for food with taxpayer dollars. The private sector can offer meeting space with snacks or light fare to facilitate collaboration and problem solving among community groups and public and private sector responders
6. Self-deploying is not the answer and can adversely impact the response. Be part of the organized response. Many business headquarters feel they must rush to the impacted area to check on their satellite offices but this may take hotel rooms and other resources from locals displaced from the disaster. Think before responding and ensure any company representatives check in with local emergency management to share information and verify credentials to access their facilities and employees
7. Connect with the local behavioral health agency or state mental health authority to participate in exercises and training opportunities such as psychological first aid (PFA). PFA training courses are available online at no cost (The National Child Traumatic Stress Network, n.d.)
8. Understand the phases of disasters (See Fig. 9.1) and assure that efforts fit the phase

(Continued)

(Continued)

9. Bring like minds together to share best practices and build personal relationships
10. Limited funds, if any, are available for public employees to attend conferences in- or out-of-state. Corporate sponsorship of state teams comprised of public and private sector leaders to attend trainings could prove valuable to building state resiliency
11. Capture, integrate, and share lessons learned
12. Sustain and maintain the institutional knowledge
13. Always practice self-care
14. Maintain a sense of humor, but know it has its place and that inappropriate humor can create endless problems
15. Be grateful. Say thank you

Making Integration Work

Behavioral health, emergency management, and the private sector continue to enhance their collaborations and response effectiveness through shared technology, mutual goals and objectives, and integrated testing and exercising. Although these formal alignments enhance the effectiveness and practice of disaster response and recovery, exceptional programs are still personality- and leadership-driven with networking being a key component. Who has the best model to emulate? That does not exist. Like all disaster response, leaders must customize the framework to fit the event and the community.

As a means to organize and operationalize the major point made in this chapter, Table 9.4 is provided. It is designed to provide a framework that can be customized by various locations and organizations. It should be used as a "place to start," a format intended to help users develop a matrix helpful in their own communities. It is hoped that the *process* of populating cells in a document such as this will prove to be valuable in promoting collaborative efficacy and preparedness that least to and strengthens integration.

TABLE 9.4 Sample Matrix to Foster Integration of Behavioral Health, Emergency Management, and the Private Sector

Goal	Key Activities	DBH	EM	Private Sector	Collaboration
Promote integration	Identify stakeholders and begin networking meetings				
Integration needs assessment	Identify the Federal, State, and Local rules for each key player (i.e., legislation, policy, regulatory agencies, and politics)				
	Review community landscape: what has/is being done				
	Review service delivery models, programs, and funding resources				
	Identify gaps—individual and collective				

(Continued)

TABLE 9.4 (Continued)

Goal	Key Activities	DBH	EM	Private Sector	Collaboration
Develop structure and plan	Identify roles, resources, priorities				
	Risk communication				
	Media				
	Community outreach and engagement				
	Public information and education				
	Develop and implement an engagement plan				
Testing/ exercise	Develop scenarios across systems to test strengths and identify problems				
Sustain & maintain	After action reports and trainings for continuous modification based on events, successes, challenges, and stakeholder change				

REFERENCES

Alive and Well. (n.d.). *About alive and well.* Retrieved from http://www.aliveandwellstl.com/learn-more/about-alive-and-well/.

American Red Cross. (n.d.-a) *What we do.* Retrieved from http://www.redcross.org/what-we-do.

American Red Cross. (n.d.-b). *Nursing and health education.* Retrieved from http://www.redcross.org/tn/knoxville/disaster-services/dat/mental-health.

Anheuser Busch. (n.d.) *Disaster preparedness and relief.* Retrieved from http://anheuser-busch.com/index.php/our-responsibility/community-our-neighborhoods/natural-disaster-relief/.

Behavioral Health Network. (n.d.). *Bridges to care & recovery.* Retrieved from http://www.bhnstl.org/current-initiatives/bridges-to-care-and-recovery/.

Bogan, J., & Moskop, W. (2015, March 16). *What's going to happen to Canfield Green apartments?* Retrieved from http://www.stltoday.com/news/local/metro/what-s-going-to-happen-to-canfield-green-apartments/article_a358dfaf-2e7f-51b1-941b-69d542c0f5a0.html.

Commonwealth of Virginia. (2004). *Community Resilience Project Regular Services Grant Final Report.* Department of Mental Health, Mental Retardation and Substance Abuse Services, P.O. Box 1797, Richmond, VA. 23218-1797.

Community Resilience Project of Northern Virginia. (2004). *Helping to heal: A training on mental health response to terrorism.* Retrieved from http://www.vdh.virginia.gov/oep/pdf/CWD-HelpingToHealFieldGuide.pdf.

Dutton, P. Personal communication, October 10, 2015.

Federal Emergency Management Agency. (2011). *A whole community approach to emergency management: Principles, themes, and pathways for action.* Retrieved from https://www.fema.gov/media-library/assets/documents/23781.

Federal Emergency Management Agency. (2013a). *National response framework.* Retrieved from http://www.fema.gov/media-library-data/20130726-1914-25045-1246/final_national_response_framework_20130501.pdf.

Federal Emergency Management Agency. (2013b). *Emergency preparedness resources for businesses (20).* Retrieved from http://www.fema.gov/media-library/resources-documents/collections/357.

Federal Emergency Management Agency. (2016a). National response framework. Retrieved from http://www.fema.gov/national-response-framework.

Federal Emergency Management Agency. (2016b). *Whole community.* Retrieved from http://www.fema.gov/whole-community.

Federal Emergency Management Agency. (2016c). *How to apply to courses at EMI.* Retrieved from http://www.training.fema.gov/apply/.

Ferris, C. Personal communication, February 24, 2016.

Ferris, C. Personal communication, n.d.

Flanigan, S. Content compiled from author participation in ongoing conference calls and Resilience Coalition report to Missouri Mental Health Commission Dec. 11, 2014.

Fuller, S. S. (2004). *The economic impacts of a major terrorist attack on the greater Washington Metropolitan area.* Retrieved from http://cra.gmu.edu/pdfs/researach_reports/other_research_reports/NVBIA_report_2007/The_Impact_of_a_Major_Terrorist_Attack_on_the_Greater_Washington_Area.pdf.

George Mason University Center for Regional Analysis. (2011). *Population change in Northern Virginia.* Retrieved from http://cra.gmu.edu/pdfs/CRA_census_report_series/Population_Change_in_Northern_Virginia.pdf.

Good360. (n.d.). *Good360.* Retrieved from https://good360.org/flint-michigan/.

Hutson, N. G. (2013, July 7). *GE Capital team deemed Newtown's "fairy godmother."* Retrieved from http://www.ctpost.com/local/article/GE-Capital-team-deemed-Newtown-s-fairy-godmother-4651560.php.

Myers, D. (1994). In B. Hiley-Young (Ed.), *Disaster Response and Recovery: A Handbook for Mental Health Professionals.* Washington, DC: SAMHSA.

Omaha Metropolitan Medical Response System, Health Care Coalition. (2015). *Behavioral health Plan.* Omaha, NE: Behavioral Health Committee. http://www.regionsix.com/services/disaster-behavioral-health-services/ (accessed 8/13/2016).

Rallo, E. Personal communication, October 30, 2015.

Redlener, I. (2016, February 17). *We need to resettle the children of Flint.* Retrieved from https://www.washingtonpost.com/opinions/get-flints-families-out-of-harms-way/2016/02/17/e8553e40-d4dc-11e5-be55-2cc3c1e4b76b_story.html?utm_term=.9561c795fae4.

SAMHSA (2000). *Training manual for mental health and human services workers in major disasters* (2nd ed.). Washington, DC: SAMHSA.

The Ferguson Commission. (2015, October 14). *Forward through Ferguson: A path towards racial equality.* Retrieved from http://3680or2khmk3bzkp33juiea1.wpengine.netdna-cdn.com/wp-content/uploads/2015/09/101415_FergusonCommissionReport.pdf.

The National Child Traumatic Stress Network. (n.d.). *Psychological first aid online*. Retrieved from http://learn.nctsn.org/course/index.php?categoryid=11.

The Partnership, Missouri State Emergency Management Agency http://sema.dps.mo.gov/programs/thepartnership.php (link verified 5/23/16).

United Way. (n.d.). *2-1-1*. Retrieved from http://www.unitedway.org/our-impact/featured-programs/2-1-1.

Voluntary Organizations Active in Disaster. (2015). *Points of consensus: Disaster emotional care*. Retrieved from http://www.nvoad.org/mdocs-posts/poc-disaster-emotional-care/.

Walmart. (2015, August 21). *Walmart and the Walmart Foundation mark 10th anniversary of Hurricane Katrina with $25 million commitment to support disaster recovery and resiliency efforts worldwide*. Retrieved from http://corporate.walmart.com/_news_/news-archive/2015/08/21/walmart-and-the-walmart-foundation-mark-10th-anniversary-of-hurricane-katrina-with-25-million-commitment-to-support-disaster-recovery-and-resiliency-efforts-worldwide.

Walmart. (2016, January 26). *Walmart, Coca-Cola, Nestle, and PepsiCo come together to provide Flint, Mich., public schools with water for students through the end of the year*. Retrieved from http://news.walmart.com/news-archive/2016/01/26/walmart-coca-cola-nestle-and-pepsico-come-together-to-provide-flint-mich-public-schools-with-water-for-students-through-the-end-of-the-year.

Section III

Special Opportunities to Enhance Integration

If integration is to achieve optimal benefit for professions as well as the victims and survivors of disasters, opportunities must be exploited and challenges must be overcome. The goal of this section is to explore several areas in which exciting opportunities occur as well as where challenges have arisen.

For example, integration in command and control operations and integration of public communication efforts have yielded exciting and constructive results. At the same time, navigating challenging legal, ethical, and political elements of preparedness and response have challenged integration activities. The success of integration rests upon the long-term ability of both professions to persist in adapting to emerging opportunities and challenges as described in this section.

Chapter 10

Integration in the Emergency Operations Center (EOC)/Emergency Communications Center (ECC)

John J. Brown, Jr.[1], Chance A. Freeman[2], Brian W. Flynn[3], and Ronald Sherman[4]

[1]*Office of Emergency Management, Virginia, VA, United States,* [2]*Texas Department of State Health Services, Austin, TX, United States,* [3]*Uniformed Services University of the Health Sciences, Bethesda, MD, United States,* [4]*Independent Consultant, FEMA Federal Coordinating Officer (Retired), United States*

Through an Emergency Management Lens

John J. Brown, Jr.

First responders, such as police, fire, and emergency medical services (EMS) personnel witness disasters up close and are certainly affected by the trauma of horrific incidents. At the same time, those in emergency operations centers (EOCs) and emergency communications centers (ECCs) must also deal with the stress of such events, albeit without the associated sights, sounds, smells, and physical hazards of on scene operations.

In many instances, the firefighters, Emergency Medical Technicians (EMTs) or police officers are able to track the survivors to determine the ultimate outcome of their efforts in the field. EOC and ECC staffs often never learn if a patient survived or died as the result of an incident. Critical incident stress management (CISM) programs have been established in most public safety agencies nationwide, but they do not necessarily include EOC or ECC staff. The complexities in CISM programs will be discussed later in this section of the chapter.

AN EMERGENCY MANAGER'S FIRST HAND ACCOUNT OF ENCOUNTERING BEHAVIORAL HEALTH EFFECTS

As a 44-year public safety professional, I have been both a first responder for 34 years and an emergency manager with responsibility for a 9-1-1 center the past 10 years in a densely populated, urban community near Washington D.C. I have responded to large-scale emergencies in the National Capital Region, throughout the United States, and overseas deployments as a Task Force Leader and Planning Officer for the Fairfax County Urban Search and Rescue Task Force, known as Virginia Task Force One (VATF-1). In 1973, I responded to a building collapse that killed 14 construction workers in the Bailey's Crossroads area of Fairfax County. In 1998, I responded to the U.S. Embassy bombing in Nairobi, Kenya and to an earthquake in Taiwan in 1999, both involving hundreds of deaths and serious injuries. I worked in our operations center in Fairfax during two separate Task Force deployments for earthquakes in Turkey in 1999. On 9/11, I responded to the Pentagon at the request of the Arlington County Fire Department to set up a planning section under the incident command system. In 2005, I deployed to New Orleans to establish an EOC and 9-1-1 center. Between 2003 and 2007, I spent 15 months in Iraq, Kuwait and the United Arab Emirates as assistant to the DHS Attaché in Baghdad. Each of these experiences, including the traumatic death of a fellow firefighter, provided different perspectives about how we deal with stressful situations and the cumulative effects of exposure to traumatic events.

I was diagnosed with traumatic brain injury and posttraumatic stress disorder (PTSD) when I returned home from Iraq in 2007. I do not consider myself a victim, but rather a survivor. I have continued to stay engaged in the business of public safety, with the help of some awesome health-care providers in the military and the Washington D.C. Veterans Administration Hospital, as well as support from family, friends, and colleagues.

Realizing that one has a problem is the first step of recovery. I look at recovery as a journey, not a destination, and continue to work on staying healthy, physically, and emotionally. Being on the front lines can be dangerous and stressful, but I maintain that being in a support position can be just as stressful. Much of this theory is based on my own experiences where my field staff was engaged in a dangerous operation while I was at a command post and not able to see them or take an immediate action that would keep them safe. I also had many discussions with 9-1-1 center staff who were truly the first of the first responders on 9/11/2001, fielding dozens of frantic phone calls, and coordinating police, fire, and EMS radio traffic for those who were going into an environment unfamiliar to them until that fateful day. The after-effects of sleeplessness, agitation, second guessing, and sorrow seemed all too familiar comments relayed to me by staff who were on duty on 9/11.

EMERGENCY OPERATIONS CENTERS

An EOC is a multiagency coordination center where staff from different disciplines, and in some cases, jurisdictions work together to support field operations and acquire resources, typically those outside of established and normal mutual aid agreements. EOCs can be elaborate facilities designed and equipped for one specific purpose or as rudimentary as a large classroom or conference room that is rapidly activated with phones and computers as needed.

An EOC is responsible for the strategic overview, or "big picture," of the disaster and does not normally directly control field assets, instead making operational decisions and leaving tactical decisions to subordinate commands. The common function of all EOCs is to collect, gather, and analyze data; make decisions that protect life and property, maintain continuity of the organization, within the scope of applicable laws; and disseminate those decisions to all concerned agencies and individuals. In most EOCs, there is one individual in charge—the EOC Manager (Definitions.net, n.d.).

The most critical component of an EOC is the individuals who staff it. They must be properly trained and have the proper authority to carry out actions that are necessary to respond to the disaster. Also, they must be capable of thinking outside the box and creating lots of "what if" scenarios. The local EOC's primary function during an emergency is to support the incident commander.

The second critical component of an EOC is its communications system. These systems can range from simple word of mouth operations to sophisticated, encrypted communications networks. Regardless of sophistication, it must provide for a redundant pathway to ensure that both situational awareness information and strategic orders can pass into and out of the facility without interruption. The EOC does not command or control on-scene response efforts, but does carry out the coordination functions through:

1. Collecting, evaluating, and disseminating incident information.
2. Analyzing jurisdictional impacts and setting priority actions.
3. Managing requests, procurement, and utilization of resources.

The decisions made through the EOC are designed to be broad in scope and offer general guidance on priorities. Information is disseminated through the EOC Manager and tactical decisions are coordinated from field response personnel. The EOC serves as a coordinated link between the chief elected official of each jurisdiction and the field personnel coordinating the execution of event priorities (Department of Homeland Security, 2016).

EMERGENCY COMMUNICATIONS CENTERS

An ECC is a call center responsible for answering calls from the public to an emergency telephone number for police, firefighters, and ambulance services. Trained telephone operators are also usually responsible for dispatching these emergency services.

LINKING THE EOC, ECC, AND FIELD OPERATIONS

Most EOCs are activated for emergencies that stress the response capabilities of a jurisdiction or for training, drills, and exercises. EOC staff may include personnel from the local or state emergency management agency and from partner disciplines and jurisdictional stakeholders. Staying current with technology, changing threats, hazards, and vulnerabilities are important skill sets for those who work in an EOC. The ability to collaborate in a fast-paced stressful environment is necessary for the team to be successful.

The ECC is an around-the-clock operation which takes calls from the public at a time when the caller may be experiencing the worst situation of their life. The job of a public safety telecommunications specialist involves receiving calls, quickly analyzing information, dispatching appropriate resources, and providing life-saving information to the caller, while assistance is responding. Often, the ECC call-taker stays on the phone with the caller until a first responder is on scene and verifies that the phone call can be terminated.

The EOC obtains much of its situational awareness from the ECC monitoring radio transmissions, such as the dispatching of calls and on scene communications voiced by first responders. In some instances, EOC staff may actually respond to the scene as liaisons to get first-hand information and to support the incident commander with resources that are outside normal operations. These resources include sheltering needs, food, water, and portable toilet facilities for responders during long-term incidents. A public safety telecommunications specialist might be sent to a scene to handle radio, phone, and computer communications at a command post. Although it is not routine for EOC and ECC staff to deploy to an incident, long-duration, complex incidents may call for such technical assistance in the field.

> By its nature, working in an ECC or an EOC is a stressful activity. Life in these settings is characterized by information overload, multitasking, task conflict, or simply dealing with many life and death situations. ECC and EOC leadership and staff are subject to work-related stress throughout their careers. This stress can, in turn, affect productivity and the quality of work. As a result, it is in the interest of senior public safety officials to place a high priority on monitoring employees routinely for signs of stress.

INTEGRATING BEHAVIORAL HEALTH IN THE EOC AND ECC
Understanding the Stresses

Stress for emergency and disaster response workers comes in many forms and is the result of complex factors. Stress can build from the everyday nature of work of this type, and it may be difficult to pinpoint a single cause. Addressing these types of stress is a key task of leadership and should involve consultation from behavioral health professionals. If there is not a disaster behavioral health (DBH) presence in the EOC or ECC, the public health representative should make sure they have the capability to readily contact and activate such DBH resources.

This type of work also entails stress generated from what is commonly called *critical incident stress*. Critical incidents are traumatic events that cause powerful emotional reactions in people who are exposed to those events. The most stressful of these incidents are line-of-duty deaths, coworker suicide, multiple event incidents, delayed intervention, and multicasualty incidents. Every profession can list their own worst-case scenarios that can be categorized as critical incidents. Emergency services organizations, for example, usually list the *Terrible Ten* (Pearson & Thomas, n.d.).

They are as follows:

1. Line of duty deaths
2. Suicide of a colleague
3. Serious work related injury
4. Multicasualty, disaster, or terrorism incidents
5. Events with a high degree of threat to the personnel
6. Significant events involving children
7. Events in which the victim is known to the personnel
8. Events with excessive media interest
9. Events that are prolonged and end with a negative outcome
10. Any significantly powerful, overwhelming distressing event

Although any person may experience a critical incident, conventional wisdom says that members of law enforcement, firefighting organizations, and EMS are at great risk for (PTSD). However, less than 5% of emergency services personnel will develop long-term PTSD symptomatology. This percentage increases when responders endure the death of a coworker in the line of duty. This rate is only slightly higher than the general population average of 3–4%, which indicates that despite the remarkably high levels of exposure to trauma, emergency workers are resilient, and people who join the field may self-select for emotional resilience. Emergency responders tend to portray themselves as "tough," professional, and unemotional about their work. They often find comfort with other responders and believe that their families and friends in other professions are unable to completely understand their experiences. Humor is used as a defense mechanism. Alcohol or possibly other drugs or medications may be used to self-medicate in "worst case" situations (Pulley, 2005).

It is important for leaders in all public safety organizations to understand the stressors of the jobs that their employees face day in and day out, as well as during large complex emergencies. Many EOC's are staffed with nonpublic safety personnel from allied disciplines and municipal departments covering areas such as public works, parks and recreation, and public information.

Although public safety staff may be conditioned to working in stressful environments, many of our partners are not. Long hours, conflicting information, changing priorities, and emergent issues can cause anxiety, fear, and feelings of helplessness. A rapid operational tempo may be just another day at the office for first responders, but certainly not the norm for administrative and support personnel. Emergency managers must be mindful of those unaccustomed to this type of work situation and monitor them for signs of stress to provide the help that may be needed to stay engaged or reengage if they need to be relieved.

Addressing the Stress: What to Do?

There is nearly universal consensus that severe stress, with multiply causes, is a factor that must be dealt with in EOCs and ECCs. Yet, there is a striking lack of consensus regarding what models are most efficacious in reducing stress and reducing long-term negative consequences. Conducting credible research in this area is difficult. Even when, or if, a model is selected, assuring consistent application and fidelity to the model is a continuing challenge.

As noted earlier, most jurisdictions have adopted some type of CISM program. CISM is typically described as an adaptive, short-term psychological helping-process that focuses solely on an immediate and identifiable problem. It can include preincident preparedness to acute crisis management to postcrisis follow-up. Its purpose is to enable people to return to their daily routine more quickly and with less likelihood of experiencing PTSD. However, evidence-based reviews have been highly critical of a central element of CISM, namely critical incident stress debriefing (CISD). There is a growing consensus that CISD is not effective in the long-run and may sometimes be harmful (Bisson, 2003; Litz et al., 2002; McNally et al., 2003), and the popularity of CISD is declining. Evidence that, it can prevent PTSD is lacking. There are numerous articles regarding CISM for readers who want additional information (Everly et al., 2002; Raphael & Wilson, 2000; Watson, 2004).

So, what should be done for emergency personnel and what appropriate roles can and should DBH play? The aftermath of the 9/11 attacks on the Pentagon may point to a worthwhile approach as evidenced by the author's own personal experience. As firefighters, EMTs, and law enforcement officers responded to the Pentagon on 9/11, Dodie Gill, a licensed professional counselor, who managed the Arlington County Employee Assistance Program, worked tirelessly at the scene to assist the responders in dealing with the horrors they had experienced. Ms. Gill's compassionate, yet straightforward demeanor, earned the respect and trust of the first responders, and in particular, the firefighters. After 9/11, she left county service to establish a private practice that Arlington County firefighters and dispatchers use today. Her innovative approach is different than those of traditional CISM programs.

Ms. Gill has been a pioneer for the Traumatic Exposure Recovery Program (TERP). TERP is designed to support first responders from recruitment, through their careers, and into retirement. TERP evolved from promising elements of CISM and enhanced through practical, hands-on knowledge and experience gained during and after the terrorist attack on the Pentagon.

The model is based on several key elements:

1. Application of information about the brain and brain trauma.
2. Understanding how repeated exposure to traumatic events can erode mental and physical resilience over time.
3. Understanding individual variations of impact.
4. Methods of empowering individuals to manage their own symptoms.

This model includes the components necessary to put together a successful "TERP team." All TERP team members, including peer and credentialed mental health professionals, are trained to meet the specific needs of the organization they serve (Gill, 2005). Methods such as those suggested in TERP have a calming effect that help to keep us grounded and focused on the important tasks we must perform.

CONCLUSION

Regardless of the program chosen and strategies employed to deal with stress, leaders of EOCs and ECCs have a responsibility to monitor and attend to the physical and emotional welfare of those who do this type of work. Planning, training, and communicating beforehand will result in building the trusted relationships needed to keep all personnel psychologically prepared for the disasters we may face in the future.

Although the approaches for preparing for and responding to stress of emergency personnel require ongoing development and evaluation, the stress of the work remain a constant part of life in the EOC and ECC. Only through integration of the knowledge and experience present in both Emergency Manager (EM) and DBH, nurtured in an environment of mutual respect and confidence, can the needs of those who give so much be met.

REFERENCES

Bisson, J. I. (2003). Early interventions following traumatic events. *Psychiatric Annals, 33*, 37–44.

Definitions.net. (n.d.) Retrieved July 30, 2016, from http://www.definitions.net/definition/emergency operations center.

Department of Homeland Security. (2016). Retrieved July 30, 2016, from http://www.dhs.gov/fusion-centers-and-emergency-operations-centers.

Everly, G. S., Eyler, V. A., & Flannery, R. B. (2002). Critical incident stress management: A statistical review of the literature. *Psychiatric Quarterly, 73*(2), 171–182.

Gill, D. (2005). Guide: Caring for public servants. *Journal of Aggression, Maltreatment & Trauma, 10*(1–2), 591–592.

Litz, B. T., Gray, M. J., Bryant, R. A., & Adler, A. B. (2002). Early intervention for trauma: Current status and future directions. *Clinical Psychology: Science and Practice, 9*(2), 112–134.

McNally, R. J., Bryant, R. A., & Ehlers, A. (2003). Does early psychological intervention promote recovery from posttraumatic stress? *Psychological Science in the Public Interest, 4*(2), 45–79.

Pearson, A. & Thomas, K. (n.d.). Psycho-social needs of responders during a mass casualty event. Retrieved from https://www.ncjtc.org/CONF/Ovcconf/AttMat/Psycho-Social%20Needs%20of%20Responders%20During%20a%20Mass%20Casualty%20Event_Pearson%20Thomas.pdf.

Pulley, S.A. (2005). Critical incident stress management. Retrieved July 30, 2016 from http://web.archive.org/web/20060811232118/http://www.emedicine.com/emerg/topic826.htm.

Raphael, B., & Wilson, J. (2000). *Psychological debriefing: Theory, practice and evidence.* Camrbridge: Cambridge University Press.

Watson, P. (2004). Early intervention for trauma-related problems following mass trauma. In R. J. Ursano, C. S. Fullerton, L. Weinsaeth, & B. Raphael (Eds.), *Textbook of disaster psychiatry* (pp. 121–137). New York: Cambridge University Press.

Through a Disaster Behavioral Health Lens

Chance A. Freeman

During response and recovery operations, there are times when a disaster occurs within a disaster. In 2003, during the space shuttle Columbia search and recovery operation, a Texas emergency operations center (EOC) was faced with this exact situation. It serves as a good example of how some emergency managers view the topic of disaster behavioral health (DBH).

> Warning: There is a price to pay when integration fails.
> "I always thought you people were a bunch of mumbo jumbo! But now that I need you, I am glad that you are here." Those words were spoken by an incident commander who had just suffered the loss of two team members and five others that sustained serious injuries following a helicopter crash while searching for shuttle debris. While on a conference call with two of his grieving team members, he handed the phone to me and said, "Here, you help with this because I do not know what to say and we need help." In that instant, he recognized the need and benefit of having DBH responders onsite. In that moment, the emotional well-being of his work-family, flight crew, and himself became the most important mission objective. It was clear to him that tending to the emotional and psychological reactions was just as important as the on-going search and recovery operations. This disaster within the disaster had a ripple effect throughout the entire operation, and behavioral health providers were able to mitigate the psychological impact by working within the EOC. They coordinated the DBH response while helping first responders and EOC personnel understand, process, and cope with their reactions so that they could continue to work.

WHY IT IS IMPORTANT FOR DBH PERSONNEL TO BE LOCATED IN AN EOC AND WHAT ROLES CAN THEY PLAY?

This integrated structure is crucial for several reasons. First, being embedded within the EOC ensures that DBH personnel and incident command remain current on the goals and objectives of the operation. DBH personnel can assist in decision-making processes involving response planning, public information, demobilization of personnel, and the operation's transition into the recovery phase. It is important for DBH personnel to learn as much about the operation and mission objectives as possible while keeping EOC informed of DBH encounters and patterns as well as current or future operational and planning needs. For example, DBH personnel with experience responding to natural disasters understand their role and how they support the EOC, but do they understand how their role and support changes in the event of a human-caused or criminal event?

> Warning: Integration only works when all parties benefit.
> "Send all of the DBH personnel home. They are creating more problems within the operation than they are solving!" This statement was made by incident command during shelter operations following a criminal event. DBH personnel provided support to the victims as they had been trained to do while incident command focused on the on-going criminal investigation, health, and safety within the shelter operation, and the pending relocation of the children and mothers. The confusion and additional stress could have easily been prevented if DBH personnel had integrated into EOC where mission goals and objectives were developed.

Situational Awareness

It is the role of DBH to maintain situational awareness and adjust DBH operations as needed. With that awareness, DBH personnel can develop a plan that meets the needs of the EOC, responders, and those directly impacted. Elements of these plans may include where DBH assets will be staged and deployed, type and levels of staffing needed, and strategies to support the EOC (Figs. 10.1–10.3). DBH professionals can support the EOC by reporting data about encounters with victims and survivors, tracking response expenses, and monitoring current or developing impact on response personnel and the impacted community.

Briefing

It is also the role of DBH professionals to ensure that the EOC leadership is briefed on completed tasks, current operations, and provider capacity. While in the EOC, DBH personnel may conduct a needs assessment to determine the impact on the behavioral health provider network. This may include answering, at minimum, five basic questions:

1. What is the impact on existing consumers of behavioral health services?
2. What is the impact on behavioral health staff?
3. What is the impact on behavioral health facilities?
    a. Are they damaged?
    b. If so, to what extent?
4. Are behavioral health provider organizations able to continue to provide services to existing clients? If not, what services are suspended and for how long?
5. If required and requested, can staff be diverted to assist in the provision of DBH services? If so, what types of staff might be available, where, and for how long?

Resource:		Behavioral Health Operations Team (BHOT)		
CATEGORY		Public Health and Medical (ESF #8)	KIND:	Team
MINIMUM CAPABILITIES:		TYPE I	TYPE II	TYPE III
COMPONENT	METRIC			
Overall Function	Primary Mission	Team will be able to assess the need for and **provide a limited range** of disaster behavioral health services. Team will support the local IC, and/or RHMOC by serving as a liaison between all entities capable of providing disaster behavioral health services. Experienced and trained to evaluate disaster behavioral health status and community needs.	Team will be able to assess the need for and **coordinate** the operations for the provision of disaster behavioral health services. Team will support the local IC, and/or RHMOC by serving as a liaison between all entities capable of providing disaster behavioral health services. Experienced and trained to evaluate disaster behavioral health status and community needs.	Team will be able to assess the need for and **provide a limited range** of disaster behavioral health services. Team will support the local IC, and/or RHMOC by serving as a liaison between all entities capable of providing disaster behavioral health services. Experienced and trained to evaluate disaster behavioral health status and community needs.
	Capabilities	1. Develop incident/population specific plans for community and first responder support 2. Team will evaluate impact on DSHS contracted behavioral health provider facilities and services. 3. Assessment and operational planning 4. Referrals to formal behavioral health services 5. Referrals to additional disaster response and recovery resources 6. Collect and report encounter data and costs related to DBH response activities	1. Develop incident/population specific plans for community and first responder support 2. Team will evaluate impact on DSHS contracted behavioral health provider facilities and services. 3. Assessment and operational planning 4. Referrals to formal behavioral health services 5. Referrals to additional disaster response and recovery resources 6. Collect and report encounter data and costs related to DBH response activities	1. Develop incident/population specific plans for community and first responder support 2. Team will evaluate impact on DSHS contracted behavioral health provider facilities and services. 3. Assessment and operational planning 4. Referrals to formal behavioral health services 5. Referrals to additional disaster response and recovery resources 6. Collect and report encounter data and costs related to DBH response activities
Mobilization	Operationally Capable	< 6 h < 2 h of arrival at the scene	< 6 h < 2 h of arrival at the scene	< 6 h < 2 h of arrival at the scene
Capacity	Number of Crew	Team of 4 comprised of: (1) BHAT Liaison (1) CISM Peer (1) Spiritual Care Liaison (1) Support Personnel	Team of 5 comprised of: (1) BHAT Liaison (1) CISM Peer (1) Spiritual Care Liaison (2) Support personnel	Team of 2 comprised of: (2) Behavioral Health Coordination Specialists
	Personnel Standard	Experience as part of a BHAT in large-scale disaster situation in home State. 1. IS-100, IS-200, IS-700.a, IS-800.b, ICS 300, Psychological First Aid, Criminal background check (ICS-400 is recommended, but not required) 2. Two years of professional experience, worked at least one disaster or emergency event 3. Any required licenses or certifications must be current from a recognized Texas behavioral health licensing board 4. CISM personnel must be trained in accordance with and adhere to the ICISF's CISM training model Pastoral care personnel will adhere to NVOAD points of consensus	Experience as part of a BHAT in large-scale disaster situation in home State. 1. IS-100, IS-200, IS-700.a, IS-800.b, ICS 300, Psychological First Aid, Criminal background check (ICS-400 is recommended, but not required) 2. Two years of professional experience, worked at least one disaster or emergency event 3. Any required licenses or certifications must be current from a recognized Texas behavioral health licensing board 4. CISM personnel must be trained in accordance with and adhere to the ICISF's CISM training model 5. Pastoral care personnel will adhere to NVOAD points of consensus	Experience as part of a BHAT in medium-scale disaster situation in home State. 1. IS-100, IS-200, IS-700.a, IS-800.b, ICS 300, Psychological First Aid, Criminal background check (ICS-400 is recommended, but not required) 2. Two years of professional experience, worked at least one disaster or emergency event 3. Any required licenses or certifications must be current from a recognized Texas behavioral health licensing board 4. CISM personnel must be trained in accordance with and adhere to the ICISF's CISM training model 5. Pastoral care personnel will adhere to NVOAD points of consensus

FIGURE 10.1 Behavioral health operations team (BHOT).

Resource:		Public Health and Medical (ESF #8)		Behavioral Health Operations Team (BHOT)		
CATEGORY				Kind:		Team
MINIMUM CAPABILITIES:						
COMPONENT	METRIC	Type I		Type II		Type III
Equipment	Computers	Laptop & Internet connectivity		Laptop & Internet connectivity		Laptop & Internet connectivity
	Phone	Cell phone		Cell phone		Cell Phone
	Transportation	Self-sufficient		Self-sufficient		Self-sufficient

	Supplies	1. Office of Management and Budget Data Gathering forms
		   a. Individual Encounter Form
		   b. Group Encounter Form
		   c. Weekly Tally Sheet
		2. Psychological educational materials
		3. DBHS Expense Tracking Form
				1. Office of Management and Budget Data Gathering forms
				   a. Individual Encounter Form
				   b. Group Encounter Form
				   c. Weekly Tally Sheet
				2. Psychological educational materials
				3. DBHS Expense Tracking Form
						1. Office of Management and Budget Data Gathering forms
						   a. Individual Encounter Form
						   b. Group Encounter Form
						   c. Weekly Tally Sheet
						2. Psychological educational materials
						3. DBHS Expense Tracking Form

Safety	PPE	Appropriate PPE as determined by the safety officer and/or the needs of the specific deployment				
	Immunizations	Recommended all personnel should have current influenza, tetanus/diphtheria, or as recommended by the TDMS Responder Safety & Health Workgroup.				
COMMENTS:		1. Team will be composed of individuals with the skills and abilities necessary to work with the affected population based on the S.T.A.R. 2. Team composition, management, membership and governance varies, but can include para-professionals, psychologist, psychiatrists, social workers, spiritual care provider, licensed professional counselors, substance abuse/chemical dependency specialists, and first responder peers. 3. Team will be deployed anywhere in the state within 6 h of notification for a 7 day deployment. This includes 2 days of travel and 5 days of direct service. 4. ICISF – International Critical Incident Stress Foundation. DBHS – Disaster Behavioral Health Services (DSHS), NVOAD – National Voluntary Organizations Active in Disasters				

FIGURE 10.1 (Continued).

CRITICAL INCIDENT STRESS MANAGEMENT (CISM) Team

RESOURCE: Public Health and Medical (ESF #8)   **KIND:** Team

COMPONENT	MINIMUM CAPABILITIES: METRIC	TYPE I	TYPE II	TYPE III
Overall Function	Primary Mission	Team will be able to conduct needs assessment, coordinate and provide CISM services to members of first responder agencies. Team is responsible for the prevention and mitigation of disabling stress among emergency responders in accordance with the standards of the International Critical Incident Stress Foundation (ICISF).	Team will be able to conduct needs assessment, coordinate and provide CISM services to members of first responder agencies. Team is responsible for the prevention and mitigation of disabling stress among emergency responders in accordance with the standards of the International Critical Incident Stress Foundation (ICISF).	Team will be able to provide CISM services to members of first responder agencies. Team is responsible for the prevention and mitigation of disabling stress among emergency responders in accordance with the standards of the International Critical Incident Stress Foundation (ICISF).
	Capability	The following is a list of CISM interventions this team can provide: • Crisis Management Briefings • Defusing • Individual Crisis Intervention • Group Debriefings	The following is a list of CISM interventions this team can provide: • Crisis Management Briefings • Defusing • Individual Crisis Intervention • Group Debriefings	The following is a list of CISM interventions this team can provide: • Crisis Management Briefings • Defusing • Individual Crisis Intervention • Group Debriefings
Mobilization	Operationally Capable	>24 h	>24 h	>24 h
		< 2 h of arrival on the scene	< 2 h of arrival on the scene	< 2 h of arrival on the scene
Capacity	Number of Crew	Team of 8-15 comprised of: • (2) Team Leads • Mental Health Professional(s) • First Responder Peer(s) • Chaplain(s)	Team of 5-8 comprised of: • (1) Team Lead • Mental Health Professional(s) • First Responder Peer(s) • Chaplain(s)	Team of 2-5 comprised of: • Mental Health Professional(s) • First Responder Peer(s) • Chaplain(s) *See Comment 3 Below
	Personnel Standard	Experience as part of CISM Team in large-scale disaster situations in home and other States. Has extensive experience in CISM administration and knowledge of ICISF standards. • IS-100, IS-200, IS-700.a, IS-800.b, ICS 300, Psychological First Aid, Criminal background check (ICS-400 is recommended, but not required) • Two years of professional experience, worked at least one disaster or emergency event • Mental health professionals with current licensure/certification from a recognized Texas behavioral health licensing board. • Trained in accordance with and adhere to the ICISF's CISM training model.	Experience as part of CISM Team in medium-scale disaster situations in home and other States. Has extensive experience in CISM administration and knowledge of ICISF standards. • IS-100, IS-200, IS-700.a, IS-800.b, ICS 300, Psychological First Aid, Criminal background check (ICS-400 is recommended, but not required) • Two years of professional experience, worked at least one disaster or emergency event • Mental health professionals with current licensure/certification from a recognized Texas behavioral health licensing board. • Trained in accordance with and adhere to the ICISF's CISM training model.	Experience as part of CISM Team in small-scale disaster situations in home State. • IS-100, IS-200, IS-700.a, Psychological First Aid. • Two years of professional experience, worked at least one disaster or emergency event • Mental health professionals with current licensure/certification from a recognized Texas behavioral health licensing board. • Trained in accordance with and adhere to the ICISF's CISM training model.

FIGURE 10.2 Critical incident stress management (CISM) team.

CRITICAL INCIDENT STRESS MANAGEMENT (CISM) Team

Resource:				
Minimum Capabilities:	Public Health and Medical (ESF #8)	**Kind:**		Team
Component	**Metric**	Type I	Type II	Type III
Equipment	Phone	Cell phone	Cell phone	Cell phone
	Transportation	Self-sufficient	Self-sufficient	Self-sufficient

	Supplies	1. Office of Management and Budget data gathering forms
		  a. Individual Encounter Form
		  b. Group Encounter Form
		  c. Weekly Tally Sheet
		2. Psychological educational materials
		3. DBHS Expense Tracking Form
			1. Office of Management and Budget data gathering forms
			  a. Individual Encounter Form
			  b. Group Encounter Form
			  c. Weekly Tally Sheet
			2. Psychological educational materials
			3. DBHS Expense Tracking Form
				1. Office of Management and Budget data gathering forms
				  a. Individual Encounter Form
				  b. Group Encounter Form
				  c. Weekly Tally Sheet
				2. Psychological educational materials
				3. DBHS Expense Tracking Form

Safety	PPE	Appropriate PPE as determined by the safety officer and/or the needs of the specific deployment		
	Immunizations	Recommended all personnel should have current influenza, tetanus/diphtheria, or as recommended by the TDMS Responder Safety & Health Workgroup.		
Comments:	1. Equipment and supplies will be determined by number of personnel deployed with team. 2. Willing to deploy anywhere in the state for a 2–4 day deployment. 3. Team Leads will be selected from the team. For Type III teams, the Team Lead may also serve in the role of mental health professional, peer or chaplain. 4. Number of team members based on size of incident and effect on emergency responders. 5. Team is responsible for the prevention and mitigation of disabling stress among emergency responders in accordance with the standards of the International Critical Incident Stress Foundation (ICISF). 6. Team composition, management, membership and governance varies, but can include psychologist, psychiatrists, social workers, and licensed professional counselors and first responder peers. 7. Source: International Critical Incident Stress Foundation			

FIGURE 10.2 (Continued).

RESOURCE:		BEHAVIORAL HEALTH ASSISTANCE TEAM (BHAT)			
CATEGORY:	Public Health and Medical (ESF#8)	KIND:		Team	
MINIMUM CAPABILITIES:		TYPE I	TYPE II	TYPE III	TYPE IV
COMPONENT	METRIC				
Overall Function	Primary Mission	A BHAT is deployed to support local community response and recovery. Through communication, BHAT will provide coordinated early psychological interventions that include but are not limited to, crisis counseling services, spiritual care, Critical Incident Stress Management and referrals to additional resources.	A BHAT is deployed to support local community response and recovery. Through communication, BHAT will provide coordinated early psychological interventions that include but are not limited to, crisis counseling services, spiritual care, Critical Incident Stress Management and referrals to additional resources.	A BHAT is deployed to support local community response and recovery. Through communication, BHAT will provide coordinated early psychological interventions that include but are not limited to, crisis counseling services, spiritual care, Critical Incident Stress Management and referrals to additional resources.	A BHAT is deployed to support local community response and recovery. Through communication, BHAT will provide coordinated early psychological interventions that include but are not limited to, crisis counseling services, spiritual care, Critical Incident Stress Management and referrals to additional resources.
		Personnel operate in teams of two and are experienced and trained to provide disaster behavioral health services including identifying current and future needs as well as potential gaps.	Personnel operate in teams of two and are experienced and trained to evaluate disaster behavioral health services including current and future needs as well as potential gaps.	Personnel operate in teams of two and are experienced and trained to evaluate disaster behavioral health services including current and future needs as well as potential gaps.	Personnel operate in teams of two and are experienced and trained to evaluate disaster behavioral health services including current and future needs as well as potential gaps.
		Assist with coordination and communication of field teams. Provide additional response workers to supplement local responses.	Assist with coordination and communication of field teams. Provide additional response workers to supplement local response.	Assist with coordination and communication of field teams. Provide additional response workers to supplement local response.	Assist with coordination and communication of field teams. Provide additional response workers to supplement local response.
	Capabilities	The following is a list of services team can provide: • Develop incident/population specific plans for community and first responder support • Early psychological intervention • Crisis counseling • Stress management • Referrals to formal behavioral health services • Referrals to additional disaster response and recovery resources • Collect and report encounter data and costs related to DBH response activities	The following is a list of services team can provide: • Develop incident/population specific plans for community and first responder support • Early psychological intervention • Crisis counseling • Stress management • Referrals to formal behavioral health services • Referrals to additional disaster response and recovery resources • Collect and report encounter data and costs related to DBH response activities	The following is a list of services team can provide: • Develop incident/population specific plans for community and first responder support • Early psychological intervention • Crisis counseling • Stress management • Referrals to formal behavioral health services • Referrals to additional disaster response and recovery resources • Collect and report encounter data and costs related to DBH response activities	The following is a list of services team can provide: • Develop incident/population specific plans for community and first responder support • Early psychological intervention • Crisis counseling • Stress management • Referrals to formal behavioral health services • Referrals to additional disaster response and recovery resources • Collect and report encounter data and costs related to DBH response activities

FIGURE 10.3 Behavioral health assistance team (BHAT).

Resource:	BEHAVIORAL HEALTH ASSISTANCE TEAM (BHAT)				
Category:	Public Health and Medical (ESF#8)	Kind:		Team	
Minimum Capabilities:	Metric	Type I	Type II	Type III	Type IV
Mobilization		>6 h	>8 h	>6 h	<6 h
Capacity	Operationally Capable	<2 h of arrival at the scene	<2 h of arrival at the scene	<2 h of arrival at the scene	<2 h of arrival at the scene
	Number of Crew	A team of 16 comprised of: (2) Team Leads (to maintain span of control) (7) BHAT personnel (7) Spiritual Care Providers	Team of 8 comprised of: (1) Team Lead (3) BHAT personnel (3) Spiritual Care Providers	Team of 5 comprised of: (1) Team Lead (2) BHAT personnel (2) Spiritual Care Providers	Team of 2 comprised of: (1) BHAT personnel (1) Spiritual Care Provider
	Personnel Standard	Specialists trained in a broad level of disaster behavioral health interventions and direct disaster experience. Team composition will be determined based on community demographics, needs of the impacted populations and incident type. Personnel work in teams of two	Specialists trained in a broad level of disaster behavioral health interventions and direct disaster experience. Team composition will be determined based on community demographics, needs of the impacted populations and incident type. Personnel work in teams of two	Specialists trained in a broad level of disaster behavioral health interventions and direct disaster experience. Team composition will be determined based on community demographics, needs of the impacted populations and incident type. Personnel work in teams of two	Specialists trained in a broad level of disaster behavioral health interventions and direct disaster experience. Team composition will be determined based on community demographics, needs of the impacted populations and incident type. Personnel work in teams of two
Equipment	Computers	None	None	None	None
	Phone	Cell Phone	Cell Phone	Cell Phone	Cell Phone
	Transportation	Self-sufficient	Self-sufficient	Self-sufficient	Self-sufficient
	Supplies	1. Office of Management and Budget Data Gathering forms a. Individual Encounter Form b. Group Encounter Form c. Weekly Tally Sheet 2. Psycho-educational materials 3. DBHS Expense Tracking Form	1. Office of Management and Budget Data Gathering forms a. Individual Encounter Form b. Group Encounter Form c. Weekly Tally Sheet 2. Psycho-educational materials 3. DBHS Expense Tracking Form	1. Office of Management and Budget Data Gathering forms a. Individual Encounter Form b. Group Encounter Form c. Weekly Tally Sheet 2. Psycho-educational materials 3. DBHS Expense Tracking Form	1. Office of Management and Budget Data Gathering forms a. Individual Encounter Form b. Group Encounter Form c. Weekly Tally Sheet 2. Psycho-educational materials 3. DBHS Expense Tracking Form
Safety	PPE	Appropriate PPE as determined by the safety officer and/or the needs of the specific deployment			
	Immunizations:	Recommended all personnel should have current influenza, tetanus, tetanus/diphtheria, . or as recommended by the TDMS Responder Safety & Health Workgroup			
Comments	1. Team will be composed of individuals with the skills and abilities necessary to work with the affected population based on the STAR				
	2. Team composition, management, membership and governance varies, but can include para-professionals, psychologist, psychiatrists, social workers, spiritual care provider, licensed professional counselors, substance abuse/chemical dependency specialist, and first responder peers.				
	3. Team will be deployed anywhere in the state within 6 h of notification for a 7 day deployment. This includes 2 days of travel and 5 days of direct service.				
	4. Number of team members based on size of incident, effect on the population and availability of local/regional resources.				
	5. Type II and Type III Team Leads are leading the activities of 3 or 2 duos (respectively) comprised of a BHAT and a Spiritual Care person.				
	6. CISM Type II Team Lead will support BHAT CISM Type teams when deployed				
	7. ICISF – International Critical Incident Stress Foundation, DBHS – Disaster Behavioral Health Services (DSHS), NVOAD – National Voluntary Organizations Active in Disasters				

FIGURE 10.3 (Continued).

A quick and ongoing assessment asking questions such as these provides essential information to EOC leadership regarding the local organizations' capacity to respond and provide services to vulnerable populations. Note that this information is essential for potential state and federal disaster declarations and will help to determine if and when external resources need to be requested.

This type of information needs to be updated regularly and be provided to emergency management (EM) leadership so they have a current snapshot of available resources and resource gaps. In this regard, a benefit of having DBH personnel within the EOC is they will have immediate access to time-sensitive information and data that is essential to assess the need for and successful development of the Federal Emergency Management Agency (FEMA) funded Crisis Counseling Program (CCP) grant (Substance Abuse and Mental Health Services Administration, 2009).

The CCP is a program funded by FEMA and administered by the Substance Abuse and Mental Health Services Administration (SAMHSA). States are eligible to apply for the immediate services program portion of the CCP in a federal disaster declaration that includes Individual Assistance. It is due 14 calendar days from the date of federal disaster declaration. DBH personnel working within the EOC should provide periodic briefings on the need for the CCP, and, if pursued, the areas served, number of teams, and budget. In addition, the EOC allows for direct access to disaster-specific data gathered by other groups working in the EOC that is often required by the application and can hold regular conference calls with state and federal partners to maintain situational awareness and receive technical assistance for the timely submission of the state's application. There are numerous advantages to having DBH integrated within the EOC, but there are challenges in making it happen.

Assure Provision of Behavioral Health Services

A key role of DBH personnel is to identify and coordinate services to vulnerable populations who immediately require, or develop a need for DBH services, and therefore mitigating the potential for life-threatening health emergencies. To accomplish this, DBH personnel must have a presence in the emergency management structure, as well as linkages with key community partners.

Case example of assuring behavioral health services:
   Following hurricane Katrina, personnel in the Joint Field Office began receiving daily calls from an evacuee staying in a local hotel. She stated that she was about to have a baby. Repeatedly, first responders were dispatched to the scene.

(Continued)

> **(Continued)**
>
> After multiple calls, the State Coordinating Officer stopped by the crisis counseling and DBH office and jokingly talked about this lady who was about to have a baby. He said, "This lady is not going to have a baby anytime soon, or ever. She's 63 years old. She is becoming such a nuisance that the hotel manager is about to kick her out and we do not have any other place to send her." Immediately, personnel from the mobile crisis outreach team from the local mental health authority were dispatched. The lady was admitted to the local clinic, and within three days was stabilized and returned to her hotel. The following week, the lady made baked goods for the first responders out of appreciation for their kindness. It turned out the lady was being treated for paranoid schizophrenia and had been off her medications since being evacuated from Louisiana.

CHALLENGES IN INTEGRATING WITHIN THE EOC

EM and DBH personnel have encountered challenges as a result of emergency response being formalized with the adoption of the incident command system (ICS) (FEMA, n.d.). Prior to the formal adoption of ICS, state-level DBH programs were often activated as a key stakeholder within their state's EOC. This allowed unparalleled access to information, enabled DBH professionals to be involved with decision making processes and created and maintained professional working relationships with partners within the EOC. With implementation of ICS, the role of DBH within the EOC has significantly changed.

In ICS, DBH is typically placed in the Operations Section under public health (PH) as a component of Emergency Support Function #8 Health & Medical (FEMA, 2008). In the EOC, DBH personnel are replaced with a PH representative, which for some programs, created a loss of coordination, lack of information sharing, and DBH resources being left out of planning, response, and recovery operations. Within this structure, it is important for PH professionals to understand the services and resources available through DBH providers and that DBH should not be limited to only the Operations Section. Rather, DBH has a roll in all aspects and areas of the ICS structure. To encourage this, partnerships through training, drills, and exercises should be established and maintained. DBH providers need to be trained in and complete FEMA sponsored ICS training programs. At minimum, these should include ICS, 100, 200, 300, 700, and 800 (FEMA, n.d.) (Figs. 10.4 and 10.5). This will establish a clearer understanding of the disaster response structure and how the system operates.

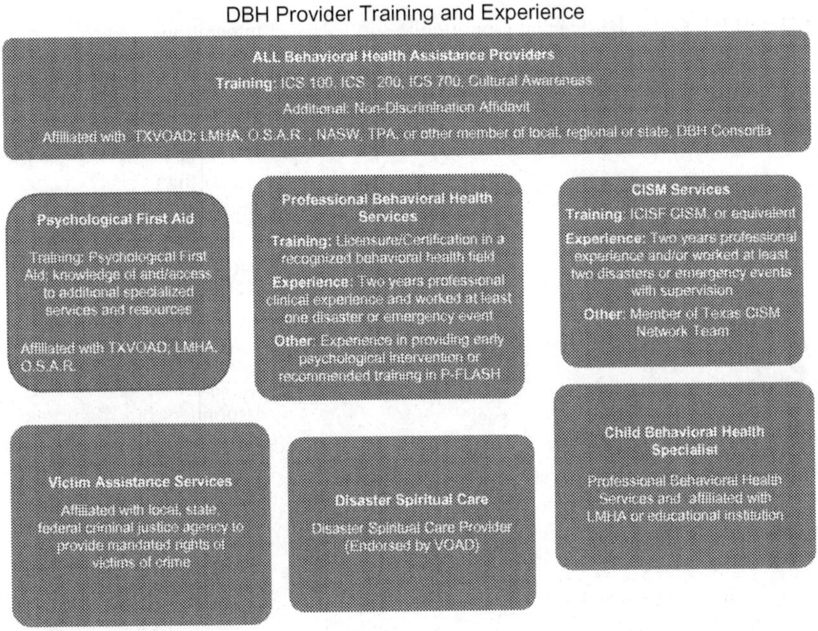

FIGURE 10.4 DBH provider training and experience. *Source: Texas Department of State Health Services' Disaster Behavioral Health Consortium and the Texas Disaster Medical Services (2015).*

Disaster Behavioral Health Qualifications Checklist

ALL DBH RESPONDERS/SEVICE PROVIDERS	Training Source	✓	Agency	Date Completed
ICS-100 - Introduction to the Incident Command System (see note 1 & 2)	http://training.fema.gov/is/nims.asp			
ICS-200 - ICS for Single Resources and Initial Action Incidents (see note 1&	http://training.fema.gov/is/nims.asp			
ICS-700.a - National Incident Management System (NIMS), An Introduction (see note 2)	http://training.fema.gov/is/nims.asp			
Affiliation with Local, Regional or State DBH Consortium (e.g. LMHA,VOAD, OSAR, State CISM Network, Criminal Justice, Emergency Management, Public Health, NASW, TPA, etc.)				
Psychological First Aid (see note 2)	http://training.fema.gov/is/nims.asp			
	http://pfa.naccho.org/pfa/pfa_start.html			
	http://learn.nctsn.org			
	http://learn.nctsn.org/course/category.php?id=11			
	http://www.ptsd.va.gov/professional/manuals/psych-first-aid.asp			
	www.redcross.org			
Criminal background check through your approving organization				
One year of professional experience as a behavioral health provider and/or experience volunteering, under supervision at a disaster or emergency event.				
Non-Discrimination Affidavit				
1 There are multiple options for the IS-100 and IS-200. IS-100HCB and IS-200HCa are for Health care workers. There are IS-100 options for Law Enforcement and schools as well 2 Online ICS and PFA courses are available free of charge				
SPECIALIZED DBH RESPONSE/SERVICE PROVIDERS		✓	Agency	Date Completed
Texas CISM Network Team Members: Individual and Group CISM	www.ICISF.org			
Spiritual Care Providers: NVOAD Points of Consensus	http://1www.nvoad.org/library/cat_view/9-points-of-consensus			
Victim Assistance Services: Affiliated with local, state, federal criminal justice agency to provide mandated rights of victims of crime				
PROFESSIONAL BEHAVIORAL HEALTH SERVICES		✓	Agency	Date Completed
Current licensure/certification from a recognized Texas behavioral health licensing board.				
Two years professional clincial experience and worked at least one disaster or emergency event				
Principles of Early Crisis Intervention. Recommended training in Practical Front Line Assistance & Support for Healing (P-FLASH)				
Additional Training Suggestions		✓	Agency	Date Completed
IS-808 - Emergency Support Function (ESF) #8 - Public Health and Medical Services (Currently under revision, but still available)	http://training.fema.gov/IS/crslist.aspx?page=1			
Think Cultural Health	https://www.thinkculturalhealth.hhs.gov/content/continuinged.asp			
Cultural Competency Curriculum for Disaster Preparedness and Crisis Response	https://cccdpcr.thinkculturalhealth.hhs.gov/			
National Organization for Victims Assistance (NOVA) "Basic Training" Providers	http://www.trynova.org/help-crime-victim/nacp/nacp-pre-approved-training/			
Functional and Access Needs - Resources (recommended to be familiar with resources available.)	https://www.preparingtexas.org/preparedness.aspx?page=32137bc8-eed7-42bb-ad7e-2765fd8abdb9			
Centers for Disease Control - Coping with Disaster - Resources	http://www.bt.cdc.gov/mentalhealth/index.asp			

FIGURE 10.5 Disaster behavioral health qualifications checklist. *Source: Texas Department of State Health Services' Disaster Behavioral Health Consortium and the Texas Disaster Medical Services (2015).*

DEFINING TERMS AND MUTUAL EDUCATION

Believe it or not, DBH, PH, and EM personnel do not always speak the same language. Each profession has their own terminology and acronyms that, if not understood, create confusion and frustration. Take the term "debriefing" for example. When "debriefing" is used within the EM setting, it means to gather or report information about an operation. However, to a DBH provider, "debriefing" can mean a phase-specific, small-group, supportive crisis intervention process. Note that in this case, the same word has two significantly different meanings and applications.

It is equally important for DBH personnel to inform and educate both PH and EM professionals regarding the populations, services, and programs behavioral health (BH) providers are responsible for on a daily basis, and how the scope of BH services changes during disaster response. These two professions, and many others, are likely to not have real time knowledge of the nature, scope, and challenges within BH systems. For example, many will not be aware that in many (if not most) locations, BH provider organizations are already operating a full capacity and have waiting lists, thus limiting their ability to redirect resources in times of emergency.

A key strategy to optimize mutual education is to integrate DBH personnel into local-, regional-, and state-level drills and exercises. In this way, DBH personnel learn the EOC structure, protocols, and language. EOC leaders and other participants can become aware of the multiple significant roles BH can play within the EOC structure. These types of experiences help BH specialists optimize their contributions through establishing strong working and trusted relationships with those in PH, EM, and other fields. In addition, it is an opportunity to illustrate the positive impact of DBH services and how including DBH providers can often make the emergency manager's job much easier. These are opportunities for all groups to work together in a manner that is educational while building trusted working relationships.

ESTABLISHING THE PARTNERSHIP

How should DBH groups get into the mix? One way is through strategic networking and coalition building.

This can be accomplished by taking a look at local, regional, and state partners and organizations who are responsible for emergency management planning, response, and recovery, such as offices of emergency management, public safety, law enforcement, and PH departments. Next, identify groups that may have a BH health role during response and/or recovery operations. For instance, consider voluntary organizations active in disaster, crime victim's compensation programs, first responder peer support programs, providers of formal behavioral health services (which include mental health and substance use treatment providers), health and human services departments, and universities. Through these relationships, groups can learn how the emergency management system works, who the key players are, and how the response framework operates. Furthermore, emergency managers and PH will learn more about the capabilities and limitations of each BH responder group as well as potential gaps in planning that may need to be addressed. Also, these linkages create credibility for BH responders within the EM and PH areas. As they will be exposed to opportunities to participate in training, exercises, and engage in dialog in nondisaster response situations, these linkages could provide as a proving ground that would ultimately ensure that BH providers are invited into an EOC.

REFERENCES

Federal Emergency Management Agency (FEMA). (2008). *Emergency support function #8—Public health and medical services annex*. Retrieved from http://www.fema.gov/media-library-data/20130726-1825-25045-8027/emergency_support_function_8_public_health_medical_services_annex_2008.pdf.

Federal Emergency Management Agency (FEMA). (n.d.) ICS Resource Center. Retrieved June 10, 2016, from http://training.fema.gov/emiweb/is/icsresource/index.htm.

Federal Emergency Management Agency (FEMA). (n.d.). Training Program. Retrieved June 10, 2016, from http://training.fema.gov/emiweb/is/icsresource/trainingmaterials.htm.

Substance Abuse and Mental Health Services Administration (SAMHSA). (2009). Crisis Counseling Assistance and Training Program (CCP).

Making Integration Work

Brian W. Flynn and Ronald Sherman

Integrating disaster behavioral health (DBH) into the context of emergency operations centers (EOC) and emergency communication centers (ECC) is not an easy task, as documented by the authors of this chapter. The organizations housed in these facilities are often highly structured and may not easily adapt to having new players at the table. At the same time, they are venues in which many important positive outcomes can take place that foster the work of both EM and DBH.

As consistently noted throughout this book, the task of mutually beneficial integration is best accomplished when based upon sound understanding of and respect for each profession by the other. Consistently, the authors of this book have stressed the benefit of establishing these relationships in the preparedness phase. In the case of EOC/ECC, this is especially true because of the formalized structure and environment.

Thus, the first task becomes mutual education in at least three areas:

- Mutual education/understanding of the roles played and contributions made by each profession.
- Mutual education/understanding of the contributions each can make to support the goals of the other.
- Exploration of the structural and resource options available for integration in these settings including identification of obstacles.

There are a number of strategies that can be implemented, as identified by the chapter authors, including:

- Identification of and linking with local, regional, and state partners and organizations who are responsible for emergency management planning, response, and recovery.
- Identification and linking with groups (including volunteer groups and faith-based organizations) that have behavioral health (BH) roles and resources during response and/or recovery phases.
- Implementation of opportunities to participate in on-site or on-line training (for both EM and DBH), as well as drills and exercises.
- Creation of strategies to monitor, evaluate, and modify the nature of integration within these settings based on experience in drills, exercises, and actual disaster situations
- Jointly efforts to develop a resource guide for DBH services in the local/regional area.

Chapter 11

Risk and Crisis Communications

Brian W. Flynn[1] and John P. Philbin[2]
[1]Uniformed Services University of the Health Sciences, Bethesda, MD, United States,
[2]Crisis1, LLC, Reston, VA, United States

Through a Disaster Behavioral Health Lens

Brian W. Flynn

In a crisis, effective communication *is* a behavioral health (BH) intervention, regardless of its content. Effective communication among leaders of all types, especially emergency management (EM) leaders, provides instruction and direction to promote desired behaviors, reduces undesired behaviors, and promotes a sense of confidence, understanding, as well as self- and collective-efficacy. This can diminish psychological arousal, which otherwise could have resulted in distressing and counterproductive cognitive, emotional, social, and behavioral consequences. Unfortunately, historically, BH professionals have had limited collaboration with emergency managers in the communications arena, especially on-site in the immediate response phase.

There are a number of enhanced and expanded ways in which BH professionals can assist emergency managers. In order for this integration to occur around communication issues, it is essential to attend to several helpful elements, including:

- Mutual understanding of roles and skills
- Mutual respect and trust
- Increasing understanding of victim priorities
- Developing anticipatory guidance
- Fostering communication with victims
- Assisting in crafting messages
- Monitoring and managing stress of EM personnel

MUTUAL UNDERSTANDING OF ROLES AND SKILLS

BH professionals can benefit from understanding not only the occupational roles and professional culture of emergency mangers but the types of communications challenges they face. These include communicating with an impacted and often highly emotional public, political forces within their own organization and outside, and a workforce that represents many organizations, work cultures, and disparate responsibilities. Increasingly, it will be necessary for BH professionals who have specialized training in communicating in effectively high stress situations to work with emergency mangers who also have specialized training in communicating effectively in high stress situations.

BH professionals may need to make EM aware of the specialized knowledge and skills the field that BH can offer that may not typically come to mind. BH professionals are often knowledgeable about how people process information and communicate in high stress situations, how to observe and assess the psychological state of individuals and groups, how to manage escalating negative interactions, and how one can express positions and opinions in ways that facilitate goals.

MUTUAL RESPECT AND TRUST

There may be some level of respect granted among parties as a result of acknowledgement of the prestige of certain academic degrees and organizational positions. However, this only goes so far unless there is both an organizational and professional recognized value of all involved. Even in cases where this integration is valued and well-documented, there is no substitute for highly personalized relationships that promote trust and confidence. People trust people, not organizations (Flynn, Bushnell, & Lurie, in press).

The ways emergency managers and BH professionals can integrate their efforts are limited only by their imaginations and shortcomings in establishing understanding, respect, and trust. Following are examples of how BH professionals might help integration in the communications areas.

WORKING WITH EMERGENCY MANAGERS TO INCREASE UNDERSTANDING OF VICTIM PRIORITIES

Anyone who has been involved in disaster preparedness, response, and recovery is aware that victims and survivors have different priorities, needs, and receptivity depending on many factors, especially the phase of the event. Emergency managers are well aware of this and historically have developed and sequenced their activities and priorities accordingly. Emergency managers may be less aware of some of the pressing (and less pressing) psychosocial priorities depending on event stage. For example, in an unexpected and life threatening event, people are likely to be concerned exclusively about their own safety and status and well-being of those they love. At this point, most other information provided to them may be ignored. Later, once personal threat is past, they may be more interested in what caused the event and how it could have been prevented or better handled. In later stages, people may be interested in recovery strategies, litigation, the cost of recovery, etc. BH professionals experienced in understanding the needs and priorities of victim and survivors, as a function of event phase, can be very helpful to emergency managers as they craft messages and prioritize communication. Following is an example.

> Not long after 9/11, two colleagues and I were asked by a prominent federal department to advise on communication strategies following a very wide-spread, no-notice event. They wanted advice on what should be said in a frightening and widely experienced event occurred and when somebody in authority needed to say *something* immediately and *before* the established mechanisms for existing government plans for organized and integrated communications had enough time to come online. The event that prompted their concern was a massive electrical power failure that covered a large portion of the Northeast Unites States. The President spoke quickly in an attempt to reduce fears. This address was widely perceived as less than optimal and this department wanted to make sure that future such communications were improved.
>
> We provided a number of suggestions, including a reminder that in situations people see as serious threats to life and health, they first want to know only three things:
> 1. Am I OK?
> 2. What about my loved ones and what is their status (alive, injured, or dead)?
> 3. What should I do?
>
> At this stage, nothing beyond that (with the exception of expressions of compassion, commitment, and optimism (Covello, 2011) are likely to be heard. Other important issues, like cause and prevention, are important but can come later.
>
> BH professionals skilled in understanding what people in crises want to know and when can be significant assets to emergency managers at every stage of disasters and emergencies.

WORKING WITH EMERGENCY MANAGERS TO DEVELOP ANTICIPATORY GUIDANCE

It has been long understood in both the BH and EM professions that being able to anticipate and prepare for challenges that might occur helps assure an appropriate and/or improved response and reaction when problems do occur. This is a core foundation of emergency and disaster preparedness. In the disaster behavioral health (DBH) community, this is usually referred to as anticipatory guidance. That is, helping people anticipate what might occur, think about it before hand, and therefore be better prepared deal with it when it does occur. A great deal of effort has been expended in disaster BH in developing the fact sheets, tips, advice documents. They are easily accessible, most many are in the public domain so they can be adapted and utilized at no cost. Examples include the American Red Cross (American Red Cross, 2016) Substance Abuse and Mental Health Services Association (SAMHSA, 2015), and the Center for the Study of Traumatic Stress (2016), to name a few. These resources cover a wide range of topics such as issues for children, the frail elderly, formal and informal leaders, health care personnel, responders and disaster workers and many more subject areas.

From time to time, BH professionals have assisted emergency managers in preparing fact sheets for the general and targeted populations. On occasion, during a response, EM has called upon BH professionals to review and make suggestions regarding press releases and public appearances of emergency managers. When these interactions have taken place, it has been largely on an ad hoc basis, not routine, and a systematized role. Integrating emergency managers, especially public communications efforts, and disaster BH professionals in all event phases will benefit facilitate and enhance the performance all involved.

Early in my disaster BH work, I began going to disaster sites as part of the Federal Emergency Management Agency response team. While there, I became acutely aware of the stresses experienced by Federal Emergency Management Agency workers both while they are in the field and when they return home after their disaster work. As a result, at Federal Emergency Management Agency's encouragement, I prepared two brochures. The first was about managing stress while doing disaster work (distributed as part of worker in-processing) and the second, *Returning Home after a Disaster*, was distributed as workers out-processed. The latter focused heavily on reintegration into family life. Both were generally well-received. Subsequent (and in my view, improved) guidance has been developed and is in wide usage (American Red Cross, 2016; Center for the Study of Traumatic Stress, 2016; SAMHSA, 2015).

However, while working at a disaster site after the distribution of these brochures had begun, I was approached by a Federal Emergency Management Agency worker I have met several times before. She said, "Brian, I'm mad at you." Surprised, I asked why. She said, "You left me out of your brochure." I am sure I looked surprised and puzzled. She continued, "You didn't talk about people like me who don't go home to families. I live alone and have no close family. For me, the people I work with here are my family so I leave my family when I go home." I was stunned by her candor and guilty about my insensitive omission. I can assure you that the brochure was quickly modified.

The lesson I learned from these events is that providing anticipatory guidance is a valuable activity. But, if it is to be done well, development must be inclusive of the complete intended audience and widely reviewed before distribution.

FOSTERING EMERGENCY MANAGERS' COMMUNICATION WITH VICTIMS

Emergency manages often find themselves in a position of communicating with disaster victims. These interactions are seldom easy and victims are often highly emotional, angry, and demanding. A critical role the BH professionals can play in disasters is to serve as a consultant to leaders of all types, especially emergency managers. This role does not emerge casually or at random. It requires time, trust, confidence, and flexibility. The following, is an example in one disaster.

> When I was working in the field during disasters, I was typically assigned to what is now called the Joint Field Office (JFO), formerly called the Disaster Field Office (DFO), where federal and state officials managed the disaster. Although working as an Officer in the U.S. Public Health Service (USPHS), an element of the Department of Health and Human Services (DHHS), I was part of the Federal EM Team when on-site.
>
> The Federal Coordinating Officer (FCO) is the lead Federal Emergency Management Agency official in the field and the President's representative. Although my day-to-day work was usually outside the DFO helping to assess and organize mental health services for victims and survivors, I always tried to spend some time, usually toward the end of the day, at the DFO. It was always important to be known, visible, and perceived as part of the team.
>
> The FCO is always overworked, juggling multiple demands, and surrounded by people. I learned early that sometimes, at the end of the day, if I just wandered by the FCO's office, I might catch him/her at a relatively quiet and sometimes reflective time. Unlike most others, I was not asking for something, reporting problems, or criticizing. Instead, I was just stopping by. I was building a relationship. I was building trust. We were getting to know each other.
>
> I recall one such evening when the FCO shared that he had to attend a community meeting that evening and was not looking forward to it. The community had sustained significant tornado damage and residents were desperate and angry. They wanted the head of anybody in authority. I asked him if he would like some company and he quickly took me up on the offer.
>
> On the ride there, we talked about many things such as his expectations for the evening and his concerns. Even in the absence of solutions or answers to their passionate and intense questioning and accusations, I had a chance to share with him notions of displaced anger and catharsis. I talked about how active and respectful listening and careful responses could be helpful and reassuring. We had a chance to practice crisis communication challenges such as potential audience questions and outbursts, as well as his potential reactions and responses.
>
> *(Continued)*

> **(Continued)**
>
> We attended the meeting together and it was every bit as bad as he had anticipated. Yet, he had a better understanding of why the crowd behaved as it did. His reactions and replies were able to acknowledge their anger and frustration, while putting appropriate limits on the very hostile treatment he was receiving from members of the audience from time to time. He was able to take an admittedly small step forward in promoting of hope and optimism, which, as noted earlier, is an important element of early intervention with traumatized individuals and groups (Hobfoll et al., 2007).
>
> On the retuning ride, we are able to discuss the experience, and I had an opportunity to reinforce the positive aspects of his very difficult experience. He expressed much gratitude and indicated that he wished he could have a "mental health guy" with him all the time.
>
> This and similar involvements could not have taken place in the absence of an in-person BH presence, at least the beginnings of a relationship built on mutual trust and respect, and a resulting willingness on both people to try something a little different.

ASSISTING IN CRAFTING MESSAGES

One of the central tasks of emergency mangers is the development and dissemination of many types of messages. Effective communications for accomplishing physical management of events, as well as psychosocial and psychoeducational, goals is important. Senior emergency managers appropriately rely on the parts of their organizations that are responsible for public information aspects of response and recovery. Typically, crafting massages does not include BH professionals. However, they can add additional and important contributions in formulating and delivering messages to intended audiences.

There are many factors that contribute to the extent to which the public receives, understands, and acts appropriately upon information. All of these factors involve psychosocial elements, including:

- Who are the trusted sources of information?
- How is information received, understood, and retained?
- Is the information addressing the most pressing concerns of the recipients?
- Are cognitive, emotional, and behavioral elements reflected in message?

This list is not comprehensive. The point to be made is that there are elements of all messaging and information dissemination that can benefit from the involvement of BH professionals who are knowledge about effective communication in high stress situations.

Integration of BH professionals and those who are tasked with public information and risk communication in both the preparedness and response phases can enhance the contributions of both fields. This type of integration can be valuable in situations outside the usual types of emergencies and disasters, such as emerging and polarized public health challenges. A case example follows.

> I was asked by communication leaders at the Centers for Disease Control and Prevention (CDC) to join a multiday process to develop messages for health leaders to deliver regarding childhood vaccine hesitancy. I initially declined due to my complete lack of expertise in this topic area. The Disease Control and Prevention organizers said that they were aware of my disaster mental health work and felt that I had something to contribute based on my understanding of how people process information in high stress situations and how to work through conflict. I reluctantly agreed.
>
> CDC had determined the questions most likely to be asked and concerns raised on the topic. The goal of the gathering was to craft messages and message sequences for officials to use in response to inquiries. After an initial presentation, other experts from areas such as risk communication and infectious disease and I worked as consultants and advisors to public information professionals to craft effective, evidence-based responses to questions about and information concerning childhood vaccines. The fundamental process was the same as work I was accustomed to in a disaster setting: bring your expertise, partner with those with other expertise, keep the needs of the target audience in mind, and jointly produce products ready to be used.
>
> Apparently, the involvement of someone with a seemingly unrelated background, like me, proved quite helpful or at least interesting. I was asked to make similar presentations and participate in related processes for years following.

ASSISTING IN MONITORING AND MANAGING STRESS OF EMERGENCY MANAGEMENT PERSONNEL

In disasters, stress-related problems are not only experienced by primary victims, but also by their families and their communities. Extreme stress is also visited upon those who respond, including both first responders and those who manage response and recovery.

BH professionals in the field should attend to the psychosocial needs of all impacted. Stress will manifest itself in more ways than can be anticipated. Being present, observant, and an integrated part of the EM structure is a prerequisite for effective responder stress management of all types. Following is a case example specifically related to public information and communications workers.

> When my staff or I were invited by Federal Emergency Management Agency to be in the field with them early in a disaster response, we always tried to put our office next to the Public Affairs Office now typically called the Joint Information Center (JIC). This was very intentional. With our offices in close proximity, we got to know the communications staff and they got to know us.
>
> Relationships were formed that typically resulted in my team being asked to have input on the creation of announcements and press releases. We were then able to provide useful consultation on the psychological impact of messages and their impact on various audiences. It was clearly a win-win situation.
>
> When working next to the Public Affairs office at one disaster, I noticed that a worker there was methodically cutting out all information from newspapers, she could find on the numerous victims killed. Initially, this seemed like a good strategy to assure that emergency managers could relate more personally to those most impacted. However, I soon noticed that this worker was doing this day after day and apparently not able to perform any of the other work she had been assigned. She had become fixated on the deaths and her actions had become compulsive and nonproductive in nature. After informally talking with her and her supervisor, she was able to be referred to a mental health professional to services.
>
> Both the ability to identify this problematic behavioral and facilitate referral was greatly enhanced by both physical proximity on-site as well as fostering collaborative relationships.

CONCLUSION

Disaster BH efforts and comprehensive communications before, during, and following disasters are more closely connected than most, even seasoned workers in both fields, often appreciate. Without appreciating, understanding, and operationalizing these interdependent specialties, accomplishing the goals of each effort will be compromised. On the other hand, long-term, creative, and dynamic integration has the potential of creating a synergistic effect.

It is easy to see the differences in professional cultures, skills, and tools. Understanding and operationalizing the *shared* values and goals is a relatively underdeveloped opportunity for both

professions. Hopefully, this chapter, in the context of communication has provided an orientation to the two fields and built a compelling rational for their integration. The Making Integration Work section will explore specific ways to establish and sustain integration.

REFERENCES

American Red Cross. (2016). American Red Cross. Retrieved from http://www.redcross.org/get-help. Accessed on June 8, 2016.

Center for the Study of Traumatic Stress. (2016). Fact Sheets. Retrieved from http://www.cstsonline.org/resources/fact-sheet-search. Accessed on June 8, 2016.

Covello, V. T. (2011). Risk communication, radiation, and radiological emergencies: Strategies, tools, and techniques. *Health Physics, 101*(5), 511−530.

Flynn, B.W., Bushnell, P., & Lurie, N. (in press). Leadership issues and disasters. In R. J. Ursano, C. S. Fullerton, B. Raphael, & L. Weisaeth (Eds.), *Textbook of disaster psychiatry* (2nd ed.).

Hobfoll, S. E., Watson, P., Bell, C. C., Bryant, R. A., Brymer, M. J., Friedman, M. J., ... Ursano, R. J. (2007). Five essential elements of immediate and mid−term mass trauma intervention: Empirical evidence. *Psychiatry, 70*(4), 283−315.

SAMHSA. (2015). *Publications and Resources on Disaster Preparedness, Response, and Recovery*. Retrieved from http://www.samhsa.gov/disaster-preparedness/publications-resources. Accessed on June 8, 2016.

Through an Emergency Management Lens

John P. Philbin

Communicating with stakeholders during and following disasters is particularly complex. Victims seek relief. Friends and family members seek information about their loved ones. EM personnel seek resources and coordination to aid those in distress. Public officials seek answers, and, depending on the nature of the event, myriad stakeholders seek to blame. Combine these competing demands with technologies that allow us to observe disasters in real time, or the inability to communicate effectively during incidents, and the communication challenge would seem overwhelming.

To succeed, how must EM personnel be prepared to respond to these divergent communication challenges and how can BH professionals assist first responders?

TODAY'S COMMUNICATION ENVIRONMENT

The environment in which first responders work today has been complicated because of the changes in the news business and the proliferation of technology. Increasingly, news appears to no longer be in the business of serving the public interest; rather, it appears primarily concerned with building audiences and enhancing ratings. For any leader—including those coordinating emergency responses—relying on the traditional means of getting information to those who matter most is to put your organization's reputation, as well as one's own, in the hands of those who are motivated not necessarily by the truth or facts, but by any frame that draws more viewers and readers. Since information demands rise exponentially, especially during high interest events, anyone associated with the response will likely be asked to comment.

Getting accurate information out quickly during a disaster is a challenge. Getting accurate information out quickly by authoritative representatives who have sufficiently coordinated the desired "talking points" is nearly impossible. Bureaucracies seldom behave efficiently—and efficiency for information demand is what disasters require.

From a leadership perspective, communicating during a crisis is on one hand very simple. It involves communicating:

1. What happened?
2. When did you find out?
3. What did you do?
4. What are going to do to ensure it does not happen again?

On the other hand, it is very challenging, because seldom are things as they appear initially. In the military, this is known as the "fog of war." Early information usually proves inaccurate, and inaccuracy will create enormous problems for organizations. Institutions, and those who lead them, will be held to a higher standard than those who are reporting information if it is wrong. *None of these important questions addresses the primary concerns of disaster victims.*

Information, like water, seems to follow the path of least resistance. But why? The concepts of confirmatory bias and cognitive dissonance indicates that once we have an opinion about someone or something, it is rather difficult to change our minds and, during stress, this would seem more so. Confirmatory bias is part of the human condition. This should lead to wondering, "How do we go about communicating effectively during crises in the wake of growing skepticism, suspicion, and erosion of public trust?"

Among many things, communication depends largely on ethical, competent leadership and our ability to influence the decision-making and behavior of our organizations. No amount of positive communication will compensate for poor operational performance and decision-making that neglects the legitimacy and concerns of our stakeholders. Integrity is the only currency we have in the business of communication, which is why trust is so important to what we do. In fact, the Department of Health and Human Services (DHHS) points out the importance of crisis communication attributes in its 2012 *Crisis Emergency Risk Communication Guide*: "Be First. Be Right. Be Credible" (Reynolds & Seeger, 2012).

No amount of positive communication will compensate for poor operational performance and decision-making that neglects the legitimacy and concerns of our stakeholders. Integrity is the only currency we have in the business of communication, which is why trust is so important to what we do.

One of the central challenges facing those who perform in the public eye is understanding how to operate in a world where our principles urge us to enhance trust and credibility versus others that historically facilitate the free flow of information and appear to conform to a much different set of rules. Especially in the early phases of a disaster, emergency managers and their communications leaders find themselves in a world where speed is more important than fact; audiences are more important than public interest; and those who used to control these processes via the mass media are rapidly becoming irrelevant.

The challenge in today's communication environment may best be illustrated by Gerald Baron's depiction of what used to happen following an event and how things occur today (Baron, 2006) (Figure 11.1).

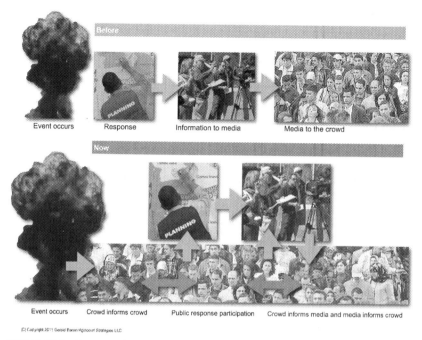

FIGURE 11.1 Representation of an event that will become or is known and has the potential to require a response and/or illicit a reaction by stakeholders or the public.

As a former Director of the Office of External Affairs for the Federal Emergency Management Agency, I understood that the reality was and remains that we have little or no control over most, if not all, forces highlighted in the preceding graphic. The space between public and private is increasingly blurred. For example, organizational decision-making and employee behaviors can become public with the click of a smart phone, a twitter feed, or a blog rant. There are no "silver bullets" in communication that magically answer all concerns—especially during disasters.

So, how might BH professionals help emergency managers shape the environment to account for this risk? If we have little ability to influence the external environment, then what are the alternatives? I believe our best options rest with our ability to make our organizations and activities as transparent as feasible. They also need to be ready to respond first to those who matter most to our organizations with simple, scalable, clear communication policies, planning and technologies that can be implemented during events. Although these attributes might seem obvious—they are difficult to achieve.

A SYSTEMS APPROACH TO EFFECTIVE COMMUNICATION DURING DISASTERS

To be prepared in today's complex information environment requires organizations examine their readiness to respond from a systems perspective.

What exactly does this mean?

From a communications perspective, it means we must:

- Be transparent in our decision-making and actions.
- Develop policies that allow and encourage our employees to communicate within established rules and boundaries—creating a "bias to communicate."
- Train our employees what the rules are and how to effectively get involved in public discourse and social media.
- Exercise our policies and procedures so that when things do go wrong, we will react instinctively.
- Leverage technology in a manner that helps manage information ethically and serves the public interest.

Together, these efforts build trust and credibility. To be successful in today's complex communication environment, EM personnel must be engaged in cultivating organizational environments that reward honesty, trust, and a bias to communicate with empathy and compassion. This is a leadership responsibility that should be shared by all who are supporting a response. Having served in nearly every Coast Guard public affairs position as a commissioned officer, the other specialists and I understood how important trust was in communicating with our stakeholders. We also understood the importance of empathy and humility.

As a communication professional, I often wonder why so many organizations fail in their efforts to communicate effectively and what can be done to enhance trust between our organizations and those who matter most. In my view, it matters not whether we are talking about public sector communication efforts—as when the National Weather Service seeks to notify those in advance of potential dangerous weather conditions or when the Federal Emergency Management Agency must get important information to disaster victims, or private sector communication efforts. These are both situations where a critical failure in quality assurance or production results in product liability risks for a company. The answer always seems to come down to leadership.

How might BH professionals "operationalize" communication leadership internally in our organizations in a way that helps increase the probability that our communication will be heard and motivates stakeholders to act in their own best interest when the inevitable occurs?

There are lots of resources to assist emergency managers in the early response phase; however, absent a framework to integrate research, policies, resources, and technologies, efforts to achieve a positive result prove elusive. This is why I believe a systems approach to communication is absolutely necessary.

For example, the risk communication research informs us that disaster victims seek information that acknowledges their affective—or emotional needs—first, followed by their cognitive and behavioral needs. The same body of research also reveals that those in distress have difficulty processing information (Covello, 2010).

Concurrently, we also must acknowledge that disasters present enormous potential for the politicization of issues that often become factors impacting leaders. Consider any number of recent incidents that include weapons, floods, or manufacturing. In many disasters, implications will emerge that affect politics, policies, planning, people, and technologies as a way to prevent or mitigate future occurrences. Even in, perhaps especially in, politicized environments, the primary concern ought to be focused on those most in need—the victims and mitigating risk of further harm.

EM personnel must routinely prepare for crises and other communication challenges by conducting regular simulations and tabletop exercises that relate to the potential risk. This will ensure that when an event occurs, reactions are instinctive and guided by simple, clear policies. Among the many lessons that emerged from Hurricane Katrina was the fact that we had a response plan. However, it was too complex and few stakeholders were aware of the plan. Planning helps build "muscle memory" for organizations, so that when a crisis emerges, those who must act are not required to pull a document off the shelf and pour through it. Simplicity is key. Exercising EM plans on a regular basis is critical. Executive action must be second nature. Business continuity and emergency response plans must be simple, clear, concise, and viable.

People who represent our organizations must be credible, trained, and *empowered*. For first responders and EM personnel, they all should be prepared to respond within their respective areas of responsibilities and expertise. Given the proliferation of technology and social media, communication policies that empower front line personnel are critical for success.

As a former Chief of Public Affairs for the U.S. Coast Guard, one of the hallmarks of the Coast Guard's public affairs principles is that *"if you own it or have responsibility for it—you can talk about it."* Questions outside this scope should be referred to the appropriate individual or office. This simple but elegant policy creates a "bias to communicate" during operations. Why is this important? Given the extraordinary demand for information during disasters, empowered first responders who are able to respond to questions within their area of responsibility are viewed more credibly. In addition, being responsive to the legitimate concerns of stakeholders will enhances trust.

Accompanied by the appropriate training, organizations can create highly capable communicators who are sensitive to the needs of disaster victims and myriad stakeholders. Given the disaster response community's subject matter expertise, BH professionals can and should play a vital role in preparing EM personnel prepare for responses.

During disasters, one wants to help stakeholders move through the series of normal reactions that generally influence the affective domain (emotions and feelings), cognitive domain (logic and thinking), and the behavioral domain (action). This is no small challenge; however, there are approaches that enhance the likelihood that stakeholders will act.

Communicating during disasters relies on trust, because without it, there is very little that leaders can do to influence emotions, thinking and behaviors. It is also one of the three underlying goals of risk communication (Covello, 2010), which include:

1. Facilitating knowledge and understanding,
2. Enhancing trust and credibility.
3. Motivating appropriate behaviors and levels of concern by stakeholders.

The ability to influence behaviors, thoughts, and emotions of key audiences is rooted in some of the key principles of risk communication and make them so important during crises. For example, it has been known for a long time that a person's ability to process information during stress declines rather dramatically. Yet, when we examine communication coming from government officials and company executives following disasters, we frequently find that their communication behavior and content ignore risk communication principles. The reality is that, during a crisis, most affected stakeholders want to know you care before they care what you know. Victims of disasters are typically seeking compassion, conviction, and optimism from highly credible sources. (Covello, 2010; Philbin & Urban, 2009).

In the various positions that I have held in government and the private sector, one of the first things I seek to understand is whether the enterprise is doing what it says it is doing. If the answer is yes, the probability of success in communicating effectively rises dramatically. If the answer is no, the challenge is nearly impossible because source credibility and integrity will undermine any communication effort. From a leadership perspective, anything that can be done to ensure that the organization is doing what it says it is doing is absolutely necessary, but not sufficient.

> Seek to understand is whether the enterprise is doing what it says it is doing. If the answer is yes, the probability of success in communicating effectively rises dramatically. If the answer is no, the challenge is nearly impossible because source credibility and integrity will undermine any communication effort.

The other important element of a systems perspective includes the *means* and *rules* by which information is communicated. On any given week during my public affairs tenures in the U.S. Coast Guard or at Federal Emergency Management Agency, we were responding to 200−1000 inquiries. To satisfy this level of demand, we needed to think strategically and systematically how to create transparency and leverage technology and processes to be responsive in a consistent, thoughtful, accurate, and timely manner.

Technology can be a wonderful enabler in communicating during disasters; however, it is important to understand that there is not a one-size-fits all approach and that there are risks as well. For example, a review of the Deepwater Horizon Oil Spill reveals some of the many challenges that *British Petroleum* (BP) had to address when unsubstantiated rumors emerging from social media began to influence operations. On the other hand, having robust technology—with simple user interfaces, that is mobile friendly, and can be operated by nonIT personnel—is critical to disseminating critical information during disasters and rapidly addressing information that is not accurate. Of course, all of the technology in world will not help if the organization's leaders and spokespeople are not considered credible and trustworthy.

SUMMARY

Although communicating effectively during disasters is extraordinarily challenging, BH professionals can look to research and best practices to inform how we communicate during a crisis. An effective communication approach requires systems thinking to address People, Planning, Policies, and Platforms (technology). I have used the "4P" model throughout my efforts to help organizations communicate more effectively. Assuming organizations have a "bias to communicate" and are "doing what they say they are doing," it then becomes a matter of ensuring that:

1. (People) Does the person in charge have clear lines of authority and responsibility?
2. (Policies) Are the communication policies simple, clear, and transparent?
3. (Plans) Are plans in place and tested regularly to create "muscle memory?"
4. (Platforms) Does your technology enable connecting anywhere and anytime? Can it be used by anyone who has authority and responsibility to release information and engage with relevant stakeholders?

The U.S. Coast Guard is considered a trusted organization by many and stands as an example of how to effectively communicate. This is largely a function of executing its humanitarian operations well most of the time. This reputation is also a function of communicating well and training their personnel. The agency does this by adhering to relatively simple public affairs guidance that is summed up as follows: *if you own it or have responsibility for it, you have an obligation to communicate about the issue.*

In addition, there are only four areas that serve as reasons not to answer questions. Known in the Coast Guard as Security, Accuracy, Propriety, and

Policy (SAPP), these include issues of (1) Security—matters involving security will not be spoken about; (2) Accuracy—information must be accurate or it will not be released; (3) Propriety—issues considered proprietary will not be released; and (4) Privacy—information considered protected under privacy regulations will not be disclosed.

In advocating this approach and when combined with lessons from research, BH professionals can create a powerful and responsive systems approach to communicating with stakeholders at any time, and especially during disasters.

REFERENCES

Baron, G. (2006). Now is too late[2]. Bellingham, WA: Edens Veil Media.

Covello, V. (2010). *RIC 2010 Risk Communication—Principles, Tools, & Techniques.* [PowerPoint slides]. Retrieved from http://www.nrc.gov/public-involve/conference-symposia/ric/past/2010/slides/th39covellovpv.pdf.

Philbin, J. P., & Urban, N. (2009). Leveraging the power of the faith based community and its critical communication role during public health emergencies. *Public Affairs in Health,* (July).

Reynolds, B., & Seeger, M. (2012). http://emergency.cdc.gov/cerc/resources/pdf/cerc_2012 edition.pdf.

Making Integration Work

Brian W. Flynn

Both portions of this chapter have described ways in which disaster BH and EM function before, during, and following disasters. There is much information that can be practically applied to optimize integration in the broad area of communication. Readers are encouraged to look at their own professions, as well as that of their counterpart, and consider actionable ways that integration can be accomplished and maintained. Table 11.1 describes what both professions might consider in several topical and phase-appropriate ways. These examples are intended only as a start in stimulating creative actions and opportunities in both professions.

TABLE 11.1 Operationalizing Communications Integration between Disaster Behavioral Health and Emergency Management

	Disaster Behavioral Health	Emergency Management
Understanding roles and skills	Assure disaster BH personnel understand disaster management structures and processes (including communications)	Include understanding of the various roles disaster BH can play in communications during all phases of disaster
	Develop a variety of disaster-related skills sets and/or develop specialized expertise (e.g., crisis communication)	Within EM (especially communications), identify individuals who can lead in establishing and sustaining integration
	Within disaster BH authorities and organizations identify individuals who can lead in establishing and sustaining integration	
	Promote inclusion of DBH in Emergency Support Function (ESF)-8 See Federal Emergency Management Agency, 2008 for more information on all ESFs, http://www.fema.gov/media-library-data/20130726-1825-25045-0604/emergency_support_function_annexes_introduction_2008_.pdf especially regarding contributions to JIC activities at all levels (local to national)	
Building respect and trust	In basic and advanced disaster BH training, promote integration with EM (and their specialized communications professionals) as a core value	In basic and advanced EM training, promote integration with disaster BH professionals as a core value that promotes formal and informal relationships
	Build formal and informal relationships with communications personnel in EM	Build formal and informal relationships with BH experts
	Build organizational relationships with EM authorities and organizations	Build organizational relationships with BH authorities and organizations
	Make presentations at EM/communications meetings and conferences	Make presentations at behavioral and disaster BH and conferences
	Author joint position papers, editorials, and articles in each other's literature base	Author joint position papers, editorials, and articles in each other's literature base

(Continued)

TABLE 11.1 (Continued)

	Disaster Behavioral Health	Emergency Management
The preparedness phase	Participate in drills and exercises	Include DBH in drills and exercises by making it part of emergency operations planning
	Assist in development of "on the shelf" messages and message sequences	Include disaster BH in message/communications plans and strategies
	Anticipate opportunities and challenges and plan accordingly	Anticipate opportunities and challenges and plan accordingly
The response phase	Be present at sites where EM takes place	Seek out and engage BH personnel in EM JIC operations
	Initiate contact and offer collaboration/services to communications leadership	Seek consultation regarding psychosocial impacts of message development and dissemination
	Assist communications leadership to target populations and/or issues where enhanced communication is needed	Enlist BH personnel in education, monitoring, and management of staff stress issues
	Locate disaster BH desk/office in physical proximity to JIC at EM sites (this is a challenge as DBH if typically part of either ESF-8 (Health) or ESF-6 (Mass Care) and will require creative approaches)	Assist BH leaders identify specific psychosocial needs/challenges when they are observed through communications processes
	Provide consultation when appropriate for development and dissemination of messages	
	Provide feedback to communications leaders when impacts of messages are observed	
The recovery phase	Share newly learned information and communications emerging needs identified through BH recovery efforts	Share emerging BH needs identified recovery efforts
	Participate in reviews/critiques of event response activities	Assure that BH is included in reviews/critiques of event response activities
	Assist communications leaders in evaluating their programs/products	Assist communications leaders in evaluating their programs/products

Chapter 12

How to Navigate External Factors: Legal, Ethical, and Political Issues

Berl D. Jones, Jr.[1] and Daniel Dodgen[2]
[1]*Federal Emergency Management Agency, Washington, DC, United States,* [2]*US Dept of Health & Human Services, Washington, DC, United States*

Through an Emergency Management Lens*

Berl D. Jones, Jr.

As a tool to facilitating integration, this section will discuss the need for mutual understanding of the respective needs and requirements of both emergency mangers and disaster behavioral health (DBH) professionals. It will attempt to address key questions such as, "Why can emergency management (EM) and cannot do certain things?" as well as "Why can disaster behavioral health (DBH) and cannot do certain things?" Understanding and respecting scope and boundary issues is key to successful integration.

SOME IMPORTANT CONTEXT: SPEED AND PRIVACY

The mission of those who assist disaster survivors is to provide assistance as quickly as possible and to make sure a referral system is in place to handle unmet needs. This drive for speed and efficiency must be balanced by the necessity to protect the personal information of disaster survivors. There are laws, regulations, and policies in place to do just that. These are often not known to DBH workers. While behavioral health (BH) professionals are

---

* It should be noted that laws, regulations and internal policies are subject to change. The author suggests checking for updates with the related organization, department, or agency.

used to working with privacy safeguards, they, especially those new to DBH work, are often unaware of these protections in the EM sector. Without a working knowledge of these privacy safeguards, DBH risks the high probability of becoming at odds with emergency managers.

We live in a world of information that has generated many new challenges to long-existing laws. The requirements to protect Personally Identifiable Information (PII) are outlined in the Privacy Act of 1974 (https://www.justice.gov/opcl/privacy-act-1974):

> "The Privacy Act of 1974, 5 U.S.C. § 552a, https://www.justice.gov/opcl/privacy-act-1974, establishes a code of fair information practices that governs the collection, maintenance, use, and dissemination of information about individuals that is maintained in systems of records by federal agencies."

PII "refers to information which can be used to distinguish or trace an individual's identity; such as their name, social security number, etc. alone or when combined with other personal or identifying information which is linked or linkable to a specific individual, such as date and place of birth, mother's maiden name, etc."

In the world of PII, there is a delicate balance between those seeking assistance and those who provide the assistance. Disaster survivors' access to funds, goods, and services requires the provision of their personal information. Protection of PII and provision of assistance can sometimes appear to be opposing forces.

At the federal level, emergency managers are directed in the ways they may collect and share information about disaster survivors. So, what does this mean for integration of EM and DHB? Here are a few considerations:

- For DBH, it means that in addition to the privacy and confidentiality safeguards they are accustomed to, they must understand and live within the requirements placed upon emergency managers
- It means that, prior to responding, DBH training must include understanding of privacy protection implemented by emergency managers
- It means that, in establishing initial integration efforts, the topic of mutual confidentially requirements, options, and expectations must be a high priority

EVOLUTION OF THE DISASTER ASSISTANCE APPLICATION PROCESS

In the early days of disaster assistance, disaster survivors visited Disaster Application Centers (DACs) and provided their personal information face-to-face to disaster service providers from nonprofit organizations and several government agencies. Information was recorded using paper applications and managed through hard copy case files. This information would then be transferred by phone, eventually by facsimile machines, or simply entered into a computer manually. The common instruction was to "press hard on the carbon paper" to create multiple copies for the sharing of information.

The head of household would sit down with the service provider and relay all of the necessary personal information for that agency to begin to build a case file and verify any loses. In the case of voluntary agencies, caseworkers would evaluate the types of needs—acute and longer-term. While the emotional state of the disaster survivor was important, it was not always at the forefront of these initial intakes. Caseworkers, staff nurses, clergy, and volunteers were often in place to provide comfort to the individuals and families.

> These physical application centers also provided an opportunity, when and if DBH workers were present, to interact with both survivors and workers to identify stress related problems, provide informal guidance and advice, and to facilitate referral to psychological services. With the evolution of such centers, as described later in this section, that early, first-hand opportunity to interact does not always exist. Other mechanisms have had to be developed and many of those are described throughout this book.

Once information was gathered, the service provider's representative would begin to find resources to meet the emergency needs. Housing is a prime example. It was imperative that agreements were in place to provide referrals. A major referral activity was the referral of disaster survivors to voluntary agencies for emergency temporary housing, also known as transient accommodations. To do this, the service provider needed very specific PII to obtain housing and to meet other emergency needs, such as food and clothing. Voluntary agencies would often obtain a release of information (ROI) from the disaster survivor so that their information could be shared among the other local agencies. After the emergency housing was taken care of, referrals for other needs such as furniture, clothing, food, medical/burial expenses, and mental health and spiritual counseling referrals could be made.

Looking back over the years, it is interesting to see what changes have occurred with respect to PII. Providers have gone from in-person paper applications, to phone calls to call centers outside of the disaster area, to internet and mobile-based applications. We now have data storage and portability through devices and methods we could never have envisioned in the early days. All of these changes have required adaptation, flexibility, and creativity for emergency managers and DBH professionals alike, not to mention survivors themselves. There is no question that integration of EM and

DBH is a dynamic process, and that change is the norm. Integration is clearly not something that can be established and then put on auto-pilot.

The advances in technology have a positive effect on disaster assistance in terms of speed of delivery, timely referrals, and availability of services. All levels of government and most agencies in the nonprofit sector have moved to electronic intake, storage, and processing. This allows for timely case and eligibility review as well as the provision of assistance. The change in technology has enhanced our ability to assist disaster survivors, but it has also elevated the concerns over and control of PII.

> Sharing of disaster survivors' information among agencies is critical to ensure the broadest possible access to services, but also to limit duplication of benefits.

The federal government is prohibited from duplicating benefits by both statute and regulations. Voluntary organizations have a keen interest in avoiding duplication as well. Nonprofit organizations rely on donated dollars, services, and volunteers to provide the assistance needed in their impacted communities. These organizations recognize that donated dollars, goods, and services are finite; therefore, they must closely coordinate with their partners to avoid wasting resources. Being good stewards of government and community resources is a shared responsibility for all disaster related activities and services.

One of the most beneficial relationships in disaster assistance is between Federal Emergency Management Agency and the National Voluntary Organizations Active in Disaster (NVOAD) (http://www.nvoad.org/). Following is a description of its history and function. It was clear after Hurricane Camille in 1969 that there needed to be better coordination among local, state, tribal, federal governments, and community organizations. Agencies were operating independently of one another, duplicating services, and gaps in service often went unaddressed. In 1970, the National Voluntary Organizations Active in Disaster was established. This organization consisted of a core of nonprofit organizations, both ecumenical and nonecumenical, committed to the principles of cooperation, coordination, communication, and collaboration. These principles exist today throughout EM.

Although the establishment of the National Voluntary Organizations Active in Disaster does not directly relate to PII, it highlights an example of establishing and sustaining a formalized system designed to ensure disaster survivors have access to disaster assistance through nonprofit, ecumenical, local, state, tribal, and federal agencies by sharing information. Today, the National Voluntary Organizations Active in Disaster continues to be the backbone of EM at the local, state, tribal, and national levels, responding to the entire spectrum of emergencies from the local level through federal presidential disaster declarations.

THE EVOLUTION OF DISASTER BEHAVIORAL HEALTH SERVICES

As understanding of the wide range of services needed by disaster survivors as well as the recognition of the magnitude and nature of BH expanded, the disaster response component organizations and systems began to develop methods to assess and meet the needs of survivors, from the point of intake for disaster assistance to DBH professionals and agencies. In the early 1970s, Federal Emergency Management Agency's predecessor agency, through the Stafford Act, initiated through the federal Department of Health and Human Services (DHHS) the Crisis Counseling Assistance and Training Program (CCP), administered through the Substance and Mental Health Services Agency (SAMHSA). The goal of this program described in more detail in later in this section as well as in other parts of this book, was to engage and train local DBH resources to meet the most pressing needs of survivors and their communities immediately following a disaster. Still today, it provides crisis counseling services to and referral to more intensive and formalized DBH services for those in need. This additional need-driven focus on DBH services expanded the distribution of a disaster survivor's PII. Where there is sound integration of EM and DBH leadership, the sharing of this information has typically gone smoothly. Where that integration is not taking place, it is not uncommon to see significant disagreement regarding sharing of PII. The result is strain on the relationship between EM and DBH as well as compromised services to survivors.

THE INTENT AND COMPLICATIONS OF PROVIDING ASSISTANCE

Government programs and assistance provided through community agencies are designed to be supplemental and cover the serious needs and necessary expenses of the disaster survivors. These programs may not be able to return an individual or family to predisaster conditions, but they will address critical needs. Insurance is intended to be the primary form of assistance. However, the reality is that not everyone can afford full or partial insurance.

Federal disaster assistance is limited through statute and regulations. Most of the assistance is based on conducting a damage assessment and identifying the seriousness and necessary expenses of an individual or household. The amount of assistance provided is determined through an inspection of an individual's property, verification of loss, and identification of needs. Information about the amount and type of assistance an individual or family receives cannot be shared unless the proper documentation is in place to do so. This has implications for DBH providers and is an important area of understanding and agreement for all involved.

The timeliness of disaster assistance is often a source of pressure on emergency managers. Disaster losses must be verified and other forms of assistance like insurance must be factored into what is provided. By statute, the federal government cannot duplicate benefits. If a household has insurance, the disaster survivor and their insurance company must share benefit information with emergency managers. Insurance companies consider their customer information to be proprietary and often require their customers' consent to release their information. It is easy to see how multiple requirements for protecting and sharing information can be a significant complicating factor in the provision of services. DBH providers should be aware of their own privacy requirements and how all of these combined "protection" factors can be stressors for service recipients, emergency managers, and service providers.

> There is no secret formula for determining "how fast is fast" when it comes to the provision of disaster assistance for longer-term items such as housing, repair assistance, and replacement of personal property. There are many factors in considering the "need for speed" of recovery assistance.

Factors impacting the speed of response include the ability to access to the damaged areas for assessing damage to property, the ability for the survivor to be present for the damage assessment, the availability of insurance information and the resources available to repair/rebuild, and the consistency of repair/rebuild plans with local applicable ordinances. Case management services, often in concert with DBH resources, often help disaster survivors identify all of their needs and prioritize their limited funds. It is in this process that many survivors realize the full extent of their losses and the limitations of their ability to recover as they had hoped. The involvement of DBH resources in this stage is often extremely helpful to both recipients and emergency mangers.

External inquiries may come to government agencies in the form of the Freedom of Information Act (FOIA) requests. These requests can be related to obtaining information on an individual or could be information on a group of individuals or the process, policies, and guidance related to disaster assistance for a specific event.

It is important to recognize that, even though there are requirements to provide information in compliance with the FOIA, factors such as the expense and the timeliness required to process the FOIA request may make compliance in whole or in part difficult or impossible. No matter where these inquiries come from, the federal government must ensure that an individual's privacy is maintained.

Congressional inquiries can be in response to reports from their constituents or simply to demonstrate to their constituents that they have their districts' best interests in mind. The same parameters of protecting information apply to congressional inquiries. Once all parties have a ROI in place, the sharing of information can occur—but not until and unless.

> Complicating factors in information sharing:
> - Frequent inquiries from many sources such as local, state, tribal governments, congressional, and the media begin once a governor of an impacted state requests federal support
> - Questions about what areas will be included, what assistance is available, and how long it will take for Federal Emergency Management Agency to help are common
> - When the President declares an Emergency or Major Disaster Declaration for a state or tribal territory, there are political pressures to get timely assistance into the impacted communities
> - In all these circumstances, the disaster survivor's personal information must be protected regardless of the external pressures to disclose their information

FRAUD AND PII

Disaster assistance programs, both governmental and community-based, are designed with the premise that the majority of disaster survivors applying for assistance are honest. However, there is also an expectation of timely assistance, as already discussed. Federal Emergency Management Agency must be able to register disaster survivors, provide immediate assistance, assess longer-term needs and provide that assistance, refer individuals and families for additional assistance, and maintain accurate, verifiable PII. In additional to automated and manual verification tools, the entire process is monitored by the Department of Homeland Security (DHS), Office of the Inspector General (OIG), the Government Accounting Office (GAO), Office of Management and Budget (OMB) and Congress, and so on.

The OIG is involved in all aspects of disaster assistance to ensure that those applying for assistance are not committing fraud. The agency's systems are set up to detect any such fraud. If fraud is detected, a process is in place to recover erroneous payments.

Examples of potential fraudulent applications include:

- Identity theft
- False addresses
- Multiple applications from the same person or family
- Redirecting financial assistance away from the head of household
- Omitting information that could impact eligibility

> **Especially Complex Situations**
> Not all disaster survivors who apply for assistance want their information shared. Two examples illustrating this are (1) undocumented members of an affected household and (2) members of households not wanting other family members to locate them.

Households where one or more members may not be US citizens are often reluctant to provide their personal information to Federal Emergency Management Agency for fear of discovery and deportation. In some cases, the US citizen is a family member under the age of 18.

Scenarios in which a disaster survivor may not want to be found and has concerns about the sharing of their information include households with a history of domestic violence and substance abuse. *There are occasions where the disaster event is the opportunity for a spouse and their children to escape an abusive home.* One member of the household seeks to find the other members, and the others do not want to be found. This becomes difficult when all involved require disaster assistance and may or may not be willing to share it.

Federal Emergency Management Agency utilizes the Routine Use mechanism within the Privacy Act that allows Federal Emergency Management Agency to share recovery information with trusted partners, provided the use of a record is for a purpose which is compatible with the purpose for which it is collected. Federal Emergency Management Agency publishes routine uses in the Federal Register, the daily newspaper of the federal government. Sharing of data with trusted partners is compatible with Federal Emergency Management Agency's overall responsibilities under the Stafford act and the purpose for which Federal Emergency Management Agency collects the data.

When a disaster survivor calls Federal Emergency Management Agency, they may be provided the National Crisis Counseling Hotline number that is staffed by Substance and Mental Health Services Agency. If the Crisis Counseling Program has not been authorized and implemented, the National Hotline will provide the impacted state's hotline number. If the program is activated, Federal Emergency Management Agency can provide callers with their specific Crisis Counseling number.

If the state requests a transfer the electronic information and an agreement is in place to share the information, Federal Emergency Management Agency can provide that information. Case by case agreements with state agencies can be established if the state requests it and Federal Emergency Management Agency approves.

It should not be difficult to see the intersection of DBH factors, and many if not all of these complicated elements, especially when victims are reluctant to share accurate information either for legitimate or nefarious reasons. These are delicate issues that require attention by both EM and DBH leaders *before an event*. Examples of areas for discussion include:

- What are the obligations of a DBH professional to inform EM if he/she suspects attempts to defraud the government?
- How are DBH professionals the legal "duty to report" requirements impacted if the potential for violence of child abuse is suspected?

These are never easy issues to confront in routine situations. They become even more difficult in a disaster environment.

IN-PERSON CONTACT

Federal Emergency Management Agency and the state or tribal governments may open Disaster Recovery Centers (DRC) where disaster survivors can go to get referrals to local assistance agencies. This should not be confused with the physical locations where survivors, in the past, had to go to *apply* for disaster assistance described earlier in this section.

It is important to remember that this may the first and only time a person goes through this kind of experience, and they may not even know who can help or what types of help are available. DRCs can be very crowded and busy places, with many agencies staffing intake or information tables. Because of close physical proximity to many people providing private information verbally, privacy challenges exist in these centers as well. It is not uncommon, and entirely appropriate, for DBH resources to be present in these locations.

All agencies must ensure proper record keeping, information exchange, and privacy for their clients. For example, to assist individuals who may have suffered a loss of a family member or friend, or are simply being overwhelmed by the event, staff is trained to identify these situations and seek help from the professional and trained paraprofessional counselors on site or nearby. In cases where the Crisis Counseling Program is in operation, funding for these types of services, as well as others, is federally provided.

THE CURRENT PROCESS: WHAT DBH NEEDS TO KNOW ABOUT INFORMATION SHARING

Once a disaster survivor calls the National Processing Service Center, the person's information is entered for initial processing. While some forms of assistance can be processed automatically through a series of algorithms, more complicated cases may be handled manually by applicant assistant specialists. The information that is gathered can be used to determine emergency needs, such as food or housing, and more long-term needs, such as a low interest loan from the Small Business Administration or Unemployment

Assistance from the impacted State's Department of Labor. With today's technology, disaster survivors can also apply for federal assistance over the internet, through smart phone applications, and on the scene of the event through field staff.

> Until agreements are in place to share information with local service providers, Federal Emergency Management Agency can only provide an applicant with referral information for emergency assistance such as counseling, food, clothing, and shelter. At that point, Federal Emergency Management Agency *cannot* share information with the local service organizations, including DBH specialists. Only referral information can be provided. Any additional information shared by applicants the entity to which they are referred will be at the applicant's discretion.

Federal Emergency Management Agency exercises extreme caution with all disaster survivor information. Typically, a few weeks after the disaster incident, local and state officials want to know the status of individual cases. They receive questions and complaints from constituents who report that they received too much or too little, or "my neighbor lied." Yes, the reporting of fraud by neighbors or family members does occur.

One of the most difficult challenges emergency managers and service provider agencies face is when media outlets, congressional inquiries, or inquiries from litigation entities begin arriving with accusations that emergency mangers and agencies are not doing as much as they need to be doing to help individuals. In cases where the accusations and claims are erroneous, it is difficult to respond without releasing prohibited information. However, on the positive side, in cases where additional usable information is provided and more review occurs, this new information opens up additional eligibility and additional assistance may result.

Inquiries from multiple family members may also create a challenge for Federal Emergency Management Agency workers. Sometimes, individuals from the same household call to find out the status of their case but they are not authorized to receive that information. This is particularly difficult in cases where a husband and wife are separated, the husband needs assistance and the wife needs assistance but the assistance will go to the head of household.

REFERENCES

National Voluntary Organizations Active in Disaster. http://www.nvoad.org/.
Privacy Act of 1974. https://www.justice.gov/opcl/privacy-act-1974.

Through a Disaster Behavioral Health Lens

Daniel Dodgen

DBH is a field that is finally coming into its own. The number of articles focusing on DBH research and practice has increased significantly in the last 20 years (Boscarino, 2015; North & Pfefferbaum, 2013), with continuing education, online courses, and conference-based training show a similar trend. Most people in the field would consider this a sign of success. There is certainly evidence that BH is much better integrated into the overall strategic thinking about disasters than it ever was before (Dodge & Meed, 2009). This growth has extended beyond the research and practice community to the policy world. In the past 10 years, BH has been included in public health and hospital preparedness grants, and state and local BH agencies are increasingly part of local healthcare preparedness coalitions (U.S. Department of Health and Human Services, 2016a). Furthermore, the White House established a national disaster mental health advisory group, which published its final report, "Integration of Behavioral Health in Federal Disaster Preparedness, Response, and Recovery: Assessment and Recommendations," in 2010 (Pfefferbaum et al., 2012). This committee focused on ways to enhance and integrate DBH at the national level. Similarly, when Congress established the office of the Assistant Secretary for Preparedness and Response (ASPR) in 2006, this new federal agency quickly founded an office focusing on BH and related issues (Dodgen & Meed, 2009; U.S. Department of Health and Human Services, 2016b).

The increased focus on BH in the context of disaster preparedness, response, and recovery has brought needed services to hurting individuals and communities. However, it has also brought new challenges that the field is only beginning to address. Many of these new challenges lay in the areas of law, policy, and ethics (Flynn & Speier, 2014; Call, Pfefferbaum, Jenuwine, & Flynn, 2012). From the perspective of these areas, there are three basic questions that anyone planning for or providing BH services in a disaster needs to ask:

1. What are the relevant current laws, policies, and ethical guidances for DBH services?
2. How does the disaster context impact these legal, policy, and ethical issues?
3. What are the long-term ethical and policy concerns for impacted individuals and communities?

This chapter will attempt to answer these questions by providing an overview of the laws, policies, and ethical concerns they raise.

WHAT ARE THE RELEVANT CURRENT LAWS, POLICIES, AND ETHICAL GUIDANCES FOR DISASTER BEHAVIORAL HEALTH SERVICES?

The Law

While many readers are familiar with common disaster response mechanisms, they may not be familiar with their legal underpinnings. The Stafford Act and the Pandemic and All Hazards Preparedness Reauthorization Act (PAHPRA) are two of the critical laws for national DBH activities in the United States. Upon these two laws rest most of the activities, particularly regarding health and BH, undertaken during national crises. A third source of federal funding for mental health services is the Crime Victims Fund (U.S. Department of Justice, 2016), which helps victims offset the cost of mental health and other expenses (Dodgen & Meed, 2008; U.S. Department of Justice, 2016). In a disaster context, this would apply to events with a criminal component such as mass shootings and terrorist attacks.

The Stafford Act outlines the process by which states and tribes can seek federal assistance in disasters (Dodgen & Meed, 2008; Federal Emergency Management Agency, 2016). If a state or tribe requests and receives a presidential disaster declaration, it becomes eligible for public assistance or individual assistance services. Individual assistance programs provide immediate direct and financial assistance to individuals, including temporary housing assistance, disaster-related unemployment assistance, and crisis counseling.

While the CCP is discussed elsewhere in this volume, the reader should know that the program is established by law and operated under regulations and guidance established by Federal Emergency Management Agency in coordination within the federal Substance Abuse and Mental Health Services Administration. The guidance from Federal Emergency Management Agency determines how CCP dollars may be used. This includes what target populations must be addressed, how needs assessments are conducted, what services are allowable, and how long services can be offered under the grant (Federal Emergency Management Agency, 2015a). Frustrations with the CCP grants can emerge because the program's parameters are established by law and may not always be as flexible as local entities wish they could be. Thus, when local BH authorities wish to use CCP funds to pay for a service that they feel is needed, they cannot do so if the service is outside the program's purview. For example, the CCP grants are designed to focus primarily on "sub-clinical" or "pre-clinical" services (e.g., crisis counseling, outreach, and education) and are not intended to pay for or replace more intensive psychiatric services, even if the state feels such services are needed.

Other legislative actions have had a significant impact on disaster related health, including BH, structures. The PAHPRA established the Office of the ASPR at the US DHHS (U.S. Department of Health and Human Services, 2016c). ASPR is tasked by Congress with coordinating the public health and medical (including BH) response to disasters, terrorism, and public health emergencies. PAHPRA not only established ASPR, it also placed the Medical Reserve Corps (MRC) and the National Disaster Medical System (NDMS) under ASPR. MRC and NDMS are two sources of deployable DBH responders during national emergencies. These two programs are authorized to exist by the legislation, and the legislation also provides broad description of their role and function.

Just as legislation provides the foundation for the existence of CCP, ASPR, NDMS, and MRC, it also provides the foundation for publicly supported BH services. At the federal level, and in most—if not all—states, the BH system is mandated to focus its efforts primarily on people with serious mentally illness (SMI) adults or seriously emotionally disturbed (SED) children. The terms vary from state to state, but the reality remains that BH authorities have to focus their minimal resources on those with chronic BH needs. These statutory requirements can restrict reallocating funding or resources during disasters, even to address emergent needs. This can impact the number of providers available during an emergency and the length of time they are available.

The final legal issue to consider in DBH service planning is the question of licensure. Licensure laws vary tremendously by discipline, with social work, master's level psychology, doctoral level psychology, and psychiatry, for example. Each field has different internship requirements, licensing exams, and licensing boards. The licensing rules also vary significantly across states. Consequently, a licensed BH provider cannot cross state lines, even to volunteer services, unless he/she is volunteering with an organization that has interstate authority to deliver services (such as the American Red Cross) or each affected state grants a waiver (Call et al., 2012). In states that have mutual Emergency Management Assistant Compacts (EMACs), mental health professionals holding a license in their home state is deemed licensed by the state requesting assistance subject to any conditions declared by the Governor of the state requesting assistance (National Emergency Management Association, 2016).

When people think about legal issues, they often are most worried about liability issues. In the famous Buffalo Creek flood in West Virginia, psychological harm was part of the successful case against the coal company (Flynn & Speier, 2014). Since the case held that a company could be held liable for causing psychological harm, the question arises whether malpractice has never been an issue in disaster mental health. To date, no appellate court has ever made a decision concerning a case involving a mental health professional providing services during the acute phase of a disaster (Call et al., 2012). However, it remains possible that someone could bring a lawsuit because of harm caused by lack of services or inappropriate services. To reduce the likelihood of such lawsuits, attention must be paid to the licensure issue as well as the ethical issues described below.

> Why do I need to know about disaster mental health and the law?
> - **Federal Laws**. Laws such as the Stafford Act and the Pandemic and All Hazards Preparedness Act determine what the federal government will pay for in a disaster
>   - For example, Crisis Counseling is only authorized when there is a disaster declaration authorizing individual assistance
> - **Federal Strategies**. These laws direct the development of the preparedness, response, and recovery strategies that dictate EM policy
>   - The National Health Security Strategy (NHSS), required by law, guides how healthcare preparedness grant dollars are used
> - **Licensure**. Licensing laws for mental health providers cover record keeping, privacy, consent to treatment, and other ethical responsibilities
>   - Most of these laws do not provide exceptions for services during disasters
> - **Liability**. Successful lawsuits after disasters have included psychological harm as part of the case
>   - No one has successfully sued individual providers for causing harm during a disaster or for failing to provide services, but the possibility exists

Policy

The laws described above provide the parameters within which DBH services must be provided. However, state, local, federal, and nongovernment agencies still have some flexibility to determine how best to implement these laws. This is why CCP grants in two states, for example, can look quite different, yet still be in compliance with the guidance from Federal Emergency Management Agency. This is where policy enters the equation.

At the national level, the National Response Framework (NRF), the National Disaster Recovery Framework (NDRF), and the NHSS provide the big picture strategic frameworks for conceptualizing DBH services. These documents are the result of a consensus development process that includes multiple government and nongovernment stakeholders, with Federal Emergency Management Agency leading the NRF, and NDRF and HHS leading the NHSS (Federal Emergency Management Agency 2015b; U.S. Department of Health and Human Services, 2016d). These strategies in these core documents offer guidance to local entities for many issues related to disaster mental health. For example, the NHSS Strategic Objective 1, "Build and Sustain Healthy, Resilient Communities," includes priorities to:

- Encourage social connectedness through multiple mechanisms to promote community health resilience, emergency response, and recovery
- Build a culture of resilience by promoting physical, behavioral, and social health; leveraging health and community systems to support health

resilience; and increasing access to information and training to empower individuals to assist their communities following incident

NHSS Strategic Objective 4, "Enhancing Public Health, Healthcare, and Emergency Management Systems," includes these priorities:

- Strengthen competency and capability-based health-security-related workforce education
- Expand outreach to increase the numbers of trained workers and volunteers with appropriate qualifications and competencies
- Effectively manage and use nonmedical volunteers and affiliated, credentialed, and licensed (when applicable) healthcare workers

These strategic objectives are accompanied by an implementation plan that provides examples of specific actions federal agencies and stakeholders can take to improve preparedness, response, and recovery. Two examples of actions relevant to readers of this volume that promote resilience are:

- State, local, tribal, and territorial (SLTT) governments and community-based organizations can cross-train public health, healthcare, and human services professionals to improve recovery service provision
- SLTT governments can work with community-based organizations to ensure that community leaders have access to BH services

The bullets above provide just a few examples of the kinds of objectives included in the NHSS and illustrate how they can influence the development of DBH services and providers. So, why is this relevant for emergency managers and planners to know? The NHSS aligns with two major grant programs for public health agencies: the ASPR Hospital Preparedness Program (HPP) and the Centers for Disease Control and Prevention (CDC) Public Health Emergency Preparedness Program (PHEP). This alignment, and the corollary alignment with other NRF Emergency Support Function #8 (ESF #8) (Public Health and Medical Services) activities, helps drive many health and BH activities. Knowledge of the NHSS can help emergency managers to understand the expectations and goals of public health and BH plans, particularly since other planning efforts then cascade from these goals.

Just as there are variations on how each state might design its CCP grant for counseling services, there are variations on how each state, tribe, and locality will implement the NHSS and utilize its HPP and PHEP preparedness dollars. Some localities design programs with robust BH capabilities, while others have more minimalist approaches. The manner in which each place chooses to implement the laws and strategies reflects that state, locality, or tribe's policy priorities. However, it also leads to questions about the ethical aspects of disaster BH.

> What do I need to know about policy?
> - The NHSS encourages public health agencies to engage with EM in several ways. For example,
>   - Government and community-based organizations are encouraged to cross-train healthcare professionals to improve recovery service provision
>   - Health agencies are encouraged to increase the numbers of trained workers and volunteers with appropriate qualifications and competencies
> - There are hundreds of healthcare coalitions across the United States that are funded to foster public health preparedness
>   - These coalitions are encouraged to integrate BH into their activities
>   - There is tremendous variability in how each community implements these laws. Not all healthcare coalitions actively involve BH

Ethics

The ethical issues in DBH are not as clearly delineated in law or policy, but certain issues tend to emerge consistently in research and in the "gray" literature. These revolve around training, professional practice, research, and allocation of resources.

The ethics of DBH professionals are not different from the ethics of other healthcare service providers, and generally remain the same during disasters as they are at other times (Call et al., 2012). Providers must adhere to laws and ethical codes regarding record keeping, competence, consent, etc. Merely being licensed is not always enough. In a disaster, interactions with individuals tend to be brief, problem-focused, and intense. BH responders need to be trained properly to provide specific disaster- or crisis-related BH services. Such training is readily available through the American Red Cross and other entities, and is usually offered by CCP grantees for their employees.

However, there is some controversy among researchers and practitioners about whether all models of psychological intervention are beneficial. So, the first ethical question emergency planners must ask themselves is: *Are the BH professionals responding to disasters in my community trained in an appropriate, research-based model?* This is not a question the emergency manager needs to be able to answer personally, but they need to ask it of the local BH authority or other entity coordinating the BH response.

Appropriate services include more than just the right intervention; they include services that are culturally competent for the impacted community. Cultural appropriateness encompasses language and national origin, but it also includes sensitivity to other issues. Without training, some BH professionals may encourage first responders to express emotions during a response in a way that interferes with, rather than facilitates, their work.

Understanding these cultural competence issues is also part of ethical practice.

The term "research-based" rather than "research-proven" is used above because the current research does not support any intervention unequivocally (Call et al., 2012). For the field to develop and improve, research is needed to evaluate the effectiveness of DBH services, pilot new tools, and improve needs assessment. However, there are many issues that must be resolved *a priori*. This raises a second ethical question: *If research is being conducted, what are the plans to ensure that it will be conducted ethically?* These include determining who is conducting the research, the purpose, and what institution has provided ethical review to insure all individuals are protected from harm (Institute of Medicine, 2015). In some cases, it will be nearly impossible to conduct ethical, meaningful research. Asking these questions in advance can prevent ethical concerns during a disaster, when some individuals could otherwise take advantage of the fluid situation to conduct research without proper safeguards. Addressing the issues in advance can also create opportunities to gather information that will improve future responses.

The final, and perhaps the biggest, ethical challenge regards the allocation of resources. In many communities, BH providers are a limited resource. Deploying them to a disaster response can mean taking them away from other necessary activities. This is partly why the CCP supports "pre-clinical" services that can be delivered by trained paraprofessionals. However, in the early stages of a disaster, CCP grants are not yet awarded and fully operational and trained paraprofessionals most likely will not be available. The question then is: *What is the best way to prioritize the use of BH professionals?* In a large-scale event, there may not be enough assets for the need and decision-makers will need to prioritize. Should the focus be on children, on first responders, or on the bereaved? Are there other vulnerable populations who have been affected? Should BH professionals be used to train local nonprofessionals in "sub-clinical" interventions as a way of multiplying the resource? What ethical issues will that raise? BH concerns are often highly visible in the early stages after a disaster, so there is a temptation to send "crisis counselors" everywhere. However, this may not be the best use of the resource. Consultation with the local BH authority and with experienced responders can help emergency managers make wise decisions about deployment of these assets.

None of the ethical questions above is unanswerable, but they are even more difficult to answer during a crisis. Emergency managers, leaders, and planners can work through them with their BH agencies before an emergency to minimize the ethical challenges during a disaster. Nevertheless, new challenges will inevitably emerge.

> What do I need to know about ethics?
> - Ethical standards for BH providers do not change in a disaster
>   - Record keeping, privacy, consent to treatment, and other ethical responsibilities remain
>   - Providers need to be trained in culturally appropriate, research-based services
> - Disaster research is needed but must be done cautiously
>   - Is anyone planning to conduct research? What is its purpose? What institution has provided ethical review to ensure protection from harm?
> - Planning for allocation of resources is valuable
>   - Are there specific groups I should focus on?
>   - Do I expect volunteer providers from outside my community to offer assistance? If so, who would they be most competent to serve?
> - Who is the lead for DBH in my community?
>   - Do I have that person's contact information with me at all times so they can answer all these questions?

HOW DOES THE DISASTER CONTEXT IMPACT ETHICAL, LEGAL, AND POLICY QUESTIONS?

As many seasoned emergency planners say, "all plans fail at the moment of first contact with the enemy." Because few disasters unfold exactly as drills, exercises, and plans predict, there is a need to consider how the context of any disaster might impact the delivery of DBH services.

One of the first challenges that emerges in DBH is defining the "victim" or "survivor." With medical services, the survivor is the one with the injuries. With BH, injuries can be caused by direct exposure to the incident (i.e., losing a family member, being dislocated, etc.) or by indirect exposure (i.e., witnessing an event on television, being afraid to travel after a terrorist incident, etc.). Different programs have different parameters regarding who can receive services and how. The CCP grants allow some latitude on whom to serve, while BH services offered through crime victim funds, as mentioned above, are restricted to direct victims (Dodgen & Meed, 2009). As the event unfolds, different groups may be identified as needing BH services. Such services can be very beneficial in the community recovery and should not be prevented. Nevertheless, policies may need to be adjusted to make sure the needs are addressed without violating law or policy.

As the community identifies the most affected populations, it may be that the needs do not map well onto the skills of the available BH responders. For example, there may be many senior citizens needing services, but few DBH workers with geriatric experience. Perhaps most of a community's BH responders are prepared for natural disasters and the community experiences a terrorist attack. Howe (2012) argues that clinicians should not be deterred from helping by fear of litigation and Call et al. (2012) cite ethical principles for psychologists that allow working outside their usual scope of practice "in order to ensure services are not denied (p 314)." Obviously, an emergency manager cannot withhold available assets, but efforts must be made to ensure that just-in-time training is made available for BH workers. Similar ethical issues emerge in all disasters and prior planning can anticipate some, but not all, of them.

Another policy challenge that emerges during a disaster is the need to advocate for additional resources. The risk here is making the case without overstating it. We know that many people experiencing psychological distress after a disaster improve over time without any intervention. We also know that many people do not. So, emergency managers and BH agencies have to make informed estimates of the long-term BH needs and request resources in line with those estimates. Clearly, there are legal, policy, and ethical implications for these decisions. This is why preevent strategic planning is so important, and why DBH and policy experts need to be engaged with emergency managers in the preparedness phase.

Does any of this change during an actual disaster response?
- What do I do if I do not have access to providers trained in DBH?
  - Can someone in my community provide just-in-time training?
  - Psychologists, for example, may provide services outside their usual scope of practice "in order to ensure services are not denied"
- How do I meet the existing need without creating unrealistic expectations for long-term services and without creating disparities in available assistance?
  - Focus on disaster-related needs
  - Stay within the program guidance for CCP and other programs
  - Utilize the expertise of your community's BH experts
  - Ensure that the BH experts and agencies are engaged in recovery planning from the beginning of the event

---

WHAT ARE THE LONG-TERM ETHICAL AND POLICY CONCERNS FOR IMPACTED INDIVIDUALS AND COMMUNITIES?

Just as the disaster context shapes the legal, policy, and ethical questions, it also shapes the long-term consequences of how those challenges are handled. Remember that communities have histories that are not erased by disasters. For some communities, disasters cause issues of social justice and resource discrepancies to play out in many ways. In Hurricane Katrina, for example, long-term inequities based on race and geography led to real and perceived inequities in response and recovery (Fussell, Sastry, & Vanlandingham, 2010). This has led to suggestions for more inclusive planning processes that consider issues including race, culture, and language (Andrulis, Siddiqui, & Gantner, 2007). Inclusive planning improves outcomes for community members, but it can also raise unanticipated concerns. For example, suppose the reader is coordinating the response to a disaster in a community with a large population of linguistically diverse children. A decision is made to apply for grants that will hire and train local paraprofessionals to do education, outreach, and counseling. Now suppose that the paraprofessionals are so successful at engaging the community that they begin to uncover many other psychological needs among the children. Since CCP grants are time limited, the existing BH system is strained, and bilingual BH professionals are few in number, what should the community do? Law and policy limit the duration of CCP grants and the focus of BH agency funds. But what are the ethics? How does a community trying to recover from a disaster handle long-term needs identified or exacerbated by the emergency that may have existed undetected before the emergency?

There is no single answer to the questions raised above, but they point to the ways that law, policy, and ethics intersect in DBH response. They also point to the ways in which decisions made during a disaster can have long-term implications for the wellbeing of a community. Uncovering and addressing previously hidden BH needs in children could be an incredibly positive thing for the community in our example. But it could also set a precedent for over-extending already taxed systems in the recovery phase, or creating an unrealistic expectation for ongoing services. Every decision can have long-term implications for policy and ethics.

These issues are raised to encourage emergency managers to include BH in their preparedness, response, and recovery planning. Excluding BH has led to serious mistakes with severe negative consequences for all involved. On the other hand, including BH entails much more than simply having a plan to deploy disaster health workers as needed. It requires a thoughtful consideration of the legal, policy, and ethical issues involved in the delivery of such services. Fortunately, there are experts across the country on these issues. The critical task is for emergency managers and planners to engage with their BH experts early and often to ensure that the emergency managers are complying with the law, correctly implementing policy, and ensuring that needs are being addressed ethically. The engagement can happen via the public health and medical support function (ESF#8) but it extends beyond that to include mass care and communications. For that reason, emergency managers need to seek out their DBH contacts just as DBH leaders need to seek out their EM officials. This engagement needs to continue throughout the preparedness, response, and recovery phases of disasters. The results of such collaboration will provide significant short- and long-term benefit for any community, even if no disaster occurs.

REFERENCES

Andrulis, D. P., Siddiqui, N. J., & Gantner, J. L. (2007). Preparing racially and ethnically diverse communities for public health emergencies. *Health Affairs*, *26*(5), 1269−1279.

Boscarino, J. A. (2015). Community disasters, psychological trauma, and crisis intervention. *International Journal of Emergency Mental Health and Resilience*, *17*(1), 369−371.

Call, J. A., Pfefferbaum, B., Jenuwine, M. J., & Flynn, B. W. (2012). Practical, legal and ethical considerations for the provision of acute disaster mental health services. *Psychiatry*, *75*(4), 305−322.

Dodgen, D., & Meed, J. (2008). Law, policy, and procedure: The role of government in addressing traumatic stress. In G. Reyes, J. Elhai, & J. Ford (Eds.), *Encyclopedia of psychological trauma*. Hoboken, NJ: John Wiley & Sons.

Dodgen, D., & Meed, J. (2009). The federal government in emergency response: An ever-evolving role. In P. Dass-Brailsford (Ed.), *Crisis and disaster counseling: Lessons learned from Hurricane Katrina and other disasters*. Thousand Oaks, CA: Sage Publications.

Federal Emergency Management Agency. (2015a). *Crisis Counseling Assistance & Training Program*. Retrieved from https://www.fema.gov/recovery-directorate/crisis-counseling-assistance-training-program.

Federal Emergency Management Agency. (2015b). *National Planning Frameworks*. Retrieved from http://www.fema.gov/national-planning-frameworks.

Federal Emergency Management Agency. (2016). *Robert T. Stafford Disaster Relief and Emergency Assistance Act, as amended, and Related Authorities as of April 2013*. Retrieved from https://www.fema.gov/.

Flynn, B. W., & Speier, A. H. (2014). Disaster behavioral health: Legal and ethical considerations in a rapidly changing field. *Current Psychiatry Reports*, *16*, 457.

Fussell, E., Sastry, N., & Vanlandingham, M. (2010). Race, socioeconomic status, and return migration to New Orleans after Hurricane Katrina. *Population and Environment*, *31*(1−3), 20−42.

Howe, E. G. (2012). What legal risks should mental health care providers take during disasters? *Psychiatry*, *75*(4), 323−330.

Institute of Medicine (2015). *Enabling rapid and sustainable public health research during disasters*. Washington, DC: National Academies Press.

National Emergency Management Association. (2016). *EMAC legislation*. Retrieved from http://www.emacweb.org/.

North, C. S., & Pfefferbaum, B. (2013). Mental health response to community disasters: A systematic review. *Journal of the American Medical Association*, *310*(5), 507−518.

Pfefferbaum, B., Flynn, B. W., Schonfeld, D., Brown, L. M., Jacobs, G. A., Dodgen, D., ... Lindley, D. (2012). The integration of mental and behavioral health into disaster preparedness, response, and recovery. *Disaster Medicine and Public Health Preparedness*, *6*(1), 60−66.

U.S. Department of Health and Human Services. (2016a). *Hospital Preparedness Program Overview*. Retrieved from http://www.phe.gov/Preparedness/planning/hpp/Pages/overview.aspx.

U.S. Department of Health and Human Services. (2016b). *At-Risk Individuals, Behavioral Health & Community Resilience (ABC)*. Retrieved from http://www.phe.gov/abc.

U.S. Department of Health and Human Services. (2016c). *Pandemic and All-Hazards Preparedness Reauthorization Act.* Retrieved from http://phe.gov/Preparedness/legal/pahpa/Pages/pahpra.aspx.

U.S. Department of Health and Human Services. (2016d). *National health security strategy.* Retrieved from http://www.phe.gov/preparedness/planning/authority/nhss/Pages/default.aspx.

U.S. Department of Justice. (2016). *Office for Victims of Crime.* Retrieved from http://www.ovc.gov/.

Making Integration Work

Berl D. Jones, Jr.

There are many complex legal and ethical issues faced by emergency managers as they implement programs and services to individuals, families, and communities following disasters. These challenges are dynamic and may change considerably over time.

To establish and sustain integration between emergency mangers and DBH professionals, these complexities must be well understood, openly and frequently discussed, and understanding and agreement must be reached. At the same time, DBH professionals are also bound by a wide variety of legal and ethical considerations that are discussed by Dr. Dodgen later in this chapter. It is imperative that emergency managers understand those factors also.

Only through ongoing discussion and information sharing can the various requirements be managed. These are not easy discussions, as problems are far more easily identified than solutions found. Complexity can easily bog down the integration process. Several suggestions for fostering integration on these topics include:

- Prior to attempting harmonizing and integrating these types of complex factors, emergency mangers and BH professionals should, within their own domains, seek to identify key legal and ethical factors and work toward agreement and consensus within their own professions. For example, for DBH, in the absence of case law on the topic, there are areas where most BH professionals (and their professional organizations) will find easy consensus and many areas where they would not. This will be an ongoing and dynamic process. Without internal consensus, forging external agreements is difficult
- As consensus within the two professions grows, it is important that the most current and accurate guidance is provided to members. For example, it is important that all appropriate federal emergency managers implement similar policies and practices regarding how and what information is shared with DBH partners. At the same time, it is important the DBH workers in various areas of the country and at the state and local levels share a common understanding of their legal and ethical responsibilities
- As progress is made on the above efforts, it is important for emergency mangers and BH professionals to work together to identify areas of easy agreement as well as those areas where there is either lack of clarity or even conflict. This process should lead to the formulation of strategies to implement agreements, explore resolution of differences, and monitor the effects of those actions

All parties will benefit from a continuing reminder of three determining factors:

1. Confidentially requirements and controls on information exist for legitimate and important reasons
2. Solutions that result from an integrated approach benefit all disaster preparedness and response and those who lead those processes
3. The ultimate value behind the need for integration and solving complex problems is to benefit disaster survivors. *This is why EM and DBH professionals do what they do.*

Chapter 13

Sustaining Integration: A Way Forward

Brian W. Flynn[1] and Ronald Sherman[2]
[1]Uniformed Services University of the Health Sciences, Bethesda, MD, United States,
[2]Independent Consultant, FEMA Federal Coordinating Officer (Retired), United States

Throughout this book, readers have examined the need for integration between disaster behavioral health (DBH) and emergency management (EM), as well as with other key stakeholders, such as the private sector. Chapter authors have shared their decades of experience and advice. Hopefully, readers now have an expanded understanding of the "why" and "how" of this integration of effort, skills, roles, and experience.

However, history is replete with noble goals, good ideas, and dedicated people that accomplish a specific goal, only to have the value of their achievements diminish with time. Why this happens probably results from a variety of causes and factors. This chapter, drawing from the authors' experiences, as well as drawing together lessons from previous chapters, will discuss why this may occur and suggest steps to optimize the possibility that integration will be sustained. Where appropriate, references to earlier chapters are provided.

WHY INTEGRATION BREAKS DOWN

In identifying why integration breaks down, it is hoped that both DBH professionals and emergency managers will be able to spot key events early and take corrective or supportive action quickly. Following are several factors to be aware of:

Change of Leadership

When system change occurs, it is often the result of key individuals and personalities. These individuals often bring passion, energy, and skills to new and/or complex tasks. These individuals are often very persistent, visible, and located relatively high on the organizational chart. They create new alliances and often accomplish much. However, too often, the sustainability of these alliances is overly dependent of these personality dependent factors. When these individuals are no longer in their roles to maintain what has been developed, integration suffers and sometimes fails.

There are a couple of strategies to reduce this risk. First, it is necessary to cultivate multiple leaders from both groups at the start of the integration process. In this way, the resignation or retirement of a single individual will be less costly to the effort when that person is no longer there to lead. Second, document, codify, and institutionalize agreements, understandings, and processes so that there is an historical record of what has been and should be done. This suggestion is made with the clear understanding that creating documentation is often among the least favorite activities of action oriented people. Yet, it is one way to promote continuity and sustainability.

Change of Authorities

As has been well documented in this book, disaster response is becoming more formalized. This formalization is reflected in existing and emerging laws, regulations, and standard operating procedures (SOPs.) This is occurring across the disaster preparedness, response, and recovery spectrum. To perhaps oversimplify, everything has rules, these are the rules, and the rules are evolving. Both EM and DBH professionals should be monitoring and influencing these changes to assure that as changes in authorities and practices change, they as practitioners will change, help foster the changes, and not impede integration.

Two examples may be helpful in making this point. Prior to the Stafford Act (discussed in several places in the book), the provision of crisis counseling services was not a legislated program option for which states could apply after presidentially-declared disasters. In this case, legislation (a law, the strongest statement of rules) enhanced integration. In other cases, emerging formalization may make integration more difficult. For example, earlier in this book, readers heard of situations where, in less formal days, behavioral health was present as a key partner in disaster operations centers. With the establishment of emergency support functions, behavioral health support is subsumed under broader health and human service missions and, as a result, may actually have less direct involvement and access in Emergency operations centers. In these cases, integration faces additional barriers and creative solutions must be sought to sustain what was already occurring.

Political Landscapes

Earlier in this book several authors state that all disasters are political events. In this context, disasters affect politics and politics affect disaster response. In the political domain, not only do key individuals change with regularity, but political priorities and attitudes can, and do, change. In some political environments, integration and collaboration are highly valued and fostered. In others, independent function, organizational, and professional competition and protectionism are more valued. Often, in these latter type environments, it is competition for limited funding that drives the professional isolationism and establishing and sustaining integration is much more difficult.

Leaders in both EM and DBH are encouraged to monitor the political landscape and take advantage of periods where integration is valued and, as suggested earlier, codify and document agreements and processes. This will help those who value integration to sustain their relationship through tough times in which other values dominate.

Changing Evidence and Practice

As noted earlier, disaster preparedness and response is becoming more formalized. There are many good reasons for this—especially the demonstrated value of consistency in response structure, such as the nearly universal adoption of the incident command system. Another factor contributing to this formalization is the accumulation of evidence about what policies, programs, and interventions are most effective. Research and evaluation is difficult in disaster situations, and there is less efficacy data in every area than is ideal. However, learning continues, data and experience bases are built and, over time, practice and policy changes to reflect best or promising practices.

Sustainability can be enhanced if both EM and DBH contribute to the evidence base and apply what is learned. There will always be situations where new approaches need to be tried in the absence of information and data. Yet, all parties must be vigilant to assure that practice and policy changes when information bases change.

As an example, mentioned earlier in Chapter 6, Not All Disasters Are the Same: Understanding Similarities and Difference, and Chapter 10, Integration in the Emergency Operations Center (EOC), critical incident stress debriefing (CISD) emerged as a popular and, for many, an intuitively sound approach to reduce adverse psychosocial consequences in first responders and others. As more research emerged, it became apparent that it did not live up to the early promises and, in fact, has the potential to make things worse for some people. It is rapidly falling out of favor nationwide. There must be continuous assessment of what is done in all aspects of disasters to assure that what is practiced, promoted, and sustained represents the best-known practices.

Visibility of Impact

Behavioral health professionals are fond of saying that behavior does not continue in the absence of reinforcement. In other words, we will do, and keep doing, what gets us what we want and we will not do what does not give us what we want. So, what does that have to do with sustainability of integration?

It means that all involved will be more willing to do the work of integration if, over time, the rewards of their efforts can be seen. Everyone likes to see the positive results of their work. The culture of EM is especially oriented toward rapid and clear results of actions, procedures, and policies. The professional culture of behavioral health often values an approach based on theory development, research that includes replication of results, publishing of research, and, over often long periods of time, acceptance and application in practice and policy. The behavioral health approach described is a sound process but one that often moves at glacial speed. It is not uncommon for a decade or more to pass between the time evidence is accepted and changes in practice are seen. The differences in professional cultures can easily lead to frustration on both sides. It is common to hear behavioral health professionals express concern that emergency managers are making decisions without adequate information and data. Similarly, it is common for emergency mangers to become frustrated when behavioral health professions cannot or will not give rapid and definitive responses to seemingly straightforward questions or requests. Again, understanding and respecting different professional cultures is a key to successful and sustained integration.

Behavioral health professionals typically have their professional roots in behavior theory. They know that behavior does not persist in the absence of reinforcement. In this case, successful integration that results in enhanced function is a powerful reinforcement. Sustainability will likely be compromised if significant time and effort is spent in developing and operationalizing integration and there is insufficient evidence of the benefits of these efforts. The more candid among us will acknowledge that, as selfless and altruistic as we like to see ourselves, there is always a "what's in it for me" question when difficult tasks are undertaken. The many answers to that question were presented in Chapter 7, What Can DBH Actually *Do* To Make Emergency Managers Jobs Easier?. All parties are encouraged to monitor the results of integration efforts and practices, as well as pay special attention to the identification and sharing of positive results.

Competing Demands and Priorities

Whether one is a BH or an EM professional, each is always faced with multiple competing, and often conflicting, demands on time, energy, and resources. To complicate the picture, many of these constraining factors are externally imposed. How many DBH or EM professionals get to determine their budget? The constraints can optimistically be seen as chances to promote creativity. There is nothing like shortage and other restrictions to produce creative problem-solving as the only alternative to failure. Unfortunately, this is part and parcel of disaster work, and few would be willing to bet on it changing any time soon.

It is easy to find rewarding efforts that are showing results but then take a back seat to new and more pressing needs and priorities. That squeaky wheel seldom seems to have enough grease. Both EMs and DBH professionals can help assure sustainability by making priorities include setting a conscious, shared process and responsibility. While there may be a need to modify priorities or shift resources, it is always better when those choices are both conscious and viable. In that way, options and strategies can be better explored.

Eight Tips to Reduce Integration Breakdown
1. Cultivate multiple leaders.
2. Put policies and procedures in place that foster integration in writing.
3. Closely and continuously monitor and influence laws, regulations, and policies that impact integration.
4. Be aware of the political landscape.
5. Anticipate and be prepared for change.
6. Track and implement improved, better, and best practices as they emerge.
7. Make positive results visible to all.
8. Monitor and openly address changing demands and priorities.

FOUR PILLARS OF SUSTAINED INTEGRATION

The previous section describes factors that can threaten and compromise sustained integration. This section will explore factors supporting sustained integration.

In considering the primary factors that establish and sustain integration, four emerge that are critical. These factors are trust and respect, demonstrated benefit, adequate resources, and adaptability. These are illustrated in Figure 13.1.

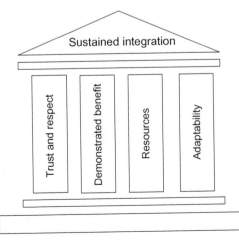

FIGURE 13.1 Four Pillars of Sustained Integration.

Mutual Trust and Respect

All sustainable relationships are based on trust and respect. The relationship between emergency managers and behavioral health professionals is no exception, and has been discussed in various sections of the book already.

Trust is based on understanding, appreciating, and acting upon shared values developed and demonstrated through mutual exposure. This exposure will be far more powerful if it is developed in person rather than through electronic or third person channels. Contrary to trends in social media, people are far more comfortable building trust in person than through some other means.

In addition, trust is not only built solely on promises and commitments. It is demonstrated in various contexts (e.g., exercises, actual disaster response, etc.) over time. As pointed out in Chapter 4, Why Is Integrating Emergency Management Essential to Disaster Behavioral Health? Challenges and Opportunities, and Chapter 5, Integration in Disasters of Different Types, Severity and Location, the middle of a response effort is not the time for introductions and exchanging business cards.

Mutual respect is the companion to trust. If effective integration is to occur, both professions must see and understand the valuable, unique, and complementary skills and training that each bring to the table. This is harder than it may seem. These are two very different professions with different professional cultures. The nature and training of each is quite different. One might suspect that each profession attracts people with potentially different interests and temperaments.

It would be easy to focus primarily on the differences. But, integration will be enhanced if each also focuses on the similarities instead. Neither emergency managers nor behavioral professionals who choose to become involved in disaster work would do this work if they did not share an uncompromising commitment to assist those who may or have experienced the tragedy of disasters. In times when integration becomes difficult or when the professional differences may be overwhelming, it is helpful to be reminded of this core value that drives both professions. The sense of a shared mission was explored Chapter 4, Why Is Integrating Emergency Management Essential to Disaster Behavioral Health? Challenges and Opportunities.

Demonstrated Benefit

As noted earlier in this chapter, integration will be sustainable if positive results can be visible. Earlier, we spoke about visibility to each other. It is also important that the positive results of this integration be visible to others as well. The impact of positive visibility pays dividends. Several examples follow:

- Survivors will be reassured when they can see an integrated and seamless response and recovery process and system.
- Political leaders who determine resources will be more likely to support both professions and their integration when they can see positive results. Remember, as stated earlier in this book, all disasters are political events.
- People who review and revise emergency operations plans will be more likely to include or expand the role of DBH in those plans.
- The media will respond positively to well-crafted messages created by EM and DBH and will present those messages in a positive way to survivors.
- Nongovernmental organizations, private sector companies, volunteer groups, and faith-based groups will be far more willing to interact with and become part of a team that is cooperatively focused on survivors' needs.

Both emergency mangers and DBH professionals will do well to consider how to make the benefits of their shared efforts visible to multiple stakeholders.

Resources

Good intentions seldom go very far if not accompanied by adequate and appropriate resources. Three types of resources are critical in this situation:

Human Resources

The likelihood of effective and sustained integration will be enhanced if adequate and appropriate resources are available. For emergency mangers, if there are insufficient human resources to prepare for and respond to disasters, it is unlikely that the important work of integrating with behavioral health will come to fruition. Other demands may overwhelm the existing EM resources. This is where incorporating volunteer groups, like medical reserve corps (MRC) and community emergency response teams (CERT), can help augment and backfill standing EM organizational resources. Both groups may have volunteers, some of whom are professionals, who are trained and skilled in at least psychological first aid.

On the behavioral health side of the equation, establishing and sustaining this relationship requires trained individuals who are interested and available to commit to this type of work. Few have this interest and training. Even if interested, commitment to a sustained effort becomes difficult, since full-time work in the field of DBH is almost nonexistent. Even those with the interest and training are often limited in the uncompensated time required for sustained effort.

If a disaster response is dependent on behavioral health resources that are based away from the impacted area, integration is even more difficult. It is imperative that DBH responders, from afar, be part of organized, recognized, and requested response organizations. As noted in other portions of this book, spontaneous, uninvited behavioral health volunteers often create more problems in a disaster response than they solve. Some of the negative results of this type of volunteer involvement were described in Chapter 4.

> Beware of the spontaneous uninvited volunteer (SUV) crisis that often accompanies disasters, especially highly publicized events. The best way to avoid problems is to have a scalable, known, and trusted source of DBH resources organized before a disaster occurs.
>
> While many SUVs are well-meaning, some are not. Their training may not be appropriate or adequate. They may generate expectations for services that the local community cannot meet when they leave. Verifying credentials in the midst of a response nearly always takes valuable resources away from the response.
>
> For guidance on how to prepare for, deal with and utilize uninvited or unaffiliated volunteers, we refer readers to a document prepared by the National Voluntary Organizations Active in Disaster, which can be found at: http://nvoad.eden.lsuagcenter.com/resource.aspx?ID=34.

Funding

There really is "no free lunch." All disaster preparedness, response, and recovery activities, especially a comprehensive and integrated approach, cost money. Historically, funding through EM channels for disaster preparedness has seldom included support for behavioral health inclusion. Likewise, seldom has public behavioral health funding included support for disaster preparation. When it has, this funding has been small and short-term, but extremely valuable.

It is incumbent and necessary for leaders in both EM and behavioral health to continue to advocate for adequate and appropriate support for this integration.

Time

Even in the unlikely event that both adequate and appropriate human and financial resources were in place, accomplishing integration take time. To be sustainable, such integration must be practiced, evaluated, and modified. It all takes time.

In a world that increasingly seeks quick fixes, emergency managers and DBH professionals alike will benefit from a coordinated and consistent voice in promoting the need for legitimate processes to take place, even if they do not yield results as quickly as all would like.

> Here are two things the editors have learned during their careers.
>
> "In most things in life, but especially in establishing and sustaining disaster behavioral health systems, I have learned one thing. Things take longer than they do." *Brian Flynn*
>
> "You can't plug a trailer into a tree. Things take longer than anybody wants." *Ron Sherman*

Adaptability

Even with good, creative, and motivated people at the table promoting integration, it is unlikely that the best answers will always be the first answers. To complicate matters, no disasters are identical. Government policies change, best and promising practices change, communities change, and perhaps most importantly, personnel change.

Even when integration is well-established, it is mandatory that agreements and practices be adaptable in changing circumstances. Change and flexibility should be anticipated in the developmental process. To borrow a disaster related metaphor: in a wind storm, it is most likely that the tree that does not bend or flex will be destroyed.

Chapter 14

Conclusion/Summary

Brian W. Flynn[1] and Ronald Sherman[2]
[1]*Uniformed Services University of the Health Sciences, Bethesda, MD, United States,*
[2]*Independent Consultant, FEMA Federal Coordinating Officer (Retired), United States*

Let us take a moment to revisit the motivation and design of this book as context for better understanding its content. The motivation to produce the book was based upon an increasingly apparent need to promote and operationalize the integration of professions of emergency management (EM) and disaster behavioral health (DBH). But why, and why now? There are two reasons:

- Experiences when integration has occurred have resulted in positive results for both professions, as well as positive impact on survivors and disaster workers.
- Experiences where integration has failed to occur have resulted in negative effects on the function of both professions, as well as diminished positive outcomes for survivors and disaster workers.

The presence of or lack of integration appears to have an effect on outcomes.

Let us also remember how the topics addressed here came to be. There was an extensive survey of ten thought leaders in both the EM and DBH fields. They were asked a wide variety of questions. Is such a book needed? Who should the target audience be? What topics are important/less important to address? How might such a book be used? Who should contribute? The list was quite lengthy.

We also did an extensive search of existing writings and publications. There was nothing in the existing literature that addressed this complex topic. The resulting sections and chapters in this book, as well as the selection of the contributors, are a result of that extensive analysis process.

The design format of the book was the result of strong editor bias toward modeling integration and making the book practical for users. A more traditional approach might have had a section approaching the topics from a DBH perspective and then another section from an EM perspective. However, in our view, that approach would have modeled a nonintegrated approach. Instead, the design of combining a contribution from each profession in each chapter was selected. Finally, each chapter ends with a section dedicated to identifying actionable steps that can be taken to address the chapter topic.

As editors, we could not be more pleased with the outcome. We also learned a great deal from all of our contributors. They were thoughtful, articulate, and pragmatic in their writing. In addition, as noted at the start of the book, they were specifically selected for not only their knowledge and ability to communicate in this medium, but for their real-life, real-time disaster experience. They were asked to talk the talk because they each have walked the walk.

KEY FINDINGS

Foundational Agreements

Writers on this book, with very few exceptions, worked independently from one another. Still, while their topical content was different, their big-picture perspectives were remarkable similar. The following themes are interwoven throughout the book:

- Integration of EM and DBH is a critical goal and worthy of serious attention for both professions.
- Integration is necessary in all event stages from preparedness, through response, to recovery.
- While ideally sharing common goals, EM and DBH are very different professions and it is important to understand and accommodate these differences while remembering common goals and values.
- Integration should focus on meeting the needs of disaster survivors (individuals, families, and communities) as well as disaster workers (including emergency managers).
- Integration can occur only if built on a foundation of shared knowledge, understanding, and respect.
- Integration must be a conscious process involving both professions. It will not happen by chance.
- Integration must be practiced through drills, exercises, and actual disasters.
- Integration is not a steady state. It must change and adapt as personnel, laws and regulations, structures, and the knowledge/experience base expands.
- Integration is most achievable within an EM culture that values collaboration, integration, shared values, and caring for survivors and workers alike.

Conclusion/Summary Chapter | 14   333

> **Editors' Note from Ron Sherman**
> During the editing of this book, and writing our own pieces, the editors observed how the differences in our professional cultures seemed to be reflected in the writing style of EM and DBH contributors, including ourselves.
>
> As emergency managers, we are used to writing short, concise statements for inclusion in incident action plans or shift transition briefings. Many of us struggled with producing the requested number of words for our sections. I recall thinking, "Three thousand words?! I can do this in three sentences." After sharing my section of Chapter 2, Where Emergency Management and Disaster Behavioral Health Meet: Through a Disaster Behavioral Health Lens, with a former co-worker, her first comment was, "You are still the king of terse verse. You need to elaborate."
>
> It did not seem that any of the DBH contributors had any difficulty in producing the requested number of words. While reviewing their submissions, I was struck by the ease with which they laid out a topic, examined its components, and then summarized their thoughts. Why the difference? Perhaps one reason is best expressed by emergency manager Nancy Dragani who wrote the lead section of Chapter 1, Where Emergency Management and Disaster Behavioral Health Meet: Through an Emergency Management Lens. In response to an email in which I mentioned our observations, she replied, "Need for decisions verses need for scientific review and discussion." I immediately understood what she meant—no elaboration needed.
>
> The importance of understanding and respecting the professional cultures of both groups is mentioned several times in this book. How we express ourselves and why is an important part of understanding and respecting different occupational and professional cultures.

Exceptional Opportunities

Interwoven in these pages are themes of concerning the extraordinary opportunities available as integration expands. These opportunities seem to cluster in two areas:

1. Integration can result in significant benefit for disaster survivors and response workers. For example, as DBH is integrated into emergency operations centers, DBH will be better equipped to more precisely target their efforts in locations with highest or emerging needs.
2. Integration, while an ongoing effort, can help both emergency mangers and DBH professions perform their respective jobs better. For example, by integrating EM public communications efforts and DBH, messages can be crafted for optimal impact and opportunities for anticipatory guidance regarding psychosocial consequences.

In addition to the above examples, throughout this book, there are specific suggestions for exciting opportunities in several specific areas. These include:

- Joint training opportunities
- Incorporating other professions in support of DBH and EM integration
- Developing on-line training accessible through Federal Emergency Management Agency's Emergency Management Institute
- Integration of behavioral health (BH) professionals and those who are tasked with public information and risk communication
- Create a list of potential stakeholders—including private practitioners and faith-based organizations
- Review of emergency operations plan to ensure DBH or EM is included in notification protocols

Significant Challenges

If integration was easy and did not require much effort, it would be functioning universally already. It is not. While convincingly identifying what can be gained through integration, contributors were also frank on the challenges facing its accomplishment.

Some Convincing is Still Needed

In both professions, some convincing regarding the value of integration is needed. Some, thankfully which is a rapidly declining amount, emergency mangers believe that behavioral health in disasters is something that comes after the response is over. Some believe that behavioral health concerns are secondary in nature and not worthy of the time, energy, and resources necessary to address these needs. Some are very reluctant to acknowledge the impact of stress, as well as policies and practices that promote stress, in the response and EM workforce.

Some DBH professionals fail to understand and appreciate the need of functioning within the EM structure of disasters. They believe their work is noble and valuable, and should proceed as they independently decide. They feel that their work should not be impacted by an overarching structure led by those in a different profession. Some have little appreciation for the special training required to provide service in a disaster environment or how different it is from more typical practice settings.

Unfortunately, integrating is complicated by a few in both professions who have had less than positive experiences with the other. Those negative experiences serve to document and support the need for true integration that results in mutually beneficially and positive experiences.

There is a need for an education and marketing task to be performed in both professions. This is best accomplished by experienced and dedicated leaders in both professions, like those who have contributed to this book, who can realistically convey to their peers the nature of the work as well as its opportunities and limits. Hopefully, readers of this book feel like they can now be an important part of this process by employing the ideas presented in the various "Making Integration Work" sections.

Taking the Time/Resources to Make It Happen

Since we were children, we have all heard that there is no free lunch. Well, there is no free integration of EM and DBH either. It will take time, effort, and resources to make it happen, as well as to sustain it. The leaders of integration need to be at the table and stay at the table. As noted earlier, whenever two professions with different histories and cultures seek to integrate, there are bound to be frustrations and setbacks. The integration process is bound to be anything but linear. Practitioners will likely need to remind each other of their shared goals and the benefits of reaching those goals.

Mutual Learning Curves

These two professions are not a natural or easy match. Yet, the contributors to this book build the case for making and sustaining this match. In addition to the elements of integration already mentioned, there is a significant learning curve facing both professions.

DBH professions must have at least a working knowledge of EM roles, responsibilities, structures, and requirements. As noted, they must also understand the professional culture that is quite different from their own. DBH professionals should be prepared, as discussed in Chapter 1, Where Emergency Management and Disaster Behavioral Health Meet: Through an Emergency Management Lens, and other places in the book, for occasionally receiving something less than full and unconditional acceptance from some emergency managers.

Emergency managers face similar challenges. Some have limited knowledge and experience with DBH professionals and may need to better understand the wide range of what DBH providers can contribute and where limitations exist. Potential roles, in addition to direct service to survivors, are provided through this book. Emergency managers will encounter many DBH professionals who have had no exposure to, or understanding of, structures such as the National Incident Management System (NIMS) and Incident Command System (ICS) and some emergency managers have not adapted to the ICS and may not know where DBH can fit in.

Understanding Changing Rules and Requirements

Throughout the book, contributors have noted changes and trends in both EM and behavioral health. For example, Chapter 1 describes the historical context of how and why EM functions as it does. Other sections, such as

Chapter 5, Integration in Disasters of Different Types, Severity and Location, and Chapter 8, Expanding the Tent: How Training and Education Partnerships with Other Professions Can Enhance Both EM and BH, discuss the changing nature of DBH. Without question, both professions are changing and it remains to be seen how they will continue to evolve. A major challenge for integration is to assure shared awareness and adaptation to a changing world.

A few trends reflected in this book are indisputable and are likely to continue:

- EM has been and will continue to become more formalized and structured at all levels of government.
- EM is embedded as part of the nation's overall homeland security structure.
- Both professions are evolving based on research, data, and documented cumulative experience. Both are becoming more evidence-based, rather than evidence-informed, resulting in changing approaches.
- Both professions are driving toward more consistent and standardized approaches. While preserving flexibility, this assures that there is consistency in approaches in diverse settings and at different levels of government.
- The range of event types experienced and anticipated is widening, requiring all involved to examine how existing and emerging approaches will need to adapt. Before the bombing in Oklahoma City in 1995, EM in the United States was not focused on large-scale domestic and foreign terrorism within the country's borders. Before the Mariel Boatlift in 1980, when 125,000 Cuban exiles arrived in mass on United States soil, EM and behavioral health had not collaborated on such a large-scale immigration emergency. The threat of pandemic has required that EM and all areas of health, including behavioral health, work together in innovative ways to deal with events that may impact many or all communities.

The Path Ahead

A clear and consistent message throughout the book is that integration of EM and behavioral health is a requirement if the needs of survivors and disaster workers are to be adequately addressed. In many ways, if both professions pledge fidelity to their professional goals, this is not a choice. It has, will, and must continue to happen.

There is not a single path forward. Each of these chapters contains a "Making Integration Work" section to suggest concrete ways to make integration around the various topics come to fruition.

Moving forward will require a multi-pronged approach. Elements to "make it work" include:

- Advocacy for integration of EM and DBH. This advocacy, to be effective, will need to occur with peers, governmental policy makers and funders, and key behavioral health and EM stakeholder organizations and individuals.
- Expanded training and educational opportunities within and between the two professions.
- Commitment of thought leaders in both fields, including the contributors to this book. As in everything, leadership matters. All of us have a professional obligation to provide leadership in every way we can.
- Cultivation of successors. One of the great challenges of innovation is that it often relies on the energy, creativity, and commitment of a few key leaders. When those leaders no longer hold these leadership roles, because of retirement or other factors, there too often remains a void. As a result, unfortunately, the gains made are too often erased. Each of us who play or have played a leadership role on this topic has an obligation to consider succession planning and prepare others to lead.

CONCLUSION

For us, as designers and editors of this book, this has been a significant milestone in a long journey. As noted at the start, the book was born out of shared values and experiences over several decades. We first met at the sites of several major disasters, and along with supportive colleagues, began thinking, talking, and experimenting with how we, from such different professions, could help each other work and serve survivors and our fellow disaster workers more effectively. Over these many years, we have continued to be close colleagues and a valuable friendship has emerged. These years of working together and friendship has sustained and inspired us, as well as through the many months of work with so many talented contributors to the book.

As the book is completed, we share the following reflections on the process and outcome:

- As the process of producing this book came to a conclusion, we realized that in this process, mostly unconsciously, we had modeled what is needed for the integration of EM and DBH.
- We proceeded with a deep and abiding respect for each other and our respective professions.
- We both realized that neither of us could complete a book of this type without each other.

- We learned from each other and our contributors every step of the way. This included explaining terms and acronyms that, even after all these years, were sometimes still not understood.
- When things got bogged down or frustrating, we supported each other and pushed through together.
- We listened to each other and to our contributors.
- We also laughed together as we learned together.

We find ourselves more optimistic and energized than ever that integration can be accomplished. We always believed it, but our wise and talented contributors reinforced that belief at every turn.

We are also sobered by the realization of how much remains to be done and how the increasing variety of events for which the nation must prepare is expanding. These challenges will test even the most committed. The road is not straight or easy, but we are confident that, in the pages of this book, readers will find guidance and support to accomplish the goal of integration. The well-being of survivors and disaster workers alike depends on us.

Index

Note: Page numbers followed by "*f*" and "*t*" refer to figures and tables, respectively.

A

After Action Report (AAR), 121
Agnes (hurricane), 76–77
All Hazards model, 14–15
All-Hazards Planning Guidance, 15
All-hazards plans, 8, 28–30
American Civic Association (ACA) Immigrant Center, 120
American Red Cross (ARC), 135–136
Andrew (hurricane), 50
AskMOVA, 169–170
Assistant Secretary for Preparedness and Response (ASPR), 307

B

Behavioral domain reactions, 161
Behavioral health practitioners, 24
Binghamton New York shootings, disaster response in, 120–121
Body responses to extreme situations, 159–165
   physical domain reactions, 161
Boston Marathon Bombing, 123–124, 141, 142*t*, 144–145
Brain responses to extreme situations, 159–165
   cognitive domain reactions, 161
Bush, President George W., 4

C

California Loma Prieta Earthquake CCP, 222–223
Carter, President Jimmy, 8
Changing Minds, 10
Cluster system for multisectoral coordinated response in disasters, 110, 112, 113*t*, 116*f*
Cognitive domain reactions, 161
Cognitive shifting, 160
Collaboration, 221
   American Red Cross and private sector Employee Assistance Programs (EAPs), 228
   basics, 221
   community support, 239
   evolution of disaster complexity and private sector, 228–239
   history and overview, 221–226
   integrating volunteers into disaster response, 240
   integration of behavioral health, emergency management, and the private sector, 248*t*
   private sector and nongovernmental organizations (NGOs), 226–228
      Flint water crisis, 229–230
      Missouri Floods, 229–230, 234–236
      9/11 terrorist attacks, 237–239
      Omaha Metropolitan Medical Response System (OMMRS)/Healthcare Coalition (HCC), 232–234, 233*t*
      Sandy Hook School Shootings, 230–232
   resilience, 235, 236*t*
   steps towards, 241–243
      education, 241–242
      live emergency management for life, 243
      preparing individuals and organizations, 242
      reach out to key partners, 242
   volunteer organizations active in disasters (VOAD), 240
Community-Based Psychological First Aid (CBPFA), 206–207
Community Emergency Response Team (CERT), 43–44

Community Mental Health Center model
    (CMHC), 78
Complex systems thinking, 101–102
Crime victim assistance agencies, 121
Crisis Counseling Program (CCP), 40–41, 46,
    76–79, 191–192, 221–222, 271
  strategies, 77
Critical Incident Debriefing Team
    development, 138–139
Critical incident stress debriefing (CISD),
    110–111, 323
Critical incident stress management (CISM),
    110–111
Critical Incident Stress Management Network,
    138–139

D

Davis, Elizabeth, 122
Deaf Mexicans, forced labor, 122
Demobilization, 136–137
*The Department of Emergency Services (DES)
    Spiderweb*, 131, 132f, 133–134
Department of Homeland Security (DHS), 5
Disaster assistance application process,
    298–300
Disaster Behavioral Health (DBH), 5t, 25–27,
    129–130. See also Emergency
    management (EM)
  action timeline, 29f
  adopting an EM framework, 28
  challenges for, 33–35, 144–145
  characteristics in terms of personality and
      work style, 35
  Concept of Operations (CONOPS), 31
  culture of, 76–78
  difference between community mental
      health care and, 15–17
  effective integration of disaster
      plan, 28–30
  emphasis on resilience, 32–33
  ethical guidances, 308–314
  future directions, 35–36
  guide to working with DBH experts, 22f
  indispensable to EM professionals, 91
  integration with EM, 87–88, 119
    adaptability, 329
    challenges, 152–153, 334–336
    change of authorities, 322
    change of leadership, 321–322
    changing evidence and practice, 323
    checklist, 126, 126t
    common and shared mission, 91
    competing demands and priorities,
        324–325
    comprehensive implementation strategy,
        188
    consultation to leadership, 155–157
    demonstrated benefit, 326–327
    example, 95
    extraordinary opportunities, 333–334
    factors hindering, 88–89
    FEMA framework for, 94
    foundational agreements, 332–333
    fundamental strategy for initiating
        change, 94–95
    funding, 328
    future of, 150–151, 336–337
    lack of, outcomes, 331
    National Biodefense Science Board
        recommendations, 95–96
    pillars of sustained integration, 325–329
    political landscapes, 322–323
    promise of, 95–96
    resources, 327–329
    time, 328–329
    tips to reduce breakdown, 325
    trust and respect, 326
    understanding culture, mission, and
        structure of EM, 89–90
    visibility of impact, 323–324
  literature, 78–80
  mass casualty event, 18–19
  methods of encouraging participation, 26
  misconceptions about, 13
  need for EM, 85–86
  as official jargon, 108
  operationalizing integration of EM into
      DBH, 74
  personal experiences from field, 26
  planners and coordinators, 28
  policy guidance, 30–32
  preparedness phase, 92
  preparing for, 19–20
  professional culture, 34–35
  professionals, 53
  providers, 110
    lessons learnt, 112–116
  readiness-for-change and systems
      perspective, 74
  recommendations for future, 21
  recovery phase, 20–21, 92–93
  response phase, 92
  roles and responsibilities within, 80

Index **341**

services
  evolution of, 301
  fraud and PII, 303–305
  information sharing, 305–306
  in-person contact, 305
  intent and complications of providing assistance, 301–303
  planning for or providing BH services, 307
  standards and practices, 34
  Super Storm Sandy response, 26–27
  systemic considerations responding to, 81
  wrap-around services, 41–42
Disaster Behavioral Health Response Network (DBHRN), 148–149
Disaster events, complexity of, 101–102
  descriptions, 224*t*
Disaster Relief Act of 1974, 75–77
Disaster response framework, 3–4
  in Binghamton New York shootings, 120–121
  in Boston Marathon Bombing, 123–124
  concept of operations' document, 107–108
  Deaf Mexicans, forced labor, 122
  deployment of DBH resources based on a mission request, 106
  for a disaster or catastrophe in United States, 107–108
  levels of, 102–112
    community level, 103
    for a crisis or emergency, 104
    for a disaster or catastrophe, 104–105, 107–108
    event management, 102–104
    events characteristics, 104–112
    example of state-and-regional level, 105–107
    at international level, 108–112
    at regional level, 106–107
    traumatic life events, 109
  liaison capacity to the HHS Incident Response Coordination Team, 108
  MHPSS applications, 111–112
  mission of the DBH Response Teams, 106
  monitoring needs for DBH services, 105–106
  New York mudslide, Yonkers response, 119–120
  Preliminary Damage Assessments (PDAs), 125
  scientific basis for interventions, 110–111
  settings for delivery of, 109–110
  for internally displaced persons (IDPs), 109–110
  variations in local, state, and federal disaster authorities, 124–125
Disaster survivors, 41–42
Disaster workers, 21
Doctors Without Borders (MSF), 110

E

Emergency communications centers (ECCs), 255, 257–258
  DBH perspective, 266*f*, 268*f*, 270*f*
  briefing, 264–271
  DBH personnel, roles of, 263–272
  DBH provider training and experience, 273*f*
  defining terms and mutual education, 274, 276
  disaster behavioral health qualifications checklist, 273*f*
  establishing partnership, 274–275
  provision of behavioral health services, 271–272
  situational awareness, 264
  field operations, 258
  integrating behavioral health in, 258–261
    addressing stresses, 260–261
    understanding stresses, 258–260
Emergency Management Assistant Compacts (EMACs), 124, 309
Emergency management (EM), 3–4. *See also* Disaster Behavioral Health (DBH)
  challenge for, 130, 144–145
  culture of, 74–76
  dynamic relationships between stakeholders and responders, 76
  resource allocation perspective, 17–18
  roles and responsibilities within, 80
  speed and efficiency, 297–298
Emergency management training and education
  allied professions, 197–198, 198*t*
  Certified Emergency Manager (CEM) program, 200
  comforter-in-chief, 194
  course syllabus, 192–196
  DBH perspective
    criminal justice programs and law enforcement, 209
    for directly affected, 203

Emergency management training and education (*Continued*)
  disaster relief directors, 203
  educating the education community, 208–209
  for indirectly affected, 203
  psychological first aid training, 209–210
  psychological support, 204
  training strategies, 207–214
  understanding of each other's fields, 210, 211*t*
  disaster drills, 212–214
  future prospects, 201
  higher education in, 199–201, 200*t*
  media approach, 196
  of mental health professionals, 211
  obstacles in, 214–216
    costs, 215
    stigma about receiving psychological support, 214–215
    time, 215–216
  practical ways to realize in, 218*t*
  professional experience in EM, 201
  rise in expectations, 194–196
Emergency managers, 18, 49–53, 76, 212
  actions to create and sustain integration, 47*f*
  checklist to conduct exercises, 44–45
  coping strategies and self-care, 64–66
  customers of, 51–53
    impassioned, 52–53
    local level customer, 51
    mitigation phase customers, 51
    preparedness phase customers, 51
    primary, 52*t*
    recovery phase customers, 51
  from disaster behavioral health perspective, 61–64
  establishing EM culture, 69
  experiences, 53–59
  first hand account of encountering behavioral health effects, 256
  focus on shared needs and challenges, 70
  identifying behavioral health resources, 67
  integration between EM and DBH, 70–71
  lessons learned, 59–60
  opportunities for promoting integration, 43–45
  orientation to world of emergency management, 68
  Population Protection Plans, 50
  provide support to DBH personnel, 86
  responsibilities, 68–69
  responsibilities of, 49, 51–53
  support and coordination mission, 51
  scope of practice, 68
  use of ESFs, 40–41
Emergency Medical Technician (EMT), 50
Emergency Operations Center (EOC), 14, 65, 85–86, 255, 257
  challenges in integration of EM and DBH, 272–273
  communications system, 257
  DBH perspective, 263, 266*f*, 268*f*, 270*f*
    briefing, 264–271
    DBH personnel, roles of, 263–272
    DBH provider training and experience, 273*f*
    defining terms and mutual education, 274
    disaster behavioral health qualifications checklist, 273*f*
    provision of behavioral health services, 271–272
    situational awareness, 264
  field operations, 258
  integrating behavioral health in, 258–261
    addressing stresses, 260–261
    understanding stresses, 258–260
  staff, 257
Emergency Support Functions (ESFs), 14–15
  areas of responsibility, 15
Emotional domain reactions, 161
Employee Assistance Program (EAP), 138
Establishment and Transfer of Command, 6
Ethical guidances for DBH, 308–314
  disaster context and, 314–315
  ethical issues, 312–314
  integration between emergency mangers and DBH professionals, 319–320
  law, 308–310
    disaster mental health and, 310
    in service planning, 309
  long-term ethical and policy concerns, 315–317
  policy, 310–312

F

Federal Emergency Management Agency (FEMA), 4, 7, 14, 50, 77, 124–125, 156–157, 191–192, 221–222
  planning guidance, 8
  preparedness guide, 137
  Routine Use mechanism, 304

Index   343

Federal Response Plan (FRP), 8
Ferguson unrest, 234–236
　Susan Flanigan's reflections, 245–247
Flint water crisis, 18
Florida's all-hazards Comprehensive Emergency Management Plan (CEMP), 105
Floyd Hurricane of 1999, 130–131
Foundation elements of any program, 134–138
Fugate, Craig, 10–12

G

Galveston Hurricane of 1900, 130
Globally networked risks, 102

H

Helbing, Dirk, 102
HHS Assistant Secretary for Preparedness and Response's (ASPR) Hospital Preparedness Program, 31–32
Homeland Security Presidential Directive 5 (HSPD 5), 4, 134–135
Homeland Security Presidential Directive (HSPD-21), 30–31
Hospital Incident Command System (HICS) Guidebook, 14
Hugo (hurricane), 50
Hurricane Sandy Recovery Improvement Act of 2013, 133

I

Incident Command System (ICS), 4–7, 13–15, 103–104, 133–134
　Communications and Information Management, 7
　core principals of, 5–6
　essential functions, 6
　Facilities and Resources, 6
　Planning and Organization, 6
　Professionalism, 7
Interventions with survivors
　benefits of incorporating DBHPs, 186–187
　EM perspective, 177–178
　　awareness and understanding of a community's complexities, 178–181
　　community leaders, role of, 179
　　emergency operations, 182
　　housing solutions, 185
　　incorporating DBH into response and recovery operations, 178–186
　　messaging, 181
　　misperceptions about government, 178–179
　　prioritization of assistance, 183
　　program implementation, 185–186
　　property acquisition and buy outs, 185–186
　　rebuilding, 186
　　redevelopment, 186
　　sheltering, 182–184
　　strategic locations, 184–185
　program evaluation, measurement, and monitoring, 173
　risk communication/media information, 165–171
　tracking and documenting behavioral health consequences, 173–174
　training and education for stress reactions, 171–172

J

Joint Field Office (JFO), 14

K

Katrina (hurricane), 3–4, 30, 33, 41–42, 50, 75, 131

L

Loma Prieta (California earthquake), 50

M

Massachusetts Office for Victim Assistance (MOVA), 169–170
Mass casualty event, 18–19
"Meals Ready to Eat" (MREs), 81
Mental Health and Psychosocial Support (MHPSS), 111–112
　in emergency settings, 111–112, 113t
　optimal for, 112
Mississippi Emergency Management Agency (MEMA), 189
Multiagency Coordination System (MACS), 4, 6

N

National Biodefense Science Board (NBSB), 31, 95, 206–207

National Center for Post Traumatic Stress Disorder (NCPTSD), 167
National Child Traumatic Stress Network (NCTSN), 167
National Crisis Counseling Hotline number, 304
National Disaster Recovery Framework (NDRF), 136
National Health Security Strategy's Implementation Plan (NHSS-IP, 2015-18), 30
National Incident Management System (NIMS), 4–7, 34, 134–135, 211, 226–227
 core principals of, 5–6
National Organization for Victim Assistance (NOVA), 110–111
National Preparedness Goal and Strategies, 31, 73–74, 135
National Response Framework (NRF), 8, 14–15, 31, 135
 Emergency Support Functions (ESFs), 135–137
National Response Plan (NRP), 141
National Voluntary Organizations Active in Disaster (NVOAD), 300
New England Hurricane of 1938, 130
New York mudslide, Yonkers response in, 119–120
9/11 terrorist attacks, 3–4, 237–239
 Diana Nordboe's reflection, 243–245
 media campaign Ad, 167–168
Northridge (California earthquake), 50

O

Ohio River, flood risk of, 10–11
Oklahoma Civil Defense Agency, 50

P

Pandemic and All Hazards Preparedness Reauthorization Act (PAHPRA), 308–309
Patient Protection and Affordable Care Act (Public Law 111–148), 73
Personally Identifiable Information (PII), 298
 fraud and, 303–305
Physical domain reactions, 161
Positioning theory, 9
Postdisaster strategies, 149–150
 DBH Coordinating Committee, role of, 150
 DBH leadership and experts, role of, 149–150
 lead mental health agency, role of, 150
 mental health experts in consulting role, 149
 Unified Command Structure, 149
Post-Katrina Emergency Management Reform Act (PKEMRA), 30
Posttraumatic stress disorder (PTSD), 18–19, 256
Predisaster strategies
 All Hazards Plan, 147
 behavioral health representatives, role of, 148
 exercises and drills, 148
 legislation, 148–149
 local and statewide disaster behavioral health assets, 147
 local and statewide disaster behavioral health plans, 147
 psychological training to emergency personnel, 148
Preparedness, development of, 131–134
 complexity of, 133
 from DBH perspective, 141
 foundation elements, 134–138
 integration barriers and strategies to enhance integration of EM and DBH, 146t
 nature of current services, 138–140
Presidential Disaster Declaration, 143–144
Presidential Policy Directive 8 (PPD-8), 73–74, 135
Privacy Act of 1974, 298
Psychoeducational information, 167, 170
Psychological First Aid (PFA), 34, 206, 209–210
 American Red Cross, 210
 CBPFA workshops for, 210
Psychological first aid (PFA), 172
Psychological reactions to disaster, 223f
Public Health core capability, 135
Public Health Preparedness Capabilities, 31
Public Information, 6
Public risk communications, 165–166

R

Recovery Support Functions (RSFs), 136
Resilience, 9–11, 31–33, 235, 236t
Responder stress, 158–159
Risk and crisis communications
 assisting in crafting messages, 282–283

Crisis Emergency Risk Communication Guide, 287
developing anticipatory guidance, 279–280
enhanced and expanded ways, 277
fosring communication with victims, 281–282
monitoring and managing stress of EM personnel, 283–284
mutual respect and trust, 278
mutual understanding of roles and skills, 278
operationalizing communications integration between DBH and EM, 295*t*
present times, 286–288
systems approach, 288–292
understanding of victim priorities, 278–279
Risk assessment, 7
Risk communication/media information, 165–171
via social media, 170–171
Risk landscapes, 102
Robert T. Stafford Disaster Relief and Emergency Assistance Act (1988), 8, 46, 50, 76–77, 94, 124–125, 144, 221–222, 308

S

Sandy Hook School shooting, 141, 142*t*, 143
Sandy (hurricane), 50, 142*t*, 144
Spiritual domain reactions, 161
State Behavioral Health Systems, 34
State Mental Health Authority (SMHA), 221–222
Stress
responder, 158–159
secondary traumatic stress (STS), 159
survivor, 157–158

Substance Abuse and Mental Health Services Administration (SAMHSA), 14–15, 77, 139–140, 173
All-Hazards Preparedness, 46
Survivor stress, 157–158
negative cognitive reactions, 161
stress indicators, 164

T

Target Capabilities List (TCL), 136–137
Threat, Hazard Identification, and Risk Assessment (THIRA), 7
Training and education for stress reactions, 171–172
A Treatise Concerning the Principles of Human Knowledge (George Berkeley), 3–4
Triggering, 168

U

Urban evacuation, 119–120
U.S. Forest Service, 4

V

Voluntary Organizations Active in Disasters (VOAD), 41, 125, 150
Vulnerability, assessment of, 7

W

"Whole Community" approach to emergency management, 8–9
Workforce behavioral health protection, 21